Springer
Proceedings in Physics 49

Springer Proceedings in Physics

Managing Editor: H. K. V. Lotsch

Volumes 1 – 29 are listed on the back inside cover

Ultrafast Phenomena in Spectroscopy

Proceedings of the Sixth International Symposium, Neubrandenburg, German Democratic Republic, August 23–27, 1989

Editors: E. Klose and B. Wilhelmi

With 245 Figures

Springer-Verlag
Berlin Heidelberg New York London
Paris Tokyo Hong Kong Barcelona

Professor Dr. Edgar Klose
Akademie der Wissenschaften der DDR, Zentralinstitut für Optik und Spektroskopie,
Rudower Chaussee 6, DDR-1199 Berlin-Adlershof

Professor Dr. Bernd Wilhelmi
Akademie der Wissenschaften der DDR, Physik, Rudower Chaussee 5,
DDR-1199 Berlin-Adlershof

ISBN-13: 978-3-642-75828-7 e-ISBN-13:978-3-642-75826-3
DOI:10.1007/978-3-642-75826-3

Library of Congress Cataloging-in-Publication Data. Ultrafast phenomena in spectroscopy: proceedings of the sixth international symposium, Neubrandenburg, German Democratic Republic, August 23–27, 1989 / editors, E. Klose and B. Wilhelmi. p. cm. – (Springer proceedings in physics; v. 49) Papers presented at the Sixth International Symposium on Ultrafast Phenomena in Spectroscopy, held in Neubrandenburg, GDR, August 23–26, 1989. Includes bibliographical references and index. ISBN 0-387-52781-8 (U.S.: acid-free paper) 1. Laser pulses, Ultrashort – Congresses. 2. Laser spectroscopy – Congresses. 3. Molecular spectroscopy – Congresses. 4. Solid state physics – Congresses. I. Klose, Edgar, 1939– . II. Wilhelmi, Bernd. III. International Symposium on Ultrafast Phenomena in Spectroscopy (6th: 1989: Neubrandenburg, Germany) IV. Series. QC689.5.L37U4682 1990 621.36'6 – dc20 90-42427

Softcover reprint of the hardcover 1st edition 1990

2154/3150-543210 – Printed on acid-free paper

Preface

This volume is a collection of papers presented at the Sixth International Symposium "Ultrafast Phenomena in Spectroscopy" (UPS '89) held in Neubrandenburg, GDR, August 23–26, 1989. This symposium brought together about 220 scientists from 18 European countries and from overseas. The participants are active in a wide range of scientific disciplines, including physics, chemistry, biology, opto-electronics and scientific instrumentation. They share a common interest in discussing problems and perspectives of ultrafast processes at picosecond and femtosecond time scales.

UPS '89 was held eleven years after the first conference of this series, which took place in Tallinn, Estonia. During this period remarkable progress has been made in this field of science and technology. Now the shortest pulses have a duration of about 5 fs, which corresponds to only about three wave periods of visible light. Other important advances concern new devices for the amplification of femtosecond light pulses to very high power.

The symposium UPS '89 demonstrated this progress in generating powerful ultrashort pulses and in applying them to a wide field of science and technology. This volume reflects this development in an impressive way. With such high-performance laser light sources, ultrafast phenomena in physics, chemistry, and biology as well as in sophisticated devices for opto-electronics and microelectronics have been studied extensively. This opens the way into the new field of ultrafast technology.

These proceedings reflect the state of the art in this field. Numerous scientists from many research groups helped to make this symposium very successful. We would like to thank all the contributors to the conference program. Especially, we would like to thank the members of the International Program Committee and of the Organizing Committee for their helpful advice and great effort in organizing the symposium.

Special thanks are due to Dr. M. Kaschke and Dr. D. Leupold from the Academy of Sciences of the GDR and Dr. W. Rudolf and Dr. U. Stamm from the Friedrich Schiller University of Jena for their assistance in preparing this volume. We are also indebted to the Central Institute of Optics and Spectroscopy of the Academy of Sciences of the GDR, the Friedrich Schiller University of Jena, the Centre for Ultrafast Phenomena in Physics (Berlin/Jena) and the Physical Society of the GDR for their sponsorship.

Berlin,
March 1990

E. Klose
B. Wilhelmi

Contents

Part I **Generation and Amplification of Ultrashort Light Pulses**

Part II Measurement, Devices and Methods

Part IV Applications to Solid-State Physics

Part V Applications to Molecular Spectroscopy

Part I

Generation and Amplification of Ultrashort Light Pulses

Towards a Laboratory X-Ray Laser

F.P. Schäfer

Max-Planck-Institut für biophysikalische Chemie, Abteilung Laserphysik, D-3400 Göttingen, Fed. Rep. of Germany

Only a few years after the demonstration of the first laser, proposals for a laser operating in the X-ray region were made [1]. Apart from the normal wish to extend the wavelength range of coherent light sources, a number of rewarding applications beckoned, like X-ray microscopy, X-ray lithography, and X-ray holography. While the early experimental work in this direction met with little success, in October 1984 the first real X-ray laser was realized at the Lawrence Livermore National Laboratory [2]. This breakthrough was only possible, however, by the use of the huge NOVA-laser as a pump source, emitting 2 kJ/400 ps pulses at 532 nm. A number of similar successes in other laboratories followed soon, using also very large laser installations with at least several hundred Joules of pump pulse energy [3,4,5]. Even though these achievements are important steps in the developmental process of X-ray lasers, one can safely say that X-ray lasers will only become useful tools for the above-mentioned applications when they have shrunk down to the size of a normal laboratory instrument.

In the following I will first point out the fundamental problems arising for the realization of X-ray lasers in general and the specific problems connected with laboratory X-ray lasers and then discuss our approach to reducing the necessary pump energy and first steps towards the realization of a laboratory X-ray laser.

A first problem of any X-ray laser is the question of a laser resonator. Even the best multilayer mirrors for the XUV and soft X-ray region exhibit high losses that make them unsuitable for laser resonators. Crystal resonators using Bragg reflections might only become useful for wavelengths of a few Ångstroms, provided the technical alignment problems could be solved in the future. For the foreseeable future the only practicable solution is a laser without a resonator, working on the principle of amplified spontaneous emission (ASE) in a single pass through the active medium. If we assume a pencil-shaped active medium of length L and diameter d, the divergence of the ASE is d/L. If the output is to be spatially

Springer Proceedings in Physics, Vol. 49 Ultrafast Phenomena in Spectroscopy
Editors: E. Klose · B. Wilhelmi

coherent across the output end, the diffraction limited divergence is λ/d. From $d/L = \lambda/d$ we obtain the optimum diameter $d_{opt} \approx \sqrt{\lambda \cdot L}$. For $\lambda = 10$ Å and $L = 5$ mm this results in $d_{opt} \approx 7$ µm, which, incidentally, is the diameter of the standard carbon fibers used for reinforcement of plastics. One could talk of a "real laser" if the single pass gain is, say, $G = \exp(\sigma \cdot N^* \cdot L) > 1000$, where σ is the cross-section for stimulated emission and N^* is the number density of the excited species. An important point to notice is that the photon transit time through the active medium would be $\approx L/c$ (with c the vacuum velocity of light), in the above example = 16.7 ps.

A second problem is the question of the active medium. Transitions in the X-ray region are either electronic transitions in highly charged ions or inner-shell transitions in atoms that have been excited by a flash of ionizing radiation. As an example of the first case, one could mention the hydrogen-like C^{5+}-ion, as an example of the second case one could offer a copper atom, in which a flash of 9 keV-radiation has created a hole in the K-shell.

A third problem is the question how to obtain an inversion in such an active medium. There are quite a few options available. The solution chosen for the Lawrence Livermore laser was the principle of the Ar^+-laser, shifted from the visible towards the XUV-region by using Se^{24+} instead of Ar^+, namely collisional excitation of the 3p upper laser levels in a hot plasma and using the 3p → 3s laser transitions at 20.6 and 21 nm, and the relaxation of the lower laser level by fast radiative transitions to the Ne-like $2s^2p^6$ ground state. This needs a delicate balance between excitation and deexcitation processes, which can only be obtained in a narrow temperature and density region and in addition needs as pump energy per ion at least the sum of the first 24 ionization energies.

Another possibility realized in several other places is a recombination laser. Taking again the example of C^{5+}, one has at first to strip the carbon atoms completely of all electrons by multiphoton ionization, i.e. create C^{6+}, and wait for the plasma to cool down in order to allow recombination to proceed, with electrons cascading down the energy levels of C^{5+}, e.g. making a transition $n = 3 \rightarrow n = 2$ at 182 Å or $n = 2 \rightarrow n = 1$ at 33.7 Å (the latter right in the middle of the famous "water window" of X-ray microscopy).

Still another possibility mentioned above with the example of the Cu-atom is the X-ray fluorescence after excitation by hard ionizing radiation, in this case most probably a transition from the L_3 to the K-shell, i.e. K_α-radiation at 1.45 Å, which, however, will be difficult to realize, as shown later.

Finally, there is the question how high the pump power and the pump energy has to be. If we look at first at the necessary pump power, we can make a quick estimate for the second example. We assume as target a piece of copper wire, 10 μm Ø, 5 mm long and want the gain $G = \exp(\sigma N^*L) \geq 1000$. Then we take from the literature the peak cross-section of the Cu K_α-line as $\sigma_m = 8 \cdot 10^{-18}$ cm^2 and calculate $N^* \geq 1.7 \cdot 10^{18}$ cm^{-3}. Multiplying by the volume of the target and by the minimum quantum energy of the pump radiation ($h\nu \geq 9$ keV) and dividing it by the lifetime ($\tau = 1.5$ fs) of this excited configuration, we obtain a minimum pump power $P_{min} \geq 26$ TW. If this pump power is used to irradiate the surface of the target uniformly, the pump intensity is found to be $I_{min} \geq 1.6\ 10^{16}$ W/cm^2. This is always assuming no losses by reabsorption or other effects. Of course, the pump power for a longer wavelength X-ray laser, operating with ions of an element with lower atomic number and as a recombination laser, will give smaller numbers, but certainly the pump power must be in the terawatt range.

For the pump energy we have to consider two different modes that give dramatically differing results. The first mode, exclusively used up to now in all X-ray laser experiments, is the simultaneous irradiation of the whole target surface. Then it is immediately evident that the minimum pulse duration must be the photon-transit time from one end to the other of the target. In our example this is $T = 16.7$ ps, so that the pump energy becomes $E = P \cdot T \approx 430$ J.

The other mode is a travelling-wave excitation, where only one small fraction of the target area is irradiated at a time and the irradiating pump beam is swept across the target with the speed of light, so that a spontaneous emitted X-ray photon starting from one end of the target simultaneously with the travelling pump wave will always experience maximum gain during its passage along the target. The minimum pump time now is the time to develop the necessary plasma conditions at any point. Theoretical simulation of the development of the plasma shows that a TW pump wave can do this in a very short time, ranging from a few tens to a few hundreds of femtoseconds, so we can take 100 fs as a typical value. This results in $E \approx 2.6$ J, which is a factor 167 smaller than in the whole-target irradiation mode.

While the peak X-ray laser power would be similar for both modes, it is clear that the travelling-wave mode will result in shorter pulses with a concomitantly smaller X-ray energy per pulse. The short pulses are very attractive for a number of applications. For other applications, however, it is the average power of the X-ray laser that counts. But even in this case it is clear that the travelling-wave excitation has an advantage, since the repetition rate of a short pulse laser with the characteristics described for the above example can have a much higher repetition rate than the huge lasers needed for whole-target irradiation, definitely greater by more than a factor 167.

It is clear from the above that a laboratory X-ray laser has to use

1. a subpicosecond pump laser and
2. travelling-wave excitation.

From earlier experiments and from plasma physics arguments another condition is

3. the shortest possible pump wavelength, which is necessary for optimum absorption of the pump light.

Our goal is a recombination laser at the 2 → 1 transition in C^{5+} at 33.7 Å, which should have an efficiency of $\geq 10^{-4}$. This would mean >100 MW peak power and >100 μJ pulse energy.

As a pump laser, a Krf-laser emitting at 248 nm pulses of 100 fs duration and 50 mJ energy plus an additional power amplifier stage with a minimum pulse energy of 1 J was chosen. While the front end is already working with the above given specifications [6] the output amplifier is still under construction [7]. Figure 1 shows the

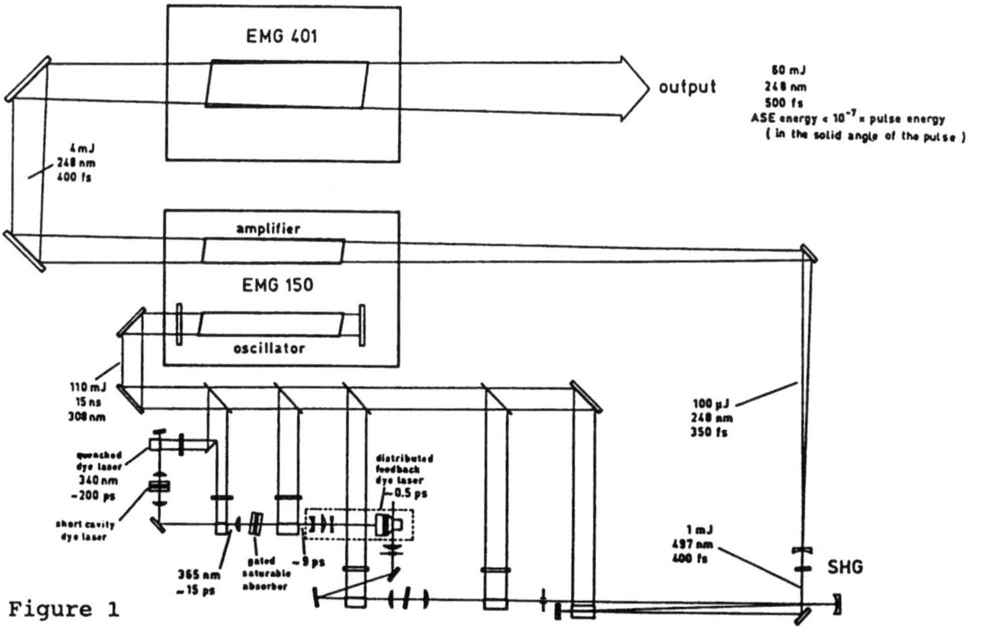

Figure 1

lay-out of the pulse generator plus preamplifier. This system is operating at 1 Hz and can be boosted to 20 Hz, if necessary, while the power amplifier presently can be operated at 1/10 Hz and can be boosted to 1 Hz. An important aspect of the laser pulse is its contrast ratio, i.e. the ratio of the short-pulse energy to the energy contained in the ASE preceding the short pulse, since this determines whether the short pulse sees the solid target surface or already a preformed plasma. After the preamplifier (50 mJ output) we have determined an energy contrast ratio of $>10^7$. Since the ASE is about 10 ns long, the short-pulse duration, however, subpicosecond, this transforms to a power contrast ratio of $>10^{11}$.

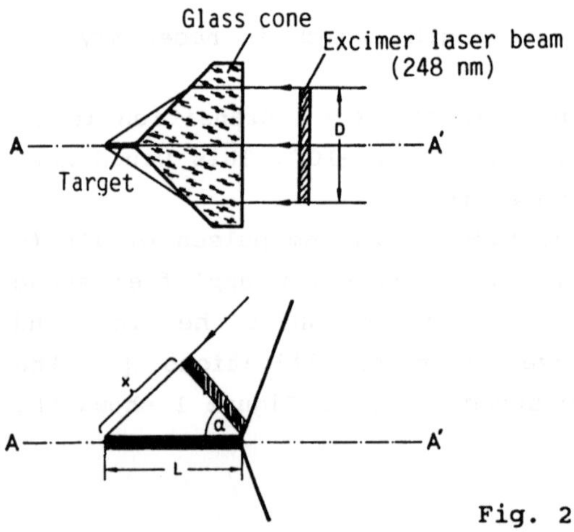

Fig. 2

The travelling-wave target irradiation was at first intended to be done using a cone of quartz-glass, as shown in Fig. 2 [8]. It is, however, directly evident from the figure that the pump wave will travel faster over the target than wanted, namely with a velocity $v = (L/x) \cdot c$, which gives a 15 % mismatch or 23 × pulse duration in this specific example. Actually, the situation is even worse than shown by simple geometrical ray-tracing, since the group-velocity in the cone is smaller than the phase-velocity, thus delaying the rays more the thicker the sections of the cone passed. This results in a pulse front where the angle α is smaller than shown here. This difficulty can only be corrected by using a two-cone system, as shown in Fig. 3, where two rays are shown and the pulse front between them at three different times. Now the length $x' = L$ and perfect synchronism between the travelling-wave and the X-ray photons is obtained [9].

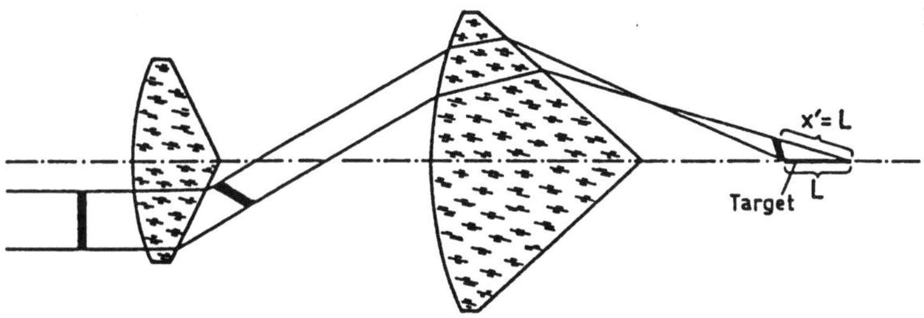

Fig. 3

While this optical system was being produced, another difficulty became apparent. We noticed by chance that even highest UV-quality quartz glass became less transmissive the higher the incident Krf-laser intensity. For example a 20 mm thick window of Suprasil® had a transmittance of only about 40 % at 120 GW/cm^2 [6]. This was soon found to be caused by two-photon absorption in our laboratory and independently elsewhere [10,11]. Systematic studies of the transmittance as a function of incident intensity at 248 nm for a number of different UV optical materials showed that only LiF, MgF_2 and CaF_2 were free of two-photon absorption, with an insignificant trace of three-photon absorption [10]. However, LiF lacks mechanical strength and MgF_2 is double-refractive, so they cannot be used for all applications, while CaF_2 is the most generally useful material up to about 100 GW/cm^2. The two-cone system will now be produced from LiF.

Preliminary tests were undertaken in an evacuated target chamber with various planar targets and focussing on a spot by a plano-convex lens, which gave the following important results:

1. At intensities of 10^{15} W/cm^2, the KrF-laser light showed maximum absorption of 60 - 80 % (dependent on the material of the target), when using p-polarization and angles of incidence for maximum absorption (between 54 and 58°, dependent again on material) [12].

2. The time duration of the X-ray pulse emitted from the focal spot was instrument-limited (<10 ps) for hard X-rays, while the long-wavelength components were longer than 100 ps [13].

3. A comparison of Al-spectra between 50 Å and 350 Å showed very distinct differences for nanosecond and subpicosecond irradiation, the latter giving much narrower line-widths and different ratios of the various line-strengths [13].

After resolving the above-mentioned difficulties in principle, the ongoing technical developments, albeit time-consuming, will certainly lead to a practically useful laboratory X-ray laser in the not too distant future.

References

1. M.A. Dugay, P.M. Rentzepis: Appl. Phys. Lett. **10**, 350 (1967)
2. M.D. Rosen et al.: Phys. Rev. Lett. **54**, 106 (1985) and D.L. Mathews et al.: ibid. **54**, 110 (1985)
3. M. Key et al.: SPIE-Conf. on X-rays from laser plasmas, San Diego, August 1987
4. P. Jaeglé et al.; J. Opt. Soc. Am. B **4**, 563 (1987)
5. T.N. Lee et al.: Phys. Rev. Lett. **59**, 1185 (1987)
6. S. Szatmári, F.P. Schäfer: Opt. Commun. **68**, 196 (1988)
7. M. Steyer et al.: Appl. Phys. B **49**, 331 (1989)
8. F.P. Schäfer: J. Phys., Coll. C6, suppl. 10, **47** C6-149/C6-150 (1986)
9. J. Jasny, unpublished
10. P. Simon et al.: Opt. Lett. **14**, 1207 (1989)
11. J.P. Roberts et al.: Opt. Lett. **13**, 734 (1988)
12. R. Fedosejevs et al.: Appl. Phys. B, in press
13. G. Kühnle et al.: Appl. Phys. B **47**, 361 (1988)

Amplification of Femtosecond Light Pulses

P. Heist[1], *F. Noack*[2], *W. Rudolph*[1], *and B. Wilhelmi*[2]

[1]Sektion Physik, Friedrich-Schiller-Universität Jena, Max-Wien-Platz 1, DDR-6900 Jena, German Dem. Rep.

[2]Zentralinstitut für Optik und Spektroskopie, Akademie der Wissenschaften, Rudower Chaussee 5, DDR-1199 Berlin, German Dem. Rep.

Abstract. Essential physical processes going on in the gain media of femtosecond pulse amplifiers are discussed. In particular, self phase modulation (SPM) and its utilization for pulse compression are investigated theoretically and corresponding experimental data on combined pulse amplification and compression are presented. Finally our approach for producing powerful femtosecond light pulses from the UV to the NIR is described.

1. Introduction

For the broad applications of femtosecond light pulses in science and technology, one is interested to have powerful pulses in various spectral ranges. Fig. 1 shows a rough view of the present state of the art in producing light pulses with durations below 200fs in different energy and wavelength regions.

The most successful concept to produce fs pulses is the mode-locked dye laser which covers the VIS and NIR, e.g./1/, and by which means it is possible to enter the UV through intracavity second harmonic generation (SHG) as well as the IR by intracavity parametric oscillation /2/. The pulse energy behind these oscillators typically does not exceed several 10pJ and can be increased up to the nJ level by cavity dumping. Pulse amplification in dye amplifiers can provide mJ pulses, where most of the work has been done around 620nm with the amplification of the pulses from the CPM dye laser. Recently, mJ pulses of 100fs duration have been obtained around 700nm after chirped pulse amplification in alexandrite /3/ and compression. Powerful pulses at certain wavelengths in the UV are produced through SHG and reamplification in excimer gain modules, e.g./4/.

In the IR the features of pulse propagation in optical single-mode fibers can advantageously be utilized. This concerns the application of the fiber compressor, for example when starting with Nd:YAG laser pulses, as well as soliton shaping. Due to the occurrence of damage and undesired nonlinearities the energy that can be handled is below 100nJ.

Femtosecond pulse amplifiers differ from each other in the achievable energy and repetition rate, where these parameters cannot be chosen independently from each other. Usually the mean power is between 1mW and 100mW and is finally limited by the pump lasers for the amplifiers which are in use today, see Fig.2.

Roughly one can distinguish between high energy / low rep rate systems such as amplifiers pumped by excimer lasers or Q-switched Nd:YAG lasers and high rep-rate / moderate energy devices. As pumps for the latter copper vapor lasers, diode pumped Nd:YAG lasers and Nd:YAG lasers with regenerative amplifiers can be used.

In designing fs pulse amplifiers one not only has to consider a high gain factor but also one has to take care to maintain the short pulse duration or even to shorten the pulses. Several schemes have

FIG.1

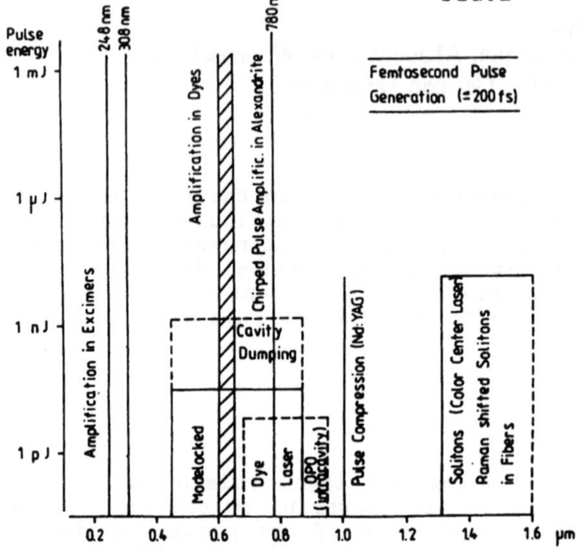

FIG 2

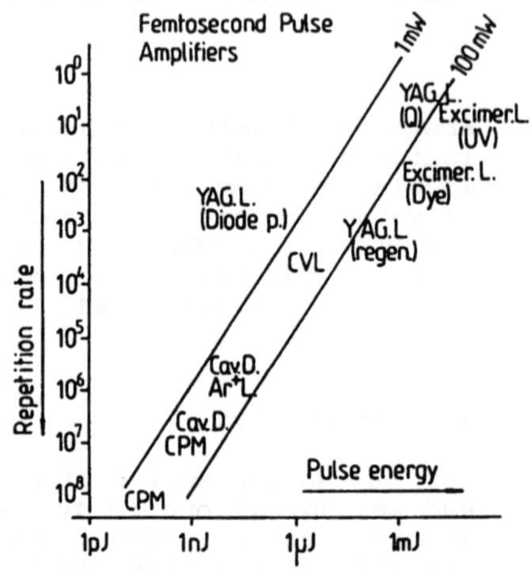

FIG.1: Femtosecond pulse sources

FIG 2: Pulse energy versus repetition rate of fs pulse amplifiers

been developed for this purpose, e.g./5/. So far the shortest pulses, which have a duration of about 6fs /6/, can be obtained by amplifying fs pulses to the nJ level and compressing them in a fiber compressor. Higher pulse energies are feasible by spectral (and temporal) broadening before amplification. This method is particularly attractive for solid state lasers/amplifiers /7/ because the peak intensities in the gain media can be decreased. Recently, this technique has also been successfully applied in dye amplifiers to produce 16fs pulses with an energy of 0.5μJ /8/.

In the first part of our contribution we want to discuss a third method - combined amplification and compression - where the spectral broadening proceeds in the gain medium itself /9/. In the second part we will describe our approach for producing powerful fs light pulses from the UV to the IR.

2. Combined amplification and self-phase modulation (SPM)

For our consideration we refer to the following simplified amplification scheme. The pulse to be amplified passes through the active material (for instance a dye cell) where it is amplified and phase modulated. The amplifier is terminated by a linear optical element (for example a prism or grating sequence). Mostly this is for compensating the group velocity dispersion (GVD) in the amplifier. The question is now whether it is possible to compress the pulses behind the amplifier to pulse durations which are shorter as compared with those after the oscillator.

At first let us have a look at the various processes that can affect pulse propagation in the gain medium.

- During the passage of the signal pulse we can neglect the pump process and the amplified spontaneous emission (ASE). The temporal behavior of the gain then is only controlled by the signal pulse through gain depletion. This influences the energy gain as well as the shaping of the pulse envelope. (Note, depending on the input pulse shape depletable gain can lead to temporal broadening and shortening of the signal pulse)
- SPM accompanied by spectral broadening can occur where we can distinguish two main sources
  * quasiresonant processes resulting from the temporal behavior of the gain
  * nonresonant processes; the most essential effect here is the intensity dependent refractive index of the host medium (dye solvent)
- The pulse propagation is influenced by GVD leading to pulse broadening and chirping without spectral changes. A characteristic sample length, the so-called dispersion length, is given by $L_D=\tau_L^2/k_L''$ (τ_L: input pulse duration, k_L'':second derivation of the propagation constant with respect to frequency) where a Gaussian pulse doubles its duration in traversing a length $L=0.6L_D$. For the solvent ethanol and 50fs input pulses this happens, for example, at $L\approx3.5$cm.

Next we want to discuss in more detail whether SPM due to gain depletion can be utilized in pulse amplifiers for an effective pulse compression.

2.1 Chirping through resonant gain depletion and compression

The resonant contribution to the chirp production rests on the change of the refractive index associated with the population numbers of the gain transition, where the sign of the change depends on whether the carrier frequency ω_L is above or below the center frequency of the transition ω_0.

The functional dependence of the chirp impressed on the pulse is given by

$$\delta\omega(t) = r(\omega_L)d/dt\{\ln[I_{out}(t)/I_{in}(t)]\} \qquad (1)$$

where I_{out} (I_{in}) is the output (input) intensity and $r(\omega_L)$ is a factor describing the dispersion of the transition which for example equals $(\omega_0-\omega_L)/\Delta\omega_0$ in the case of a homogeneously broadened two-level system. The functional dependence of I_{out}/I_{in} is controlled by the gain depletion, a measure of which is the ratio of input pulse energy and saturation energy of the transition $\epsilon=W_{in}/W_{sat}$.

As an example we want to look at what happens with a Gaussian-shaped input pulse in passing through a depletable amplifier and then through an optimum quadratic compressor, see Fig.3. The small signal gain was kept constant (10^4) and the normalized output intensity and chirp is depicted for three different ϵ. Obviously with increasing ϵ the pulse broadens and the frequency sweep around the pulse center becomes larger. A compression after a quadratic compressor (lower curves) is only possible for high depletion (solid curve) which, however, is not desirable with the view on high energy amplification. Although this type of SPM has been observed in ps dye amplifiers /10/ it does not seem very attractive in such gain media. The situation is quite different in semiconductor amplifiers where the factor r in (1) can be larger by several orders of magnitude as compared with dyes and where recently a spectral broadening and a subsequent compression by a factor of two has been observed with ps pulses /11/. For illustration the corresponding compression factors

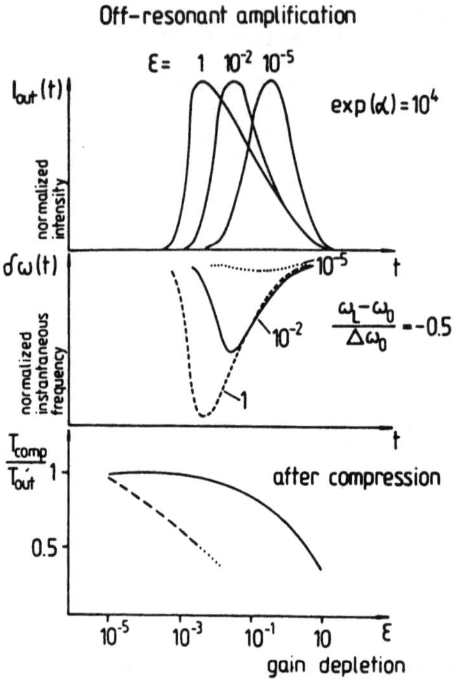

FIG.3: Shaping of pulse envelope and instantaneous frequency in a depletable amplifier traversed off-resonantly and pulse shortening after a quadratic compressor (T_{out}: pulse duration after amplification)

for a ten times larger value of r are shown in Fig. 3 (dashed curves) revealing an effective compression at smaller ϵ.

2.2 Chirping through the combined action of SPM, GVD and gain depletion

SPM through an intensity-dependent refractive index $n=n_L+n_2I(t)$ produces a frequency modulation which is proportional to

$$\delta\omega(t) \sim -n_2L\, d[I(t)]/dt \qquad (2)$$

and accordingly increases with peak intensity and sample length L. The peak intensity cannot be increased arbitrarily in an amplifier but is limited through the gain depletion. The sample length in turn controls the pulse broadening resulting from GVD. The combined action of these effects on the pulse shaping is governed by the nonlinear Schrödinger equation (NLSE) for the electric field E(t,z), which for example describes pulse propagation in single-mode fibers, supplemented by a polarization term $P_s(t,z)$ which is responsible for the gain.

$$\delta[E(z,t)]/\delta z = 0.5k_L''\delta^2[E(z,t)]/\delta t^2 - i\omega_L^2\mu_0/k_L * [\varepsilon_0 n_L n_2 |E(z,t)|^2 E(z,t)] + P_s(z,t) \qquad (3)$$

(t,z: coordinates in a frame moving with the group velocity, k_L'':

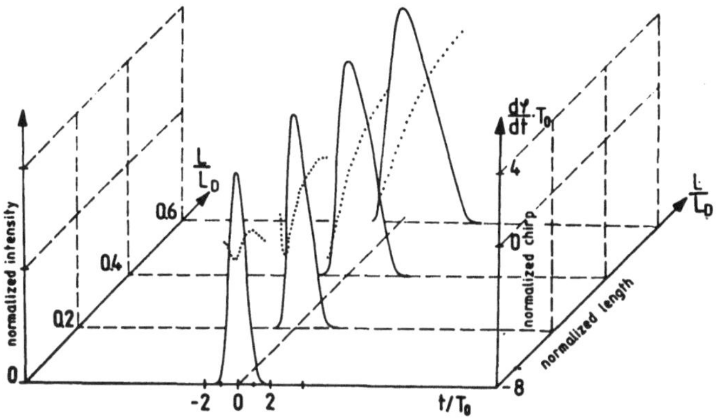

FIG.4: Envelope (solid line) and chirp (dotted curve) evolution as a function of the amplifier length

second derivation of the propagation constant with respect to frequency). Fig. 4 shows the envelope and chirp evolution (normalized quantities) in an amplifier with SPM and GVD for a Gaussian input pulse. The broadening here is due to GVD and gain depletion. The strong nonlinear behavior at the beginning develops into an almost linear behavior around the pulse center, which is most desirable for subsequent pulse compression.

Fig. 5 summarizes the calculations for a Gaussian (left) and asymmetric sech (right) input pulse. Obviously, to apply this kind of pulse compression the amplifier should satisfy the following conditions:

- the amplifier (the last stage) has to be operated near the saturation regime (which is also favorable for decreasing the influence of pump pulse fluctuations)
- the dye cell should have a certain length (in our example several cm)
- asymmetric input pulses are preferred which, for example, can be achieved by cutting the leading pulse edge in a saturable absorber in front of the amplifier stage.

Fig. 6 compares the measured spectral behavior (a) in the last stage of our amplifier (section 3) with calculated spectra (b). Different gain depletion could be adjusted by varying the input pulse energy. With increasing ϵ the spectrum broadens and a long wavelength component appears. The latter is a result of the accumulation of these components in the leading pulse edge due to GVD and of depletable gain which favors this part of the pulse.

After combined amplification and SPM the pulses are sent through a prism sequence for compression. In Fig.7 results are shown for a 100fs (left, /9/) and 140fs (right) pulse from the oscillator. In both cases a total compression of about two was observed which is in qualitative agreement with the calculations.

Now we want to discuss limits of this type of compression. First of all a limit is given by the finite width of the gain profile of the transition. Spectral narrowing was observed in /12/ where the authors tried to amplify 20fs pulses. If one takes into account the bandwidth of the gain profile, i.e. one considers a certain phase memory, the achievable compression decreases, which is shown in Fig.6 (dashed curves) for ϵ=0.5. To illustrate that on a 100fs time scale the temporal behavior of the gain described by rate equations (no phase memory) is the dominant shaping process we measured the gain

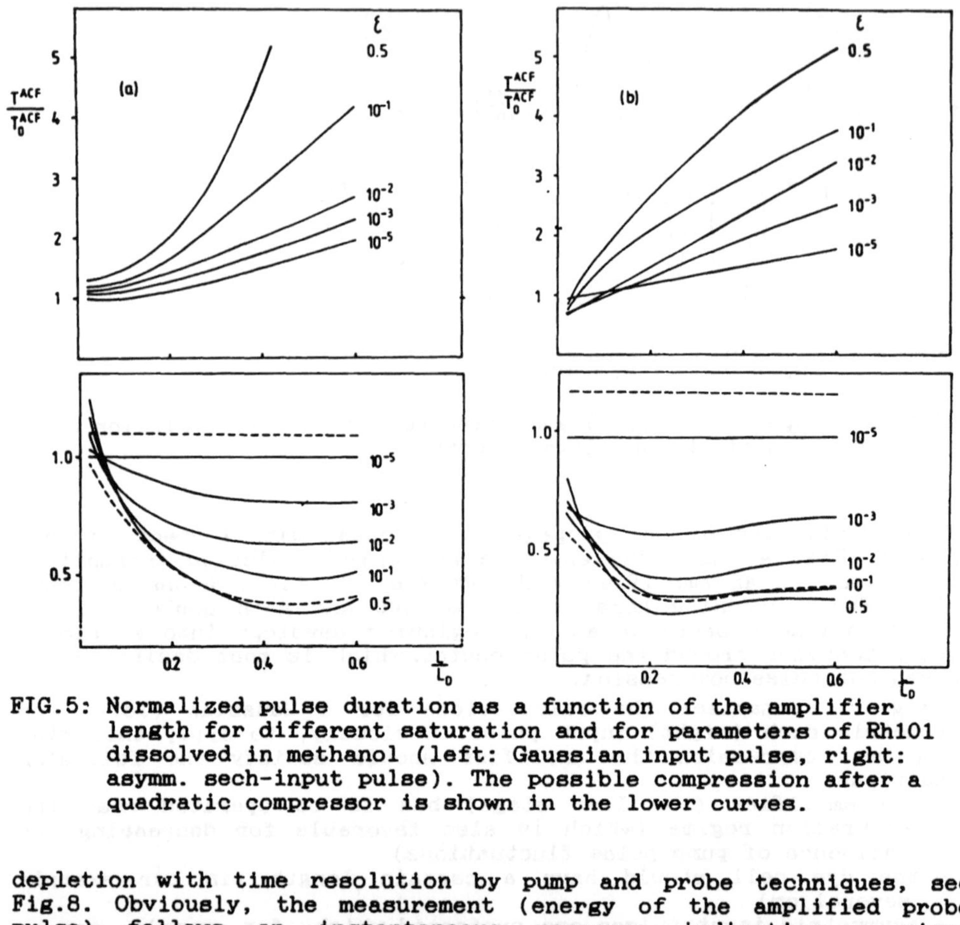

FIG.5: Normalized pulse duration as a function of the amplifier length for different saturation and for parameters of Rh101 dissolved in methanol (left: Gaussian input pulse, right: asymm. sech-input pulse). The possible compression after a quadratic compressor is shown in the lower curves.

depletion with time resolution by pump and probe techniques, see Fig.8. Obviously, the measurement (energy of the amplified probe pulse) follows an instantaneous response indicating a minor influence of the phase memory. For the calculation we also considered a fast (75fs) deexcitation of the vibronic level populated by stimulated emission, which can correspond to a vibronic relaxation or to an iso-energetic energy transfer to non-active vibronic modes. Since we could not observe a corresponding dip in the measured curve such relaxation is expected to be significantly faster than our time resolution (100fs). Such fast processes were observed in the S_1 state in /13/.

The situation is different in excimer gain media, for example in XeCl. Here the bandwidth of the transition already corresponds to the transition width of the gain medium. Consequently, the output pulse shape is strongly influenced by the characteristics of the transition profile. The pulse autocorrelation curve (recorded by means of two-photon luminescence in CsI:Na /14/) of a pulse (input spectral width corresponds to 100fs pulse duration) amplified in XeCl at weak saturation shows equally spaced (≈1.3ps) satellites /14/ revealing vibronic features in the lower electronic (X-) state which have also been observed by other means /15/.

As compared with dye and excimer media, some new solid state materials with spectral widths of several hundred nm have the potential to amplify even shorter pulses. Although, because of the larger interaction length required, GVD and undesired nonlinearities are more critical.

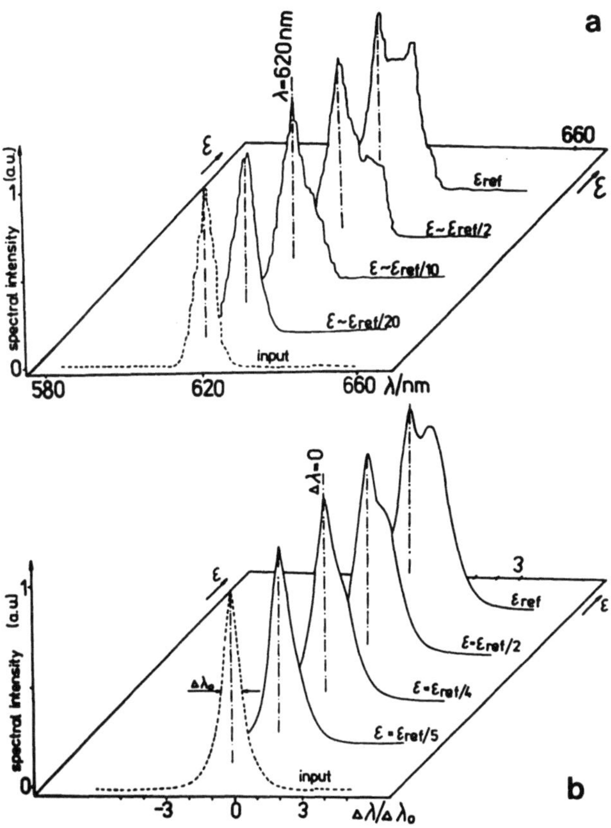

FIG.6: Measured (a) and calculated (b) spectral changes through gain depletion and SPM

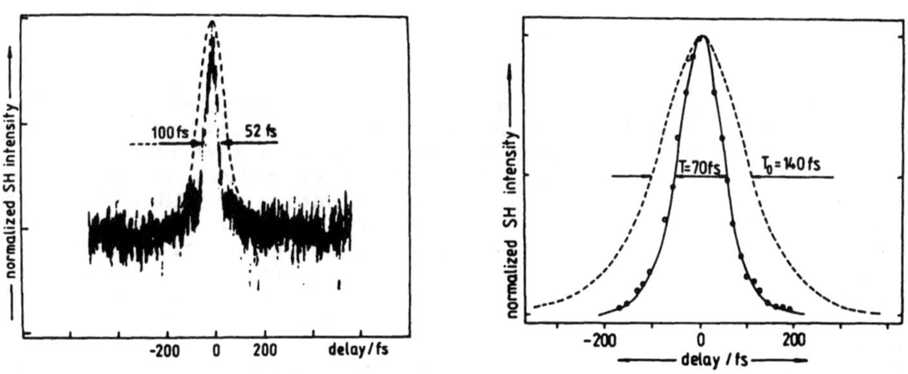

FIG.7: Pulse autocorrelations after the oscillator (dashed line) and after combined amplification and compression (solid line)

FIG.8: Time-resolved gain depletion

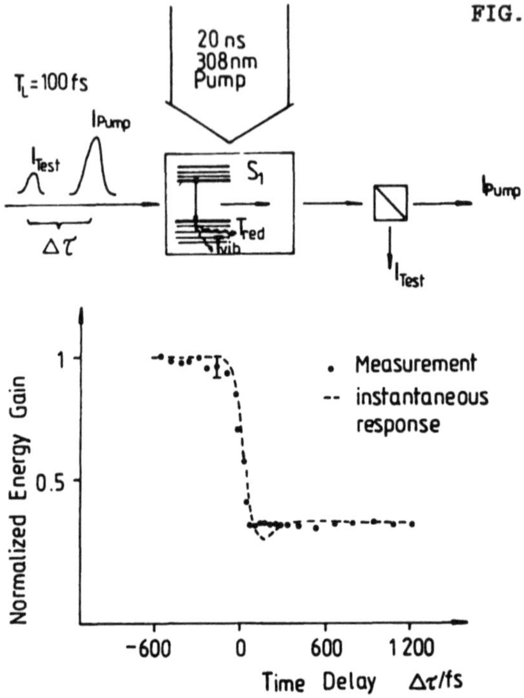

Further progress could be expected when utilizing coherent pulse amplification where the limits given by the transition width can be circumvented to some extent, e.g. /16/.

3. Femtosecond pulses from the UV to the NIR

Several concepts have been developed to produce fs light pulses in different wavelength regions.

(i) One can start with different oscillators and then amplify these pulses in different amplifiers. This large effort, however, seems only to be justified for special purposes.

(ii) The frequency is changed after amplification by nonlinear optical processes. Dealing with fs pulses the conversion efficiency usually does not exceed a few percent and the pulses must be reamplified at the respective wavelength.

A simple way to get wavelength-tunable fs pulses is continuum generation, spectral filtering and reamplification /17,18/. With two additional dye amplifier stages (possible gain factor ≈ 10^3-10^4) the initial pulse energy can be reached again in a spectral range of several hundred nm. Having in mind that with a four stage amplifier an energy gain of 10^7 is possible where the input pulse energy can be as low as some ten pJ, a tuning from the VIS to the NIR seems to be realistic.

Now we want to describe our approach to produce fs light pulses of high energy from the near UV to the near IR. The scheme of our device is shown in Fig.9 and consists of one oscillator and one pump laser for the amplifier. The femtosecond pulses are derived from a passively modelocked dye laser in the CPM regime. These low-energy pulses are amplified in four dye cells which are pumped by an excimer laser (EMG 150, 308nm) synchronized to the pulse train from

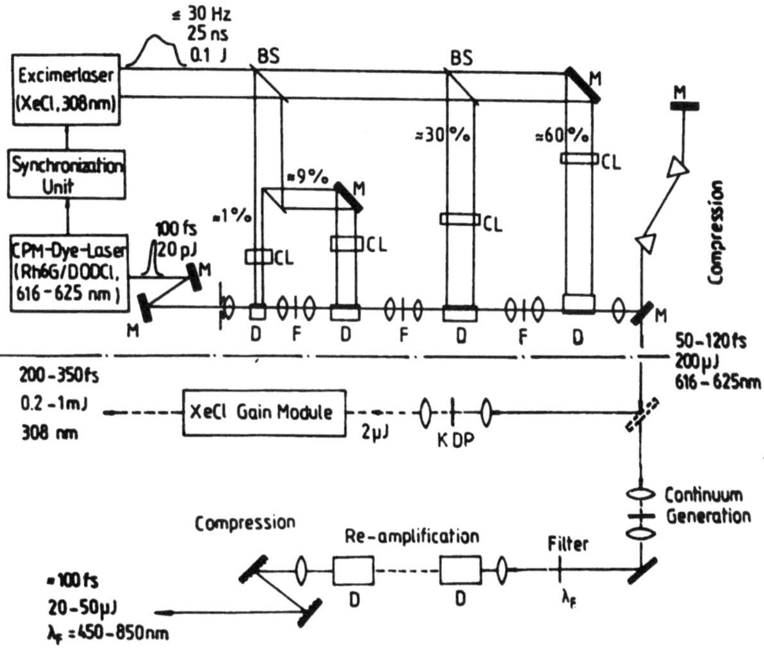

FIG.9: Set-up for producing high-power fs light pulses from the UV to the NIR

the oscillator. The first two stages are ordinary dye cells of several cm length which are transversely pumped, the last stages are designed as Bethune cells. Semiconductor-doped color glass filters (RG 5) are used as saturable absorbers to isolate the gain stages. A total energy amplification of about 10^7 was achieved varying from about 5000 in the first stage to 5-10 in the last stage. Behind the amplifier the pulses pass a prism compressor made from high dispersive SF glass where, under optimum operation conditions, the self-phase modulated pulses can be compressed by a factor of two as compared with the pulses from the oscillator.

For the generation of pulses in the UV (308nm) the red pulses are frequency-doubled in a 1mm KDP crystal (conversion efficiency ≤5%) and amplified in the second module of the excimer laser (single pass gain 50-100, double-pass gain 200-500).

Tunability from the blue to the near IR spectral region is achieved by focusing amplified (red) pulses in a suitable nonlinear optical material (CCL_4) to generate a fs broadband continuum and by filtering out the desired wavelength range. The energy loss is compensated in two dye stages which are pumped by the second excimer gain module. This gain section again is terminated by a linear optical element (prism or grating pair).

Acknowledgement

We thank Dr. C.Rempel and Dr. V.Petrov for fruitful cooperation.

References

/1/ J.Herrmann,B.Wilhelmi:Lasers for Ultrashort Light Pulses, Akademie-Verlag,Berlin 1986

/2/ a. G.Focht,M.C.Downer,IEEE J.Quant.Electr. **QE-24**(1988)431
b. D.C.Edelstein,E.S.Wachmann,C.L.Tang, Appl.Phys.Lett.**54**(1989)1728

/3/ M.Pessot,J.Squier,G.Mourou,D.J.Harter,Opt.Lett.**14**(1989)797

/4/ S.Szatmari,F.P.Schäfer,E.Müller-Horsche,W.Mückenheim, Opt.Commun.**63**(1987)5

/5/ W.Rudolph,B.Wilhelmi: Light Pulse Compression (Harwood acad. publ.,London,Paris,New York,Melbourne 1989)

/6/ R.L.Fork,C.H.Brito Cruz,P.C.Becker,C.V.Shank, Opt.Lett.**12**(1987)483

/7/ T.Damm,M.Kaschke,F.Noack,B.Wilhelmi,Opt.Lett.**10**(1985)176

/8/ G.Boyer,M.Franco,J.P.Chambaret,A.Migus,A.Antonetti,P.Georges, F.Salin,A.Brun,Appl.Phys.Lett.**53**(1988)823

/9/ a. W.Dietel,E.Döpel,V.Petrov,C.Rempel,W.Rudolph,B.Wilhelmi, G.Marowsky,F.P.Schäfer,Appl.Phys.B **46**(1988)183
b. P.Heist,W.Rudolph,V.Petrov,Appl.Phys.B **49**(1989)113

/10/ Zs.Bor,G.Szabo,F.Raksi,Proceed. UPS´87,World Scientific, Singapore 1988

/11/ G.P.Agraval,N.A.Olson,Opt.Lett.**14**(1989)500

/12/ W.H.Knox,IEEE J.Quant.Electr. **QE-24**(1988)388

/13/ G.Angel, B.Gagel, A.Laubereau, Proceed. EQEC`88, Hannover (1988)

/14/ P.Heist,F.Noack,W.Rudolph,T.Schröder,paper PTh34, UPS´89, Neubrandenburg 1989

/15/ S.Szatmari,F.P.Schäfer,J.Opt.Soc.Am.**B4**(1987)1943

/16/ V.Petrov,W.Rudolph,paper PTh5,UPS´89,Neubrandenburg 1989

/17/ A.Migus,A.Antonetti,J.Etchepare,D.Hulin,A.Orszag, J.Opt.Soc.Am.**B2**(1985)584

/18/ P.Heist,D.Lap,F.Noack,W.Rudolph,T.Schröder, paper PTh 64 UPS`89, Neubrandenburg 1989

Generation of Tunable Femtosecond Pulses of High Energy

P. Heist, D.V. Lap, F. Noack, W. Rudolph, and T. Schröder

Sektion Physik, Friedrich-Schiller-Universität Jena, Max-Wien-Platz 1, DDR-6900 Jena, German Dem. Rep.

Abstract: Starting with fs pulses from a CPM dye laser and using one XeCl-excimer laser we demonstrate the possibility of pulse amplification at $\lambda_0 \approx 616$ nm and at $\lambda = 308$ nm as well as the generation of pulses tunable from the blue to the near infrared (NIR) spectral region. The latter is achieved through continuum generation and spectral filtering.

1. Introduction

In the last years a lot of work has been done in the development of fs light pulse sources and fs pulse amplifiers. Such devices are an excellent tool in the field of ultrafast spectroscopy.

The performance characteristics of a fs-spectrometer are not only determined, for instance, by the pulse duration, pulse intensity and repetition rate, but also by the possibility of wavelength tuning. Starting from amplified pulses of a CPM dye laser at a wavelength of $\lambda \approx 616$ nm two main concepts have been applied for the generation of tunable fs pulses of high energy:

(1) The generation of a white light continuum followed by spectral filtering of the desired spectral range and reamplification in an appropriate dye [1].

(2) Amplification of seed light at a wavelength $\lambda_s > \lambda_0$ through traveling-wave pumping of a dye cell with powerful fs light pulses at $\lambda = \lambda_0$ [2].

Applying (1) we demonstrate the special capability of a XeCl-excimer laser pumped amplifier system for producing high power fs pulses in the green and blue spectral range. Moreover, this system enables the amplification of UV fs pulses ($\lambda = 308$ nm) in the excimer gain module.

2. Direct Amplification of CPM dye laser pulses

The experimental setup is given in Fig.1. Pulses from a CPM dye laser ($\lambda \approx 616$ nm, energy $E \leq 100$ pJ, duration $T \approx 100$ fs) are amplified in a 4-stage arrangement pumped by a XeCl-excimer laser (EMG 150). The amplifier is terminated by a double-pass prism arrangement. Under certain conditions the interplay of group velocity dispersion (GVD), self-phase modulation (SPM) and (depletable) amplification results in a spectral broadening which can be utilized for pulse compression [3]. The parameters of the output pulses at a repetition rate of ≤ 30 Hz are $T \leq 100$ fs, $E \leq 200$ µJ.

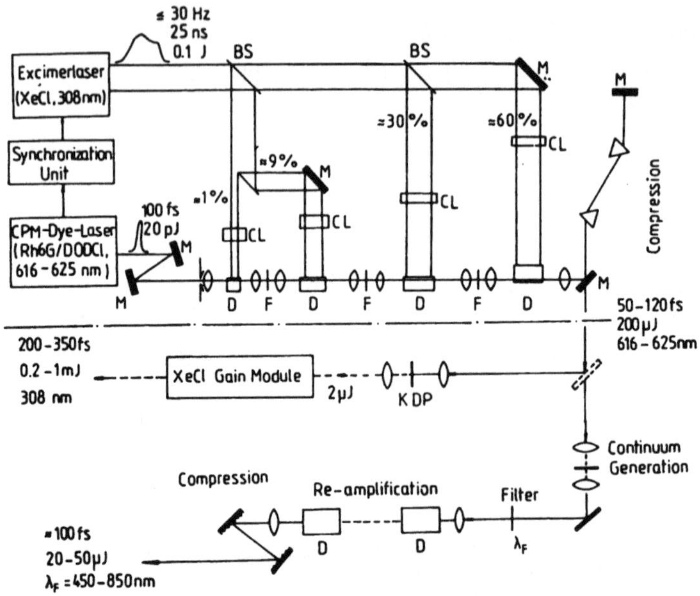

Fig.1: Experimental set-up for producing fs light pulses from UV to NIR spectral region

3. Amplification in the XeCl Excimer Module

To obtain high-power fs pulses in the UV the amplified pulses at λ_0=616 nm are frequency doubled in a thin (d≈1 mm) KDP crystal. These UV pulses (E≈2 µJ) are sent through the second excimer gain module in a single- or double-pass configuration. Typical output pulse parameters are E≈200 µJ and T≈300 fs in the case of a weak saturated amplification (single-pass) and E=1...2 mJ, T≈200 fs for stronger gain depletion (double-pass). For the pulse autocorrelation measurements two-photon luminescence in CsI doped with Na can advantageously be used [4].

4. Tunable Amplification

For the broad tunability through the VIS to the NIR (450...850 nm) an ultrafast spectral continuum is generated by focusing the amplified pulses at 616 nm into a nonlinear optical material. Figure 2 shows typical examples for various liquids.

The desired spectral range can be filtered out of the continuum spectrum by interferential filters (IF) as well as by prism- or grating arrangements. The energy loss can be compensated in a one- or multi-stage dye amplifier which is pumped by the output from the 2nd excimer gain module (E_P≤100 mJ). The number of stages which is necessary depends on the selected wavelength and desired output energy. (A limitation, of course, is given by the available pump pulse energy.) The percentage of the total pulse energy that can be used as seed signal for the reamplifier is shown in Fig.3 as a function of the wavelength.

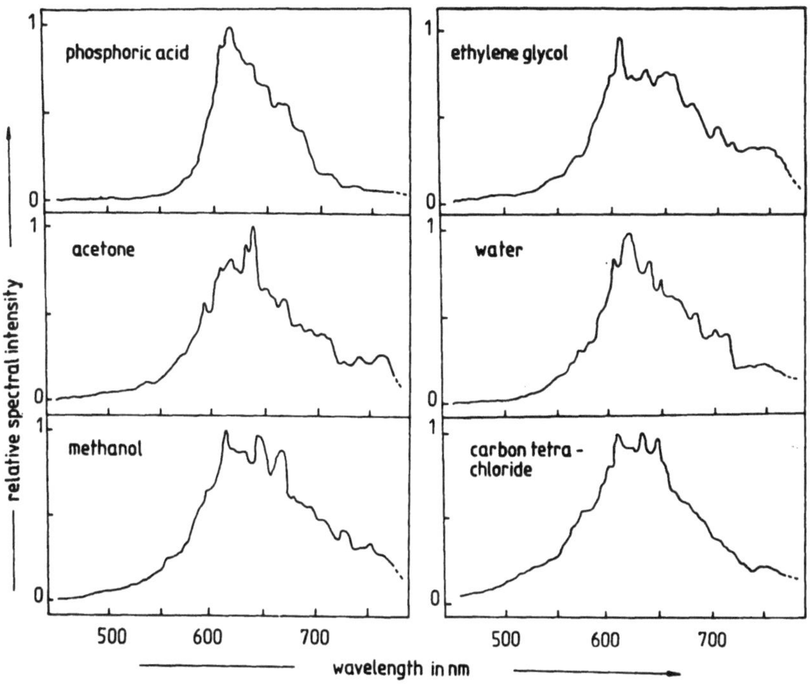

Fig.2: Examples of spectral continuum generation (T=100 fs, E=20 μJ/(100 μm)²).

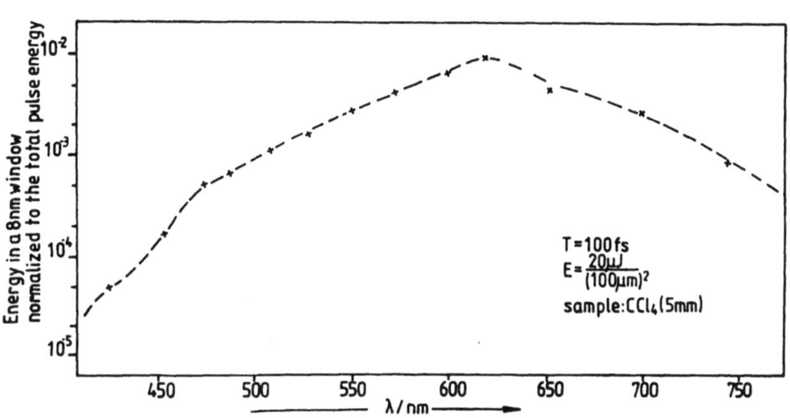

Fig.3: Distribution of energy within the continuum

As an example we want to describe the generation of fs pulses in the green spectral region through continuum generation in CCl_4 (5mm cell, E=20μJ/(100μm)²) and spectral selection by an IF ($\Delta\lambda$=8 nm) at 530 nm. The input pulse with an energy E≈50 nJ is amplified in two cells with the dye Coumarin 6 dissolved in benzene. Amplified spontaneous emission is suppressed by semiconductor doped glasses OG2. The amplification by a factor ≥1000 is accompanied by a pulse broadening

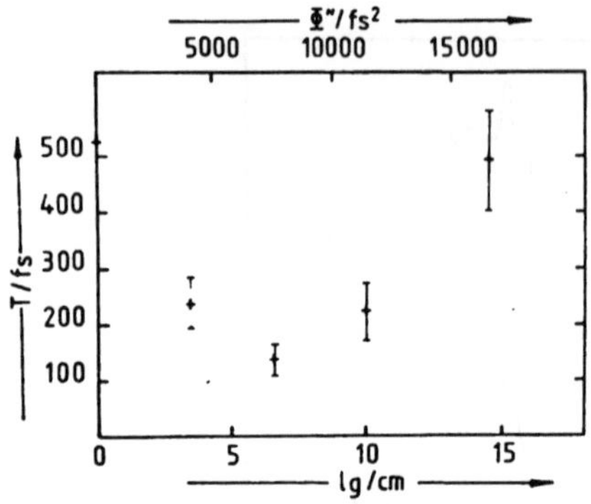

Fig.4: Pulse duration as a function of the grating spacing

of a factor 5. This can be explained by the action of GVD in the solvent (benzene, λ=530 nm: $d^2k/d\omega^2$=210 fs^2/mm) and taking into account that the input pulse is non-transform-limited. Terminating the amplifier by a grating pair, a pulse compression is possible, see Fig.4. The output pulse energy is 50 µJ and the pulse duration varies from 100...140 fs.

References

[1] A.Migus,A.Antonetti,J.Etchepare,D.Hulin,A.Orszag, J.Opt.Soc.Am.B2(1985)584

[2] J.Hebling,J.Kuhl,Opt.Lett.**14**(1989)278

[3] W.Dietel,E.Döpel,V.Petrov,C.Rempel,W.Rudolph,B.Wilhelmi, G.Marowsky,F.P.Schäfer,Appl.Phys.B**46**(1988)183
P.Heist,W.Rudolph,V.Petrov,Appl.Phys.B**49**(1989)113

[4] P.Heist,F.Noack,W.Rudolph,T.Schröder,R.Deich, W.E.Postovalov,UPS VI,poster PTh34,Neubrandenburg 1989

Analysis of Actively Mode-Locked Lasers with Intracavity Second Harmonic Generation

T. Damm[1], *U. Stamm*[1], *K.-P. Stolberg*[1], *and B. Wilhelmi*[2]

[1]Sektion Physik, Friedrich-Schiller-Universität Jena, Max-Wien-Platz 1, DDR-6900 Jena, German Dem. Rep.
[2]Zentralinstitut für Optik und Spektroskopie, Akademie der Wissenschaften, Rudower Chaussee 5, DDR-1199 Berlin, German Dem. Rep.

1. Introduction

The investigation of UV-induced dynamical processes, e.g. in biology [1], solid-state physics [2] and plasma physics [3], requires the development of ultraviolet short pulse laser sources with good beam quality.

One method of generating laser radiation in the UV-region is optical frequency doubling in nonlinear optical crystals [4]. The application of short pulses results in an increased efficiency of the frequency conversion due to the nonlinear character of the harmonic generation. A further increase of the electric field of the second harmonic is possible by inserting the crystal in a (active or passive) cavity because of the enhancement of the fundamental wave inside the cavity [5].

Here we want to discuss how second harmonic generation (SHG) in the cavity of an actively mode-locked laser affects its pulse characteristics.

2. Round Trip Model for Intracavity Second Harmonic Generation

There exist two counteracting mechanisms in the cavity influencing the developing pulses: The mode-locking process prefers the peaks in the noise of the active medium, whereas the SHG process reduces the fundamental peaks due to the quadratic conversion into the second harmonic.

In the following we give the results of a numerical treatment of pulse evolution in actively mode-locked lasers with intracavity SHG (ICSHG). For the description of the pulse formation in the laser we use a round trip model [6] which considers the development of a pulse travelling in one direction in a cavity configuration as depicted in Fig. 1.

We describe the interaction of the laser field with the active medium in the rate equation approximation. The bandwidth-limiting properties of the whole cavity are represented by an intracavity Lorentzian filter element with the bandwidth $\Delta\omega$. The mode-locker is described by the time-dependent amplitude transfer function T_A [7]. A round trip is completed by the linear losses $1 - R$ and the nonlinear losses (described by the power conversion coefficient $\varkappa = P_2/P_1^2$) and a shift in time of the laser pulse resulting from a possible mismatch between mode-locker frequency and intermode spacing of the linear laser cavity (h).

With these assumptions the system can be described by the following relation for the field amplitude E_L and the gain g of the amplifying medium:

Editors: E. Klose · B. Wilhelmi

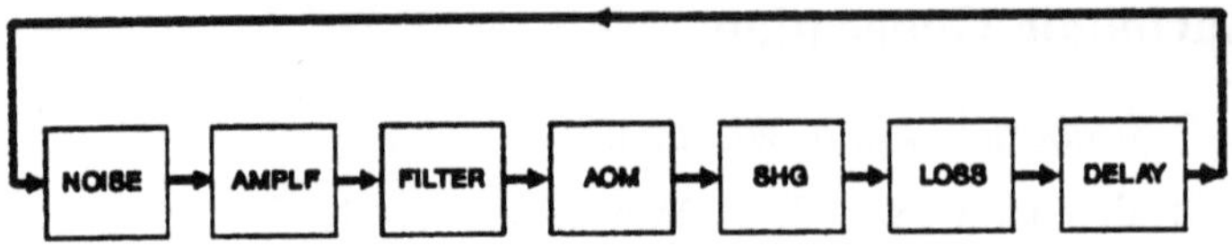

Fig. 1. Cavity configuration for the round trip model

$$E'_{\mathrm{L}}(\eta) = \sqrt{R}\sqrt{1 - \varkappa\varepsilon_0 c A\, T_{\mathrm{A}}^2(\eta+h)\,\mathrm{IFT}^2\{\sqrt{\mathrm{e}^{g(\eta+h)}}\,E_{\mathrm{L}}^2(\eta+h)\}}$$
$$\times\, T_{\mathrm{A}}(\eta+h)\,\mathrm{IFT}\{\sqrt{\mathrm{e}^{g(\eta+h)}}\,E_{\mathrm{L}}(\eta+h)\} + E_{\mathrm{sp}}$$
$$\frac{\partial g(\eta)}{\partial \eta} = \sigma_{32}\frac{\varepsilon_0}{2\pi\hbar}\lambda\, E_{\mathrm{L}}^2(\eta)\,(1+\mathrm{e}^{g(\eta)}) + \sigma_{32}Pnl - \left(P + \frac{1}{T_{32}}\right) g(\eta)\,,$$

where the notation is as follows:

η local time,
E_{sp} field amplitude of the spontaneously emitted light,
σ_{32} emission cross-section of the laser transition ($10^{-19}\,\mathrm{cm}^2$),
T_{32} energy relaxation time of the laser transition (4.8 ns),
A beam cross-section in the active medium ($1\,\mathrm{mm}^2$),
n_3 occupation density of the upper laser level,
n_0 particle density of the active medium ($2.7 \times 10^5\,\mathrm{m}^{-3}$),
n sum of all occupation numbers ($0.01 \ldots 0.001 \times n_0$),
R effective power reflection coefficient (0.8),
λ wavelength of the laser transition (514 nm),
$\Delta\omega$ homogeneous line width (40 GHz),
l length of the active medium (1 m),
ε_0 permittivity of free space,
c vacuum velocity of light,
P pumping rate,
$\hbar$ Planck's constant.

$g = g(\eta, l) = \sigma_{32}\int_0^l n_3(\eta)dz$,

IFT is an inverse Fourier transform:

$$\mathrm{IFT} = \mathrm{IFT}\{f(t)\} = \frac{\Delta\omega}{2}\int_{-\infty}^{t} \mathrm{e}^{-\Delta\omega(t-t')} f(t')dt'\,.$$

The values in parentheses correspond to the experimental data of an actively mode-locked argon ion laser with an ICSHG crystal. For an ADP crystal $\varkappa$ is about $10^{-3}\,\mathrm{W}^{-1}$.

In our simulation we drop the usual assumptions of a given (Gaussian) initial pulse shape [7–9] and self-consistency [7, 10] or periodicity [11] of the pulse envelope. The emission of statistically distributed background light (E_{sp}) in the active medium as well as phase modulation is taken into account. Therefore besides the determination of the steady-state SHG pulse characteristics the investigation of the pulse formation process and the influence of spontaneous emission becomes possible.

3. Results of Computer Simulation

The essential results of the computer simulation are:

1. The pulses evolve to their steady-state parameters within about 700 round trips (Fig. 2). Therefore the following results refer to pulse parameters after at least 700 cavity round trips.
2. The shape of the fundamental pulse becomes more asymmetric for greater conversion efficiency. Due to the lengthening of the trailing edge of the pulses ($t.2$) the pulse duration increases whereas the rise time $t.1$ remains nearly independent of the conversion efficiency (see Fig. 3). Only for negligibly small conversion values is the pulse nearly Gaussian shaped. This is in good agreement with other analytical or numerical treatments [8, 12, 13].
3. For an increasing conversion parameter the fundamental peak power converges to a lower limit. As a consequence, the average power of the second harmonic is saturated for greater conversion efficiency due to the increasing fundamental pulse width (Fig. 4).
4. Spontaneous emission cannot be neglected in the round trip model: The pulse maximum is shifted to earlier times and the peak intensity is reduced (Fig. 5).
5. The optimum temporal position of the pulse peak does not coincide with the maximum of transmission of the mode-locker. There is a considerable timing mismatch for getting the shortest pulses or the maximum pulse energy (Fig. 6).

To prove the results of the computer simulation we have performed experiments with an actively mode-locked internally frequency-doubled argon-ion laser. Different values of the power conversion coefficient could be realized by variation of the temperature of a temperature-tuned SHG crystal. A comparison of fundamental and theoretical results will be published elsewhere.

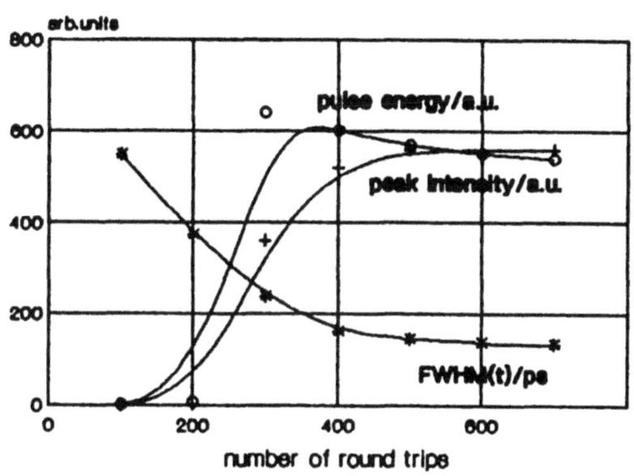

Fig. 2. Stability of solution

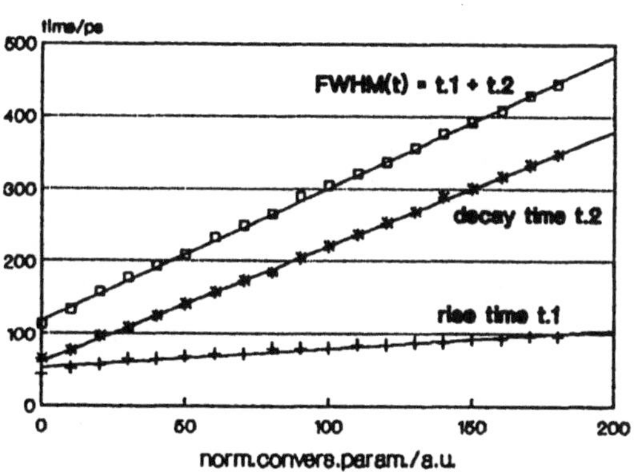

Fig. 3. Pulse duration depending on conversion

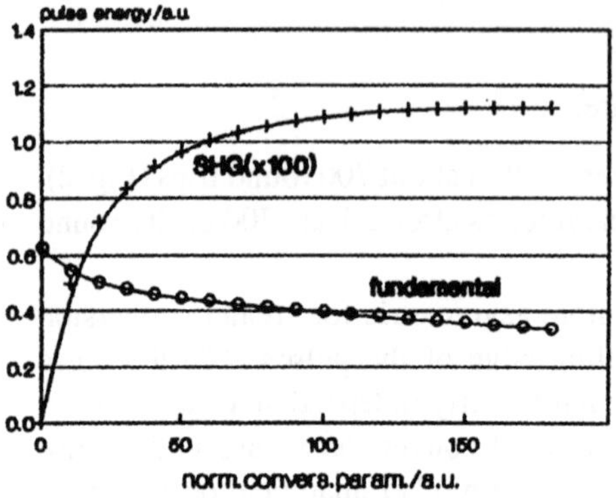

Fig. 4. Fundamental and SHG power depending on conversion

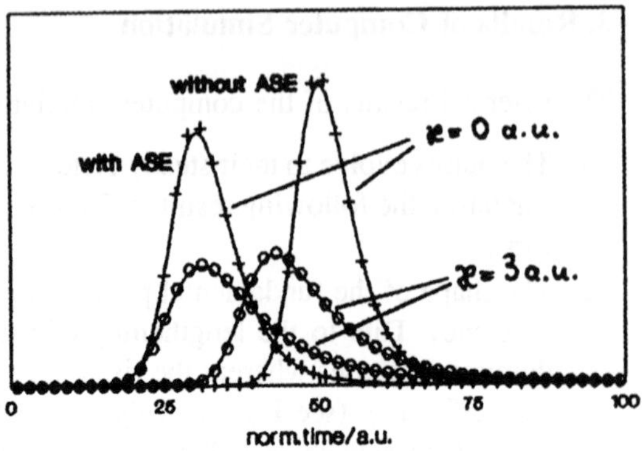

Fig. 5. Influence of ASE

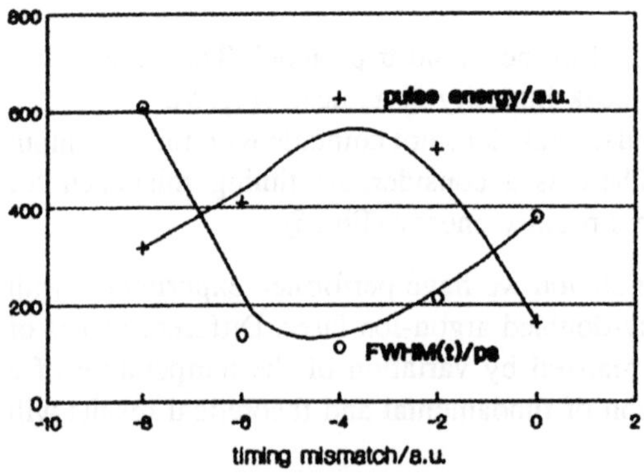

Fig. 6. Influence of detuning

References

1 H. Urabe, M. Tsukakoshi, T. Sugiyama, Y. Yazaki, T. Kasuya: Jpn. J. Appl. Phys. **25**, L937 (1986)

2 S.E. Clark, P.C. Emmony: Phys. Rev. B **40**, 2031 (1989)

3 S. Szatmári, F.P. Schäfer: In *Ultrafast Phenomena VI*, ed. by T. Yajima, K. Yoshihara, C.B. Harris, S. Shionoya, Springer Ser. Chem. Phys., Vol. 48 (Springer, Berlin, Heidelberg 1988)

4 G.D. Boyd, D.A. Kleinman: J. Appl. Phys. **39**, 3597 (1968)

5 D. Welford, W. Sibbett, J.R. Taylor: Opt. Commun. **35**, 283 (1980)

6 U. Stamm: Appl. Phys. B **45**, 101 (1988)

7 D.J. Kuizenga, A.E. Siegman: IEEE J. QE-**6**, 694 (1970)

8 J. Falk: IEEE QE-**11**, 21 (1975)

9 A.E. Siegman, J.M. Heritier: IEEE J. QE-**16**, 324 (1980)

10 T.R. Zhang: IEEE J. QE-**24**, 1877 (1988)

11 M. Piche, P.A. Belanger: Opt. Commun. **35**, 137 (1980)

12 M. Yamashita, K. Yamada, T. Sato: IEEE J. QE-**18**, 95 (1982)

13 K.P.J. Reddy: J. Appl. Phys. **53**, 7076 (1982)

Design Considerations for a Linear Cavity Femtosecond Dye Laser

J.J. Korpershoek, E.W. Castner, Jr., and D.A. Wiersma

Department of Physical Chemistry and Ultrafast Spectroscopy Laboratory,
University of Groningen, Nijenborgh 16, 9747 AG Groningen, The Netherlands

We have designed and constructed a linear cavity femtosecond dye laser. Theoretical modeling of the laser resonator was done using computer optimization based on the ABCD matrix formalism for Gaussian optics.

1. Introduction

Femtosecond pulse dye lasers have proven to be of great importance in solving problems in chemistry, biology, and solid state physics, as the other articles in this volume illustrate. We have undertaken the design and construction of a linear cavity, hybrid modelocked femtosecond dye laser to be used in nonlinear coherence spectroscopy.

One reason for choosing to build a linear femtosecond laser was the success obtained by Kubota et al. [1] in generating 29 fs pulses from a hybrid modelocked laser. The linear laser resonator is the simplest to build and optically align, and is most amenable to theoretical modeling. It is easiest to achieve the synchronous pumping condition with the linear laser. Also, the linear laser allows for simple insertion and removal of a third focusing fold. Such a third fold can be used for intracavity frequency doubling, or for a double-pass cavity dumper.

2. Theoretical Modeling of a Linear Resonator

We have designed a laser resonator using six intracavity mirrors, including two focusing folds for the gain and absorber media. A schematic diagram of our laser is shown in Figure 1A. Tunable net GVD is provided for by a four prism sequence [2]. Four prisms were used in place of two because the four prism sequence allows us to take the output beam from either end of the laser without need for spatial dispersion compensation. The ratio of the radii of curvature for the absorber fold mirrors to that of the gain fold mirrors was chosen to be 0.5 so that the saturable absorption necessary for hybrid modelocking could be increased.

A linear resonator can be modeled by an equivalent periodic lens sequence. The ABCD matrix for one complete round trip of the resonator equals the ABCD matrix for one period of the lens sequence. The criteri–

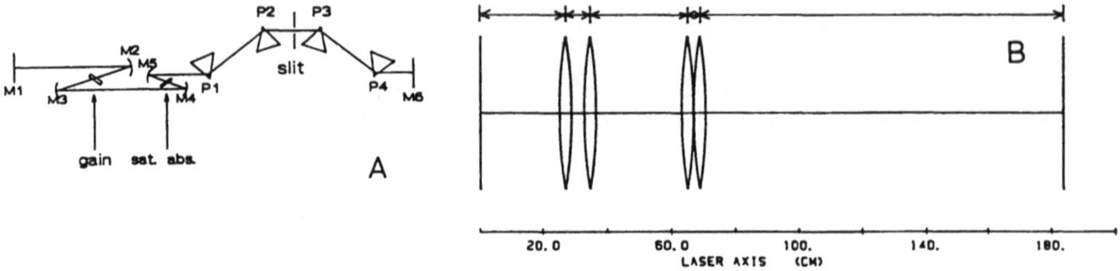

Figure 1. A. Schematic diagram of the linear resonator femtosecond dye laser. B. Equivalent lens sequence for the resonator in A.

on for a stable resonator is $|\frac{1}{2}(A+D)| \leq 1$, where here A and D are the elements of the resonator ABCD matrix.

The curvature of the wave front $R(z)$ and the beam radius (or spot size) $\omega(z)$ (1/e(max.amp.)) of the TEM_{00} mode are given by the beam parameter $q(z)$

$$\frac{1}{q} = \frac{1}{R(z)} - i\frac{\lambda}{\pi\omega^2(z)}. \tag{1}$$

The beam parameter can be determined at any position from

$$q(z_2) = \frac{Aq(z_1) + B}{Cq(z_1) + D} \tag{2}$$

where A,B,C and D are the elements of the ABCD matrix for the optical system between the points z_1 and z_2. If the beam parameter at one of the points is known, or if the optical system between points z_1 and z_2 is equal to one period of the lens sequence ($q(z_1)=q(z_2)$), then Equation 2 can be solved. The location and the spot size of a focus, or waist, can be found from the fact that the beam parameter at a focus must be purely imaginary. With the method described above, we can calculate many useful parameters of the laser beam both inside and outside the cavity: stability zones, beam contours, beam waists and divergence, and astigmatism. Overall astigmatism compensation may be achieved using the prescription of Kogelnik et al. [3] by choosing the appropriate angle of incidence on the folding mirrors. Using the model given by Magni et al. [4], the sensitivity to angular misalignment may also be calculated.

The astigmatism resulting from the four prism GVD compensator is ignored in our model; in practice this astigmatism is minimized by sending the beam through only the apex sections of the fused silica Brewster prisms. We have chosen to calculate all of the above parameters directly for all stable resonator configurations rather than use the complicated but analytic procedure of reducing the given laser cavity to an "empty resonator" [5]. This has allowed us to write a completely general com–

puter program that can be used to optimize a generic resonator. These numerical tasks are accomplished using a PC/AT type personal computer with a numeric coprocessor.

3. Theoretical Results and Discussion

We now present the results for a linear laser resonator obtained using the above model. In all of our calculations and experiments, the radii of curvature for the mirrors used was $R_2=R_3=7.5$ cm, and $R_4=R_5=3.75$ cm. The output coupler and end mirror were plane. The gain and absorber jet streams were positioned at about 30 cm and 67 cm from the output coupler. The cavity length L was chosen to be 183 cm to satisfy the length requirement for fundamental order synchronous pumping at about 82 MHz. The cavity length $L=d_{12}+d_{23}+d_{34}+d_{45}+d_{56}$ is held fixed to meet the synchronous criterion. As d_{23} and d_{45} are varied, the longest segment of the cavity d_{56} is changed to compensate. Though the results to be presented are for a specific choice of mirrors, our results are generally applicable to any linear resonator with two focusing folds.

To optimize our resonator, we vary the separation of the gain fold mirrors, d_{23}, and/or the separation of the saturable absorber fold mirrors, d_{45}. The stability zones for the laser shown in Figure 1 are plotted in Figure 2 as a contour plot of the focusing mirror separations d_{23} versus d_{45}. We have arbitrarily named the stability area on the left zone I, that on the right zone II. In Figures 3A and 3B, the beam con-

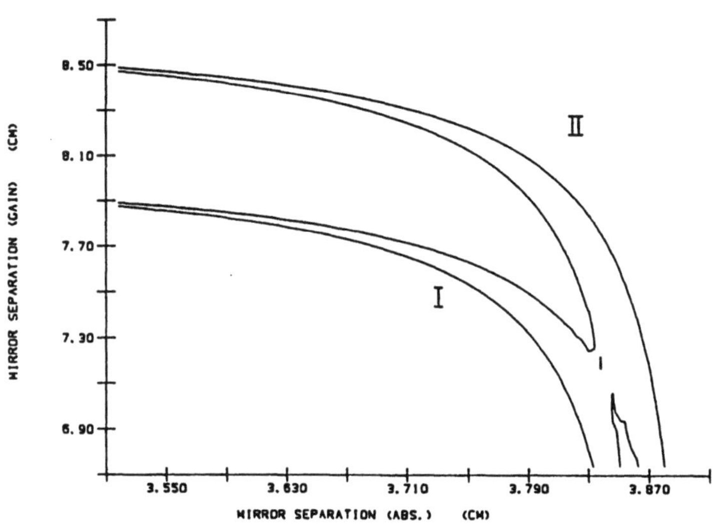

Figure 2. Stability contours for the resonator shown in Figure 1B as a function of gain mirror separation d_{23} versus absorber mirror separation d_{45}. Stability zone I is on the left, zone II on the right.

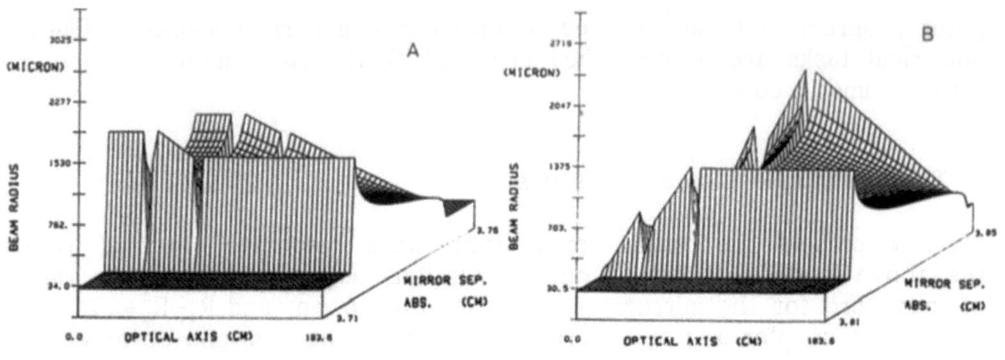

Figure 3. A. Beam contour plot in stability zone I, as a function of absorber fold separation d_{45}. B. Beam contour for zone II.

tour in the cavity is plotted as a function of the absorber mirror separation d_{45}, while the gain mirror separation d_{23} is held fixed at 7.6 cm. The sensitivity of the resonator to angular misalignment, shown in Figure 4, is several times greater in zone I than in zone II. These results are in complete agreement with those of DeSilvestri, et al. [6], in addition to our own experimental results with a single-fold dye laser.

Considering stability zone II, in Figure 3B, we find that the beam comes to a focus on one or both end mirrors. To use stability zone II, the output coupler should be placed at the other end of the cavity (at 183 instead of 0 cm), to obtain a low divergence output beam. Use of stability zone II minimizes the effects of imperfect alignment and mechanical vibrations. Also, a resonator that is less sensitive to mirror misalignment will also be less sensitive to pointing fluctuations of the pump laser, sometimes a severe problem with cw-modelocked Nd^{3+}-YAG lasers.

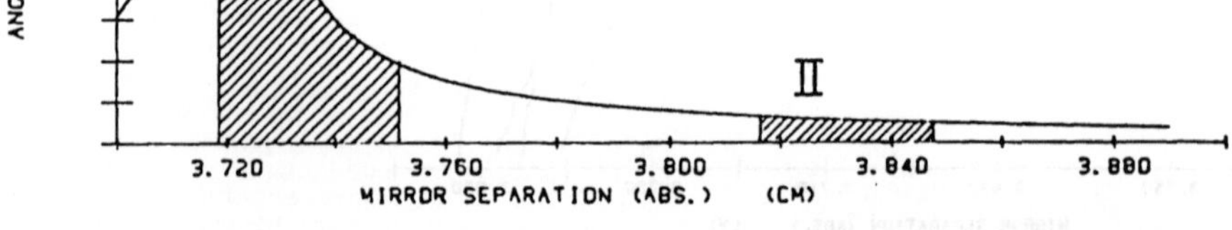

Figure 4. Sensitivity of the resonator to angular misalignment as a function of absorber fold separation d_{45}, for both zones I and II.

4. Experimental Details and Results

We have constructed the linear femtosecond laser described above. Kiton Red S was used as the gain dye and a combination of DODCI/DQOCI for the saturable absorber; both gain and absorber were in ethylene glycol solution. The gain jet nozzle was 300 μm, and that of the absorber either 50 or 100 μm. The lasing threshold was about 650 mW, with optimum performance obtained with 800–900 mW of 532 nm pump light from a frequency doubled cw-modelocked Nd^{3+}–YAG laser operating at 82 MHz. The quarter-wave (single stack) dielectric cavity mirrors were obtained from Laser-Optik, coated for 620 nm center wavelength. The laser bandwidth was restricted by varying the position of a variable slit placed between the second and third prisms in the laser, shown in Figure 1A. As explained in reference 1, this was essential to limit the laser gain to the spectral region where the saturable absorber mixture works best in order to obtain femtosecond operation. We were not able to obtain fs-laser operation at 615 nm, but rather at 635 nm. The laser pulse autocorrelation is shown in Figure 5A, with the corresponding laser spectrum in Figure 5B. With a 2% transmission output coupler, we obtained 10 mW average power output with usual pulse durations in the range of 60–65 fs (55 fs shortest), assuming the canonical $sech^2$ form. Our laser was operated only in stability zone I to date, and suffered from fluctuations of about 10% in the intensity of the output, with corresponding spectral variations. These were completely correlated with the pointing instabilities of our Nd^{3+}–YAG laser output. We are planning to improve the performance of our

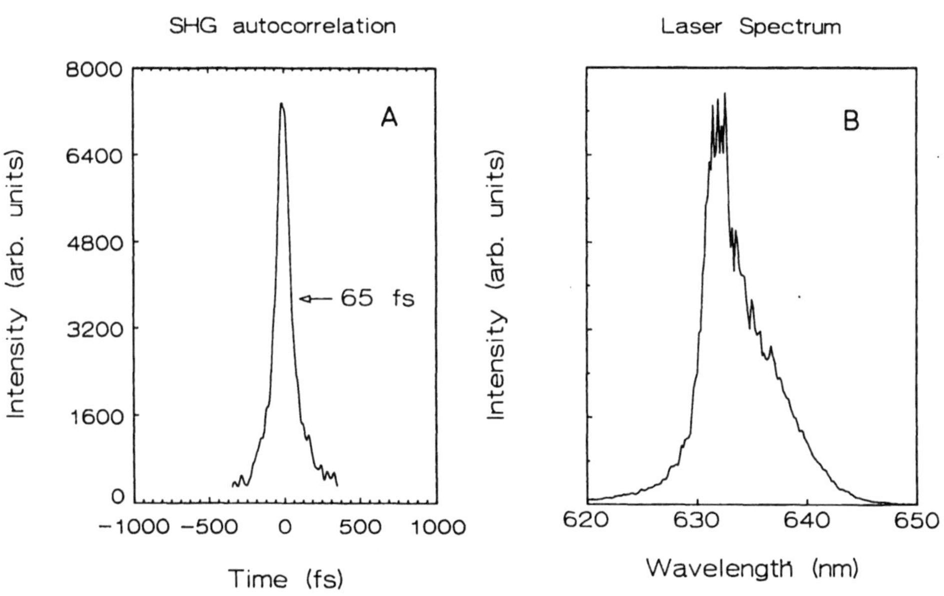

Figure 5. A. SHG intensity autocorrelation of the laser pulse, measured using a 200 μm KDP crystal. B. Corresponding spectrum of laser output.

laser by rebuilding it with R_{gain}=10 cm and $R_{absorber}$=5 cm focusing mirrors to allow us to attain the best compensation for astigmatism, and to use the laser in stability zone II. Also, active stabilization by the scheme of Chesnoy and Fini [7] will also be essential.

References

[1] H. Kubota, K. Kurokawa, and M. Nakazawa, Opt. Lett. **13**, 749 (1988).

[2] R. L. Fork, O. E. Martinez, and J. P. Gordon, Opt. Lett. **9**, 150 (1984).

[3] H. Kogelnik, A. P. Ippen, A. Dienes, and C. V. Shank, IEEE J. Quant. Electron., **QE-8**, 373 (1972).

[4] V. Magni, J. Opt. Soc. Amer. **A**, **4**, 1962 (1987).

[5] K. S. Budil, I. A. McIntyre, and C. K. Rhodes, Opt. Comm. **64**, 279 (1987).

[6] S. DeSilvestri, L. Yu-Pu, V. Magni, and O. Svelto, in Ultrafast Phenomena VI, Springer (Berlin), 33, (1988).

[7] J. Chesnoy and L. Fini, Opt. Lett. **11**, 635 (1986).

Ultrashort Pulse Generation from the Sweeping Oscillator Dye Laser

Nguyen Dai Hung, Y. Meyer, M. Martin, and F. Nesa*

Laboratoire de Photophysique Moléculaire du CNRS, Bât. 213,
Université Paris Sud, F-91405 Orsay Cedex, France
*Permanent address: Institute of Physics, Academy of Sciences of Vietnam,
Nghia do, Tu Liem, Hanoi, Vietnam

Abstract. We review recent progress in generation of high power ultrashort pulses by a new technique of pulse shortening from 10 ns to subpicoseconds. In a first step, the sweeping oscillator technique is used to obtain 100 ps. Further compression is obtained in a second step by propagating structured 100 ps pulses emitted by a microcavity in extracavity saturated amplifying/absorbing dye solutions.

1. Introduction

Femtosecond and picosecond pulsed lasers are useful light sources in ultrafast spectroscopy. High power, ultrashort laser pulses are generally produced at a few wavelengths, by using mode locking techniques. In practice these methods require the use of two expensive pump lasers (one CW, one pulsed). Recently, some other methods of generating short laser pulses have appeared, using cavity transients, distributed feedback, cavity competition or spectral sweeping processes. In this report we review and present progress in a new method for obtaining high power ultrashort laser pulses starting from a single standard ns pump laser.

In the first part of this paper we present the principle of the sweeping oscillator method and some novel results recently observed. This method, when applied to a short and low Q cavity, allows one to shorten an initial ~10 ns pump pulse by two orders of magnitude. In the second part, we describe the shaping of a structured 100 ps input pulse by extracavity propagation in amplifying and absorbing, saturable dye solutions which can lead to subpicosecond pulse generation.

An interesting application of such a high power subpicosecond pulse system is to generate in a simple way, a spectrally broad ultrashort continuum from UV to IR.

2. Spectro-temporal selection (STS) method

Y.H. Meyer and P. Flamant first presented the concept of intrinsic spectro-temporal evolution in an untuned dye laser cavity [1]. Theoretical analysis of the spectral dependence of intensity transients was studied in a multimode pulsed dye laser to describe the evolution of the laser spectra during the pulse

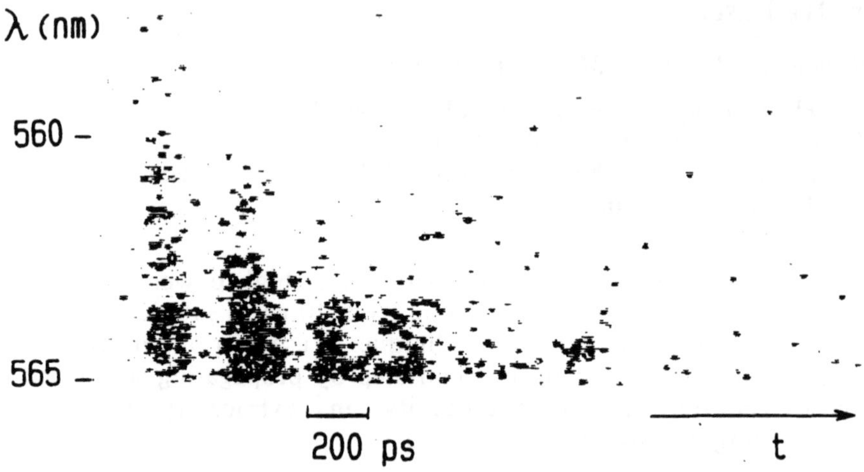

Fig. 1. Time-wavelength analysis of the 1 cm sweeping oscillator emission (two-dimension detection streak camera ARP)

emission. The relative intensity of any cavity mode (or group of modes) is a function of time. The driving force of the spectral sweep is the interaction between the intracavity instantaneous laser broadband intensity and the instantaneous broadband gain. As a result a definite spectro-temporal evolution process develops. The spectral sweep rate can be made very fast in short and low-Q cavities. This effect can be used to generate short pulses by sending the sweeping (untuned) oscillator output through a spectral filter tuned to an adequate region to obtain at once both a narrow line and a short pulse.

Spectral sweeps have been observed for broadband untuned dye oscillators with roundtrip times shorter than the pumping pulse duration. An example of spectral sweep was reported in [2] in the output intensity of a multimode 200 μm Rhodamine 640 oscillator pumped by a 120 ps (FWHM) pulse. The spectral sweep within the well resolved longitudinal modes was observed with a streak-camera after spectral filtering. A better observation of the spectro-temporal structure of the 200 μm cavity emission was measured with a two-dimension analysis in a single laser shot [3]. A single 4ps peak is emitted by a few modes.

Another example of spectral sweep is shown in fig. 1 in the case of a 1 cm cavity pumped by a 10 ns pulse. The image obtained with a two-dimension streak-camera (ARP - 50ps/mm), clearly shows that generation of a short and single pulse can be obtained by filtering in the blue wing. Such a cavity is used in the pulse shortening system described below.

3. Pulse shortening system from 10 ns to 100 ps

The complete pulse compression system is shown in fig. 2. The sweeping oscillator O_1, which is simply a standard 1 x 1 cm spectroscopy cell containing a solution of Rh6G in ethanol (4.10^{-4} M/l), is excited by a small part (less than 0.5 mJ - 10 ns - 532 nm) of the output from a standard Q-Switch 10 Hz Quantel Nd:YAG laser, through a cylindrical lens CL of 5 cm focal length. The output from the sweeping oscillator is collimated by a 20 cm lens and diffracted in the first order of a 2400 g/mm grating G_1. The spectrum is formed with a 50 cm lens on a screen with a 1 mm slit S. The transmitted beam is amplified by a dye multipass amplifier (MPA_1) pumped by the same Nd:YAG laser.

The amplifier MPA_1 is a 1 mm path cell, longitudinally pumped with 2 mJ of the Nd:YAG at 532 nm through a 10 cm f.l. lens. The path from O_1 to MPA_1 is long enough so that the green frequencies emitted at the beginning of the pump pulse are delayed by 5 ns in order to be amplified by the maximum of the pump pulse.

The effect of tuning the spectral filter G_1 is presented in fig. 3. When the grating is tuned close to the maximum of the oscillator broadband spectrum, the pulse transmitted through the slit S is several ns long. When the grating is tuned to the blue wing of the oscillator broadband spectrum, the transmitted pulse becomes shorter and around ~560 nm, a single peak of 100-200 ps is transmitted.

After amplification, the output pulse energy at ~560 nm is measured to be 20 μJ and the pulses observed with a streak-camera (50 ps/mm, resolution 4

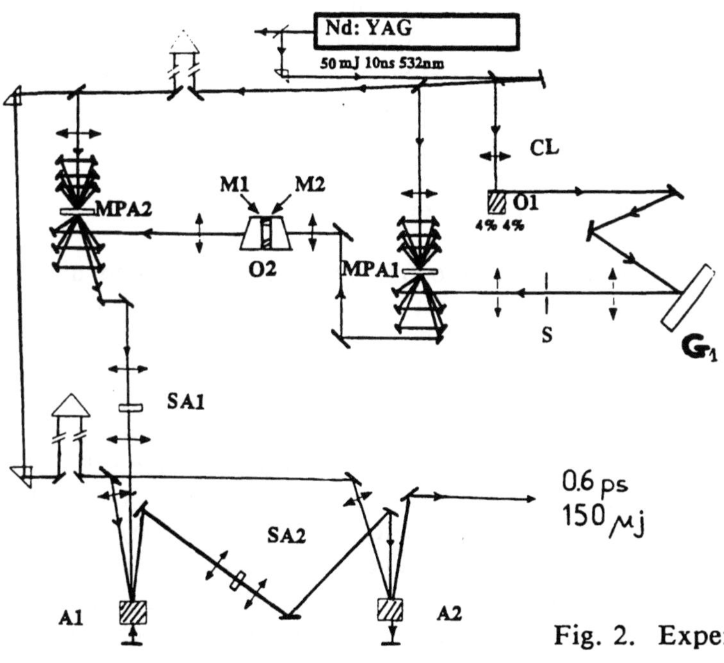

Fig. 2. Experimental set-up

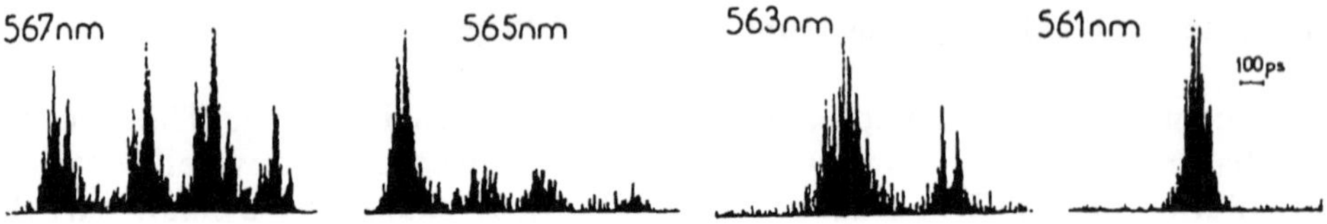

Fig. 3. Pulse shape of the filtered sweeping oscillator emission (streak camera ARP 50 ps/mm)

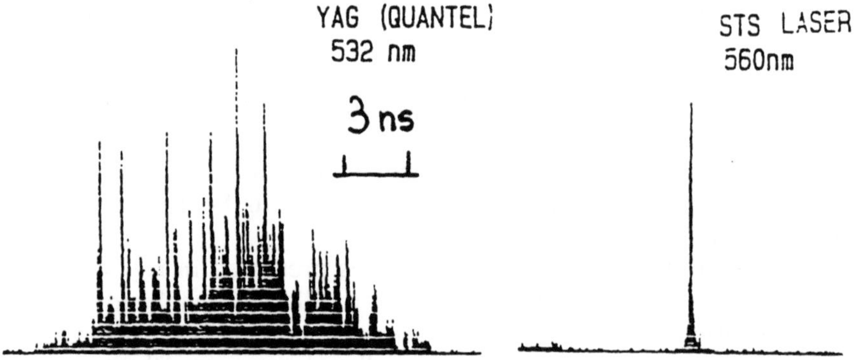

Fig. 4. The STS dye laser pulse and the Nd:YAG laser pump pulse (streak camera 1ns/mm)

ps) have durations of ~100 ps (FWHM). Fig. 4 shows the amplified single pulse as compared to the typical irregular spikes of the pump pulse at 532 nm observed with a streak-camera (1 ns/mm speed).

Thus the spectro-temporal selection (STS) method, i.e. spectral selection in the blue wing of the time-dependent broadband laser spectrum, results in production of a narrow band short pulse.

4. Pulse shortening system from 100 ps to 1 ps

In order to produce shorter pulses, the STS method was applied again to another sweeping oscillator of shorter cavity length (200 μm) pumped with the first stage output pulse (100 ps) [3] and a nearly transformed limited pulse of 4 ps was reported.

A way to produce shorter pulses was proposed recently [4] and is reported here in more detail. It is based on the shaping of a spiked input pulse by propagation in extracavity, saturable, amplifying/absorbing, dye solutions.

The 100 ps pulses obtained by the STS method at 561 μm are focused with a 10 cm-lens on a multimode low-Q microcavity O_2 (length 40 μm) filled with R640 in methanol (~2 10^{-2} M/l) (fig.2). The mirror reflectivity R_1 is ~0.1. The Fresnel reflection coefficient R_2 ~10^{-3} between the silica plate and the alcohol appears to be sufficient to initiate laser emission, well characterized

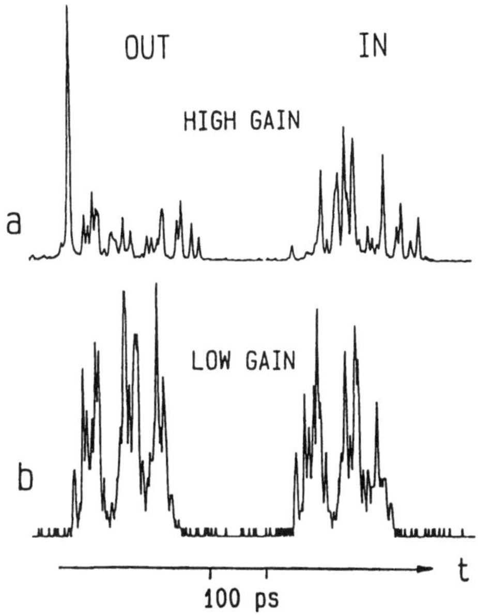

Fig. 5. a: Pulse shaping by saturated amplification (the input and output pulses are delayed to be simultaneously recorded with a one-dimension streak camera ARP, 50 ps/mm); b: Same as in a but with low gain

by the longitudinal mode pattern of the time integrated spectrum (shown in fig. 6b). The cavity mode spacing is 3.6 nm.

The microcavity output is collimated with a 30 mm f.l. lens and sent into the multipass amplifier (MPA_2) cell (1mm path with anti-reflection coatings on the external faces) which contains a recirculating solution of R640 in ethanol ($C = 5.10^{-4}$ M/l). The pump beam (532 nm - 10ns from the Nd-YAG laser) is focused with a 10 cm f.l. lens on a 1 mm diameter spot.

The role of saturable amplification in the shaping of a structured input pulse is evidenced in Fig. 5 by comparison of the pulse shapes before (in) and after (out) amplification in the MPA_2 for different pump energies of the amplifier cell. For the shot presented in trace a, which corresponds to the high gain case (3 mJ pump energy), the first spike of the amplified pulse dominates and the energy gain for the whole pulse can reach 10^3. With low gain (trace b), no gain saturation occurs.

The output beam from the MPA_2 is focused with a 5 cm lens onto a saturable absorber SA_1 (1mm path cell filled with Malachite Green (MG) in methanol). The non-exponential recovery time of the MG absorption is known to have a long decay component of 2 ps in methanol, i.e. is smaller than the interval between two spikes. Adequate focusing of the MPA_2 output in an optically thick MG saturable absorber cell ($D_0 = 7.5$) wipes out the spiked tail and a single pulse is obtained. It is to be noted that the MG absorber cell also completely eliminates the MPA amplified fluorescence background 10 ns pulse.

5. High power subpicosecond pulses

The output pulse from the MG saturable absorber cell of about 1 ps duration and 1 μJ energy can be easily amplified by two orders of magnitude by means of additional amplifiers A_1 and A_2. The duration of the amplified pulse is estimated by noncollinear intensity autocorrelation in a 1 mm thick KDP crystal. Fig. 6 shows autocorrelation traces. In trace a, the microcavity is tuned for single mode emission. The spectral linewidth recorded is 0.6 nm (FWHM). The overall autocorrelation trace width is found to be of 1 ps or less. Assuming a $sech^2$ envelope shape, pulse duration of about 600 fs is calculated.

It is possible to switch easily from the one-mode to a two-mode emission spectrum by translating the slightly wedged cavity in its plane. In trace b of fig. 6, the microcavity is tuned for a two-mode emission. The total autocorrelation trace is modulated at the cavity roundtrip time T = 340 fs. This shows that a train of 150 fs pulses is emitted with a repetition rate of 3 THz.

By focusing the subpicosecond or femtosecond-modulated pulse in a 1 cm cell of water, an intense continuum from 350 to 1100 nm is generated; this continuum is generally assigned to several nonlinear mechanisms such as self-phase modulation. The white light was collimated and its spectrum was recorded with a polychromator equipped with an SSR.OMA. Comparison between the continuum generated by a single mode pulse and by a two-mode pulse with the same energy showed that the intensity of the continuum is 2-3 times stronger with the modulated pulse [5].

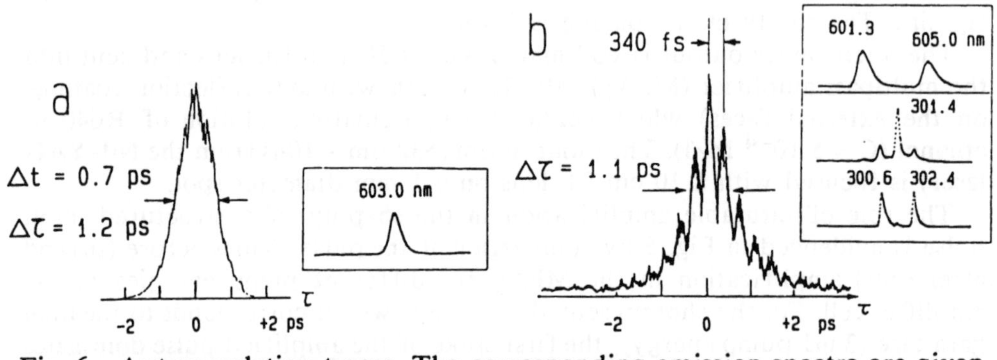

Fig. 6. Autocorrelation traces. The corresponding emission spectra are given in the inserts. **a**: Single mode microcavity emission; **b**: two mode case. The two UV spectra correspond to double and sum frequencies detected after the KDP crystal at two different delays τ separated by 170 fs

In conclusion, we report progress in the generation of ultrafast and powerful pulses by the sweeping oscillator method and pulse compression by extracavity propagation in saturable dye solution media. Such a technique presents several potential advantages. It can be applied to a non-mode-lockable laser. The STS method has a great efficiency in pulse compression.

The possibility of production of femtosecond pulses at very high repetition rate is straightforward. High power subpicosecond pulses can be obtained in a simple way with a single ns pump laser with no electronic synchronisation system.

References

[1] Meyer Y.H., Flamant P., Opt. Commun., 19 20-24 (1976).
[2] Meyer Y.H., Benoist d'Azy O., Martin M.M., Breheret E., Opt. Commun. 60, 64 (1986).
[3] Meyer Y.H., Martin M.M., Breheret E., Benoist d'Azy O., Springer series in Chem. Phys. 46, Ultrafast phenomena V, Ed. G.R. Fleming and A.E. Siegman p. 89 (1986).
[4] Martin M.M., Nesa F., Breheret E., Meyer Y.H., Springer series in Chem. Phys. 48, Ultrafast phenomena VI, Ed. T. Yajima, K. Yoshihara, C.B. Harris, S. Shionoya p. 473 (1988).
[5] Meyer Y.H., Martin M.M., Nesa F., Breheret E., J. Phys. 48, C7-12, 397 (1987).

Ultrastable Subpicosecond Nd:Glass Laser and Stretched Pulse OPOs: A New Approach to High Intensity Tunable fs Light Fields

R. Danielius, R. Grigonis, A. Piskarskas, D. Podenas, V. Sirutkaitis, and A. Varanavichius

Laser Research Centre, Vilnius University, Sauletekio Ave. 9, 232054 Vilnius, Lithuania

Abstract. We report on the generation of widely tunable light pulses with a duration 0.3-0.9 ps by optical parametric oscillators pumped by subpicosecond pulses of a passively mode-locked Nd:glass laser with inertial negative feedback.

1. Introduction

The fundamental stage in the development of femtosecond spectroscopy is the construction of femtosecond lasers with continuously tunable emmission wavelengths in a wide spectral range. Optical parametric oscillators (OPOs) can be tuned continuously over exceptionally wide intervals and occupy a special place among the tunable lasers. But femtosecond pulse generation in OPOs is connected with some difficulties and has been achieved in only a few cases. In earlier work /1,2/ the basic possibility of producing subpicosecond optical pulses using an OPO pumped by laser pulses with a duration of several picoseconds has been demonstrated . Nevertheless, the application of such sources was limited by the insufficient reproducibility of pulse characteristics. On the other hand, the parametric generation in a NPP crystal, pumped by a subpicosecond dye laser pulse , was seriously affected by the stimulated Raman scattering /3/ and was not successful. A singly resonant OPO based on a thin crystal of KTP pumped by intracavity femtosecond pulses from a standard colliding-pulse passively mode locked dye laser was reported /4/. Oscillation in this case results in a stable, continuous output of femtosecond pulses (220 fs) at a 10^8 Hz repetition rate and milliwatt average power levels in both signal and idler beams, making this a uniquely versatile femtosecond laser source.

In this paper we report on the generation of femtosecond pulses with high intensity in single pass and synchronous OPOs pumped by subpicosecond pulses of a passively mode-locked Nd:glass laser with inertial negative feedback.

2. Subpicosecond Nd:glass laser

In earlier publications complex cavity systems /5/ and saturable absorbers with an extremely fast recovery time /6/ were applied to generate subpicosecond pulses in passively mode-locked Nd:glass lasers. The main disadvantages of these methods are (i) large shot-to-shot variations of pulse parameters and (ii) pulse self-modulation due to a high peak intensity. During the past few years some groups /7-11/ have realized a new method for the control of the process of passive mode-locking in solid state lasers. The question concerns mode-locking in conditions of inertial negative feedback. The idea lies in limiting the pulse intensity to the level

Springer Proceedings in Physics, Vol. 49 Ultrafast Phenomena in Spectroscopy
Editors: E. Klose · B. Wilhelmi

of the saturable absorber saturation intensity, which extends the nonlinear regime of pulse formation and allows the generation of more stable and shorter ultrashort pulses (UPS).

It enables us to avoid self-focusing and self-phase modulation of pulses, too. We used this method of pulse stabilization and shortening in our passively mode-locked Nd:glass laser for subpicosecond pulse generation.

A conventional resonator with flat rear and output (R=50%) mirrors and a contacted dye cell for the passive mode locking were used. The laser rod was made from a Nd:phosphate glass GLS-23. The electro-optic feedback controller consisted of a thin film polarizer and a Pockels cell. The amplified output signal of fast photodiode, illuminated by part of a laser beam, was applied to the Pockels cell. As a bleachable absorber we used a solution of dye 3274 in methanol. The pulse train duration with an interpulse spacing of 9 ns depends on the initial dye absorption and ranges from 2 to 50 microseconds with the total energy of the train 0.8-0.3 mJ. The autocorrelation functions of pulses had lowest background at pulse train durations in the range 6-14 μs. The shortest pulses with duration 0.6 ps were achieved at the same pulse train durations. The standard deviation of a single pulse duration after 3.5 μs from the beginning of a train measured by the detuned autocorrelation method /12/ was 0.7 % at pulse durations less than 1 ps (Fig.1a). The pulses are transform limited up to an intensity of 100 MW/cm^2; at higher intensities some phase modulation was detected at the trailing part of the train. The maximal deviation of energy of a single pulse as well as of a whole train can be diminished to ± 6-8% by carefully choosing the value of the initial transmission of the saturable absorber and the depth of modulation of the feedback.

The experimental set-up used for the investigation of OPO is depicted schematically in Fig. 2. The train of the master oscillator (MO) is amplified in 5-stage of Nd:phosphate glass amplifier (AMP); the total amplification factor is about 1000. The electrooptic switch (EOS) isolates the master oscillator from the amplifier chain. It is activated by a laser triggered spark-gap with a delay time of 1.5-2 μs in order to cut off the leading part of the train. The amplifier stages are separated by spatial filters, a 7 mm diameter laser rod is used in the last amplifier. Up to a single pulse energy of 0.5 mJ no self-focusing was detected. It should be noted that due to gain saturation /13/ the train envelope is distorted, in particular its leading part (see Fig. 2.). The second harmonic is generated in a type I KDP crystal with an efficiency of 30%. The collinear pump beam

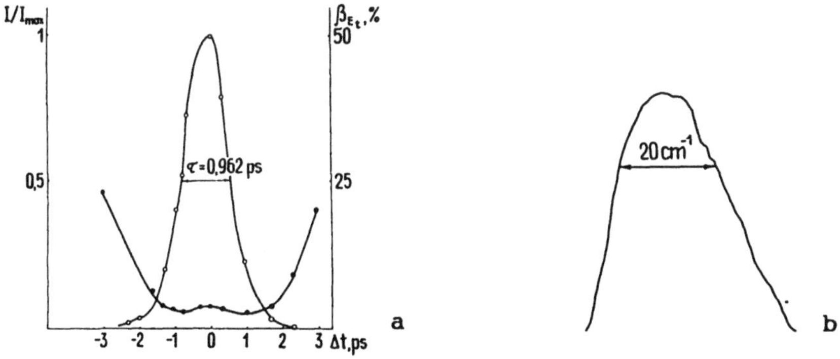

Fig.1. a) The autocorrelation function (∘) and variation coefficient of autocorrelator signal (·). b) pulse spectrum densitogram.

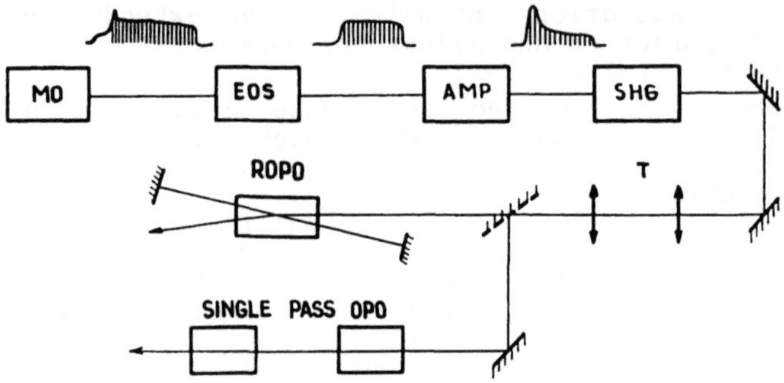

Fig.2. The experimental set up for the investigation of OPO pumped by the subpicosecond pulses.

is formed by a spherical telescope;the maximum intensity inside the OPO crystal is 100 GW/cm^2.

3. OPO pumped by subpicosecond light pulses

The resonator of resonant OPO (ROPO) consists of two flat aluminium-coated mirrors. The angle between the resonator axis and the pump beam is 2-3°. This configuration provides a high Q-factor for the resonant wave and enables one to dump the non-resonant one. Either a signal or an idler wave can be chosen as the resonant one by proper adjustment of the cavity length. $Ba_2NaNb_5O_{15}$, $LiNbO_3$, CDA, KDP and KD^*P have been tested as nonlinear crystals in this OPO. In $Ba_2NaNb_5O_{15}$ and $LiNbO_3$ crystals the parametric interaction was suppressed by large group velocity dispersion (GVD), no parametric oscillation was detected with pump intensity below the damage threshold. In the cases of both CDA and type I KDP crystals a broad-banded radiation (>600 cm^{-1} near degeneracy) was observed. For a type II phase matching KDP crystal the spectral bandwidth is <30 cm^{-1}. The tuning range limited by IR absorption in OPO is within 800-1500 nm for KDP and 750-1700 nm for KD^*P. The beam divergence of the ROPO is nearly diffraction limited. The output power of the ROPO depends resonantly on the cavity length. In general, synchroresonant curves have two peaks, corresponding to the signal and idler waves. The width of the peaks and the distance between them is proportional to the crystal length and, for the type I phase matching, depends on the operating wavelength. As for the type II phase matching, the shape of the synchroresonant curve changes a little within the tuning region.

An average energy conversion rate up to 8% for the signal wave was obtained for the 2.5 cm KDP and KD^*P crystals at a pump pulse intensity of 75 GW/cm^2, which corresponds to a 3-4 -fold threshold intensity. It should be noted that the conversion in the leading part of the train is considerably higher due to the higher intensity of the pump pulses.

Efforts have been made to find the conditions for a pulse shortening in ROPO. As shown in the work /2/, pulse shortening takes place when the group velocity mismatch and the parametric gain factor are in a definite relation. In this case the OPO output mirror of 90% reflectivity was used. We have measured the duration of the OPO resonant pulses for different crystal lengths at three cavity detuning ΔL values, corresponding to the maximum conversion rate (W_{max}) and levels of (0.2-0.3)W_{max} and (0.1-0.05)W_{max}. Fig. 3 depicts

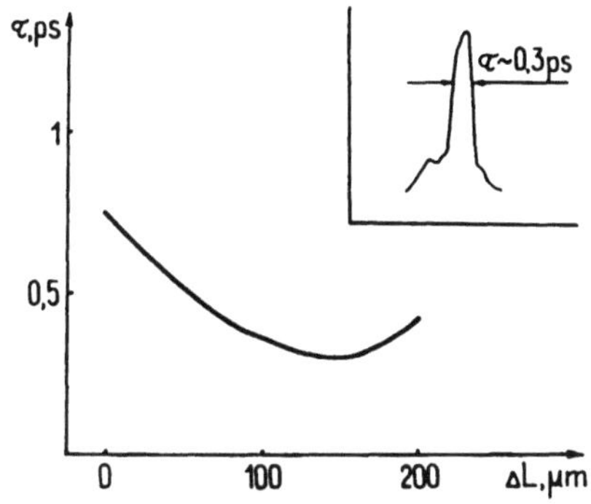

Fig.3. The OPO pulse duration dependence on cavity length detuning ΔL.

a plot of pulse duration vs resonator length detuning in the case of a 4 cm type I phase matching KDP crystal. The general trend of the pulse duration dependence on the resonator length detuning qualitatively remained the same for other crystal lengths (from 2.5 cm to 6 cm), however, the minimum pulse duration for the crystals both shorter and longer than 4 cm was larger. The pulse broadening up to 2-3 ps and its fragmentation was observed at a cavity length corresponding to the peaks of synchroresonant curve in 6 cm KDP crystal. Only slight pulse shortening, as compared with data /2/, can be explained by a larger influence of group velocity mismatch for subpicosecond pulses. The inset in Fig.3 shows the densitogram of the autocorrelation function trace of the shortest pulse detected.

In the case of a single pass OPO we used two type II phase matching KDP crystals spaced by 0.5 m. The optic axes of the two crystals were arranged in such a way that the spatial walk-off occurring in the first crystal was compensated by the second one. The pulse duration depends on both crystal length and pump intensity. The minimum pulse duration was found to be 0.3 ps using two 4 cm long crystals at a conversion efficiency < 5%. The pulse duration > 2 ps was observed in the case of saturating pump intensity and two 6 cm crystals. It was rather unexpected that a pulse shortening occurs without accurate adjustment of the signal and the pump pulse arrival at the second crystal /2/. A detailed investigation of the pulse duration dependence on the crystal length and the pump parameters should be carried out using a single pulse pump. The estimated value of the single pulse energy at the leading part of the train was ≃1 μJ. Both the beam divergence and the spectral bandwidth depend on the distance between the crystals. The bandwidth was less than 100 cm^{-1} for 0.5 m distance between the crystals. Our experiments show that in order to produce the parametric superradiance a longer crystal is needed than in the case of the ROPO. For crystals shorter than 4 cm the laser-induced breakdown threshold was lower than that of the parametric superradiance. In our case of a long train of subpicosecond pulses, the KDP crystal damage occurred at intensities higher than 1.5-2 TW/cm^2.

4. Parametric conversion of chirped light pulses

The limiting factor of parametric light conversion in the subpicosecond time domain is GVD /3/. As was mentioned earlier, no parametric oscillation was obtained in crystals with large GVD ($Ba_2NaNb_5O_{15}$, $LiNbO_3$). We verified experimentally the idea of avoiding the influence of GVD by pulse expanding (chirping) in a

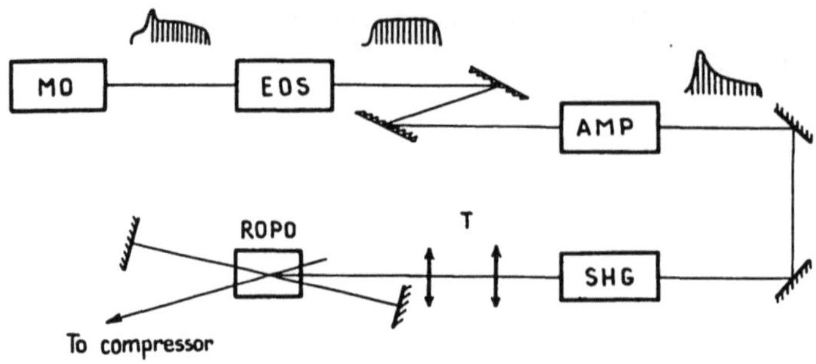

Fig.4. Experimental scheme for study of stretched pulse OPO.

linear dispersive medium before a parametric interaction. The scheme of the experiment is presented in Fig.4. The subpicosecond Nd:glass laser pulses before amplification were expanded by a pair of diffraction gratings (1800 lines/mm) from 0.7 ps to 65 ps. It should be noted that the pump chirp parameter in this case didn't depend on pulse intensity and was the same for all the pulses in the train. The ROPO with $Ba_2NaNb_5O_{15}$ was investigated. The train consisting of 50-60 parametric pulses was observed at the OPO output. The OPO energy conversion efficiency was $\simeq$ 14% at the pump intensity over 50 MW/cm^2.

It has been demonstrated that type I phase matching nonlinear crystals pumped by chirped pulses emit signal and idler radiation with chirp magnitude far exceeding that of the pump /14/. Under certain circumstances the duration of generated pulses can be reduced below the coherence time of the pump pulse by means of the conventional compression technique. In our experiment the chirp enhancement factor at signal pulse wavelength 1.01 micrometer was found to equal 9. The spectral bandwidth of the pulse linear frequency modulation measured by employing a dynamic interferometer system /15/ reached 170 cm^{-1}. This, in principle, assuming Gaussian pulse shape, allows us to compress such pulses to 90 fs duration. The positive GVD compressor /16/, consisting of two diffraction gratings (600 lines/mm) and a telescope with a unit magnification between them was used for the OPO signal pulse compression. Unexpectedly, a low compression rate was obtained and the minimum pulse duration in the middle of the train was measured to be 13 ps. An additional investigation showed that compression was limited by the presence of random phase spectral components. The compression rate of the chirped pulses can be expressed by /17/

$$N = [1 + g^2/1+(t/T)^2]^{1/2}$$

where g is the chirp parameter, t is the pulse duration before compression, and T is the pulse correlation time. The measured T value in our experiment was 0.13 ps and the calculated compression rate was in good agreement with the experimental results. In this way, in order to improve the compression efficiency, one must minimize random phase components e.g. by increasing the effective length of the parametric interaction.

The parametric amplification due to its high gain and broad bandwidth seems to be a very attractive technique to produce powerful fs pulses. As demonstrated in the above experiment pulse stretching enables one to overcome the limitations on parametric interaction efficiency imposed by GVD. A question remains whether the chirp distortion due to GVD does not make it impossible to compress the amplified stretched pulse down to the initial duration. In order to clarify this point the following experiment has been carried out. The pulses of master oscillator were expanded up to 20 ps. After amplification and frequency doubling a small fraction of remaining fundamental radiation was injected into the parametric amplifier (PA) (type II phase matching 6 cm KDP crystal) pumped by the second harmonic. Due to frequency doubling, the chirp slope ($\Delta\nu/\Delta\tau$) of the pump pulse must be larger by the factor of 2, as compared with the signal pulse. Therefore, the idler pulse chirp must be similar to the signal chirp according to the parametric conversion phase matching conditions.

The duration of the compressed idler pulse was 1.5 ps, which is close to the original master oscillator pulse duration of 1.4 ps, despite a 4.5 ps walk-off between the pump and idler pulses. The difference is attributed to the pump pulse shortening during second harmonic generation process.

5. Conclusions

We have demonstrated the parametric generation of widely tunable powerful subpicosecond pulses using the stabilized passively mode locked Nd:glass laser as a pump source. Such a system appears to be a promising base unit of subpicosecond spectrometer suitable for wide range of applications. It has been shown that appropriate stretching of ultrashort signal and pump pulse offers the possibility of efficient phase preserving parametric amplification even in the presence of large group velocity mismatch. Further applications of this method might include the pulse spectrum broadening in the fiber and subsequent parametric amplification in order to generate pulses in fs region. The combination of the fiber with the grating expander would allow the use of the (sub)nanosecond high energy pulses for PA pumping.

References

1. A.Fendt, W.Kranitzky, A.Laubereau, W.Kaiser: Optics Commun. **28**, 142 (1979).
2. B.Bareika, G.Dikchius, A.Piskarskas, V.Sirutkaitis, Y.Yasevichiute: Sov.J.Quant.Electron. **13**, 1507 (1983).
3. T.Ledoux, J.Zyss, A.Migus, J.Etchepare, G.Grillon, A.Antonetti: Appl. Phys. Lett., **48**, 1564 (1986).
4. D.Edelstein, E.Washman, C.Tang: Appl.Phys.Lett. **53**, 345 (1989).
5. A.Penzkofer: Opt. and Quantum Electron. **17**, 219 (1985).
6. B.Bareika, A.Piskarskas, G.Sinkyavichyus, V.Sirutkaitis: Sov.J. Quant.Electron. **14**, 407 (1984).
7. O.Martinez, M.Marconi, F.Diodati: J.Appl.Phys. **53**, 6673 (1982).
8. K.P.Komarov, A.S.Kuchyanov, V.D.Ugozhayev: Opt.Commun. **57**, 279 (1986).
9. A.M.Valshin, V.M.Gordienko, S.V.Krayushkin, V.T.Platonenko, V.K.Popov: Sov.J.Quant.Electron. **16**, 1125 (1986).
10. P.Heinz, W.Kriegleder, A.Laubereau: Appl.Phys.A **43**, 209 (1987).
11. K.Burneika, R.Grigonis, A.Piskarskas, G.Šinkyavichyus, V.Sirutkaitis: Sov.J.Quant.Electron. **18**, 1034 (1988).
12. A.Birmontas, R.Kupris, A.Piskarskas, V.Smil'gyavichyus, A.Stabinis: Sov.J.Quant.Electron. **8**, 1475 (1978).
13. P.C.Magnante: IEEE J.Quant.Electron. **QE-8**, 440(1972).

14. A.Piskarskas, A.Stabinis, A.Yankauskas: Sov.Phys.Usp. **29**, 879 (1986)
15. A.Piskarskas, A.Stabinis, A.Yankauskas: Sov.J.Quantum Electron. **14**, 1575 (1984).
16. O.E.Martinez: IEEE J.Quantum Electron. OE-23, 59 (1987)
17. A.Stabinis: Private communication.

Femtosecond Pulse Generation from a Titanium-Doped Sapphire Laser

P.M.W. French, J.A.R. Williams, and J.R. Taylor

Femtosecond Optics Group, Physics Department, Imperial College, London SW7 2BZ, UK

Abstract. This report concerns the optimisation and mode locking of a continuous wave titanium-doped sapphire laser. Pulses as short as 75ps duration were obtained through active mode locking by loss modulation. Upon adding a nonlinear external cavity, subpicosecond pulses were directly generated from a titanium-doped sapphire laser for the first time. Passive mode locking using nonlinear external cavity feedback, and mode locking with a linear external cavity have also been investigated.

The recent demonstration of c.w. laser operation of titanium-doped sapphire ($Ti:Al_2O_3$), tunable over ~400nm, is important for many areas of laser physics [1]. As a c.w. mode locked laser, it is a convenient spectroscopic tool for time-resolved studies of semi-conductors with a potential femtosecond time resolution. Owing to the relatively long upper-state lifetime (~3μs) and low gain cross-section (~$3\times10^{-19}cm^2$), the techniques of synchronous pumping and passive mode locking with a slow saturable absorber, which have been successfully employed to mode lock c.w. dye lasers, are not appropriate for $Ti:Al_2O_3$. Active mode locking through loss modulation, however, has generated pulses as short as 6ps [2] though pulse durations of several tens of picoseconds are more typical.

The performance of $Ti:Al_2O_3$ lasers is very sensitive to the quality of the laser crystal. For the work reported here, commercially available crystals (from Union Carbide) of nominal 0.1% concentration of Ti^{3+} were used. Initially the simple astigmatically-compensated cavity of Figure 1 was constructed. The folding mirrors (M_1, M_2) were of 15cm radius of curvature and lens L_1 was of 10cm focal length. The $Ti:Al_2O_3$ rod was 16mm long with Brewster-angled end faces and was mounted on a copper block, water cooled to 17°C. Pumping with the "all-lines" visible output of an argon ion laser, the laser threshold was 0.9W absorbed pump power and the laser exhibited a 16% slope efficiency for up to 6W pump power. For higher powers the laser efficiency dropped rapidly due to thermal lensing. In a later configuration employing folding mirrors of 10cm radius, a 22mm length rod which was cooled to 5°C maintained a 16% slope efficiency for up to 15W pump power. Figure 2 shows the tuning curve obtained

Springer Proceedings in Physics, Vol. 49 Ultrafast Phenomena in Spectroscopy
Editors: E. Klose · B. Wilhelmi

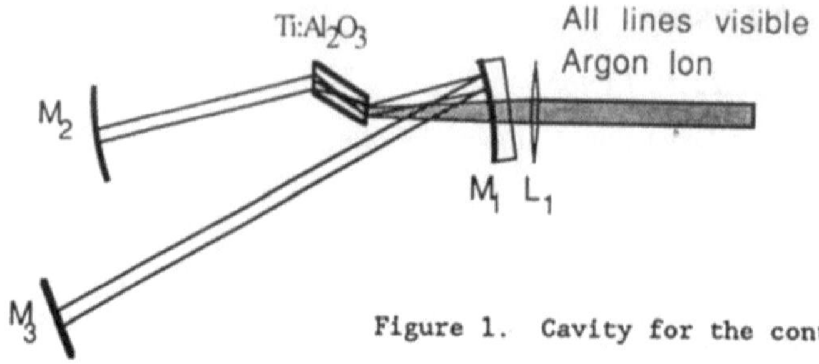

Figure 1. Cavity for the continuous wave Ti:Al_2O_3 laser.

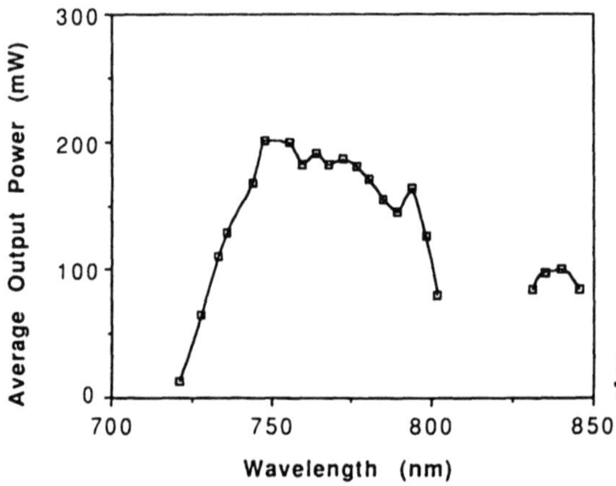

Figure 2. Tuning curve of the c.w. Ti:Al_2O_3 laser for 15W pump power.

for 15W pump power. The limited tuning range and the gap around 820nm is believed to be due to excited state absorption.

The laser was actively mode locked in the cavity shown in Figure 3. A standard Spectra Physics acousto-optic prism modulator driven with ~1W RF power yielded pulses as short as 75ps in duration at a repetition rate of 138MHz. With a 5% output coupler the laser tuned from ~720nm to 840nm (though not between 800nm and 830nm) and with a 20% output coupler the laser yielded a maximum average output power of 920mW. Pulse durations were routinely maintained below 100ps.

In order to generate subpicosecond pulses, the technique of nonlinear external cavity feedback was used, based on the theoretical and experimental work of Blow et al. [3,4]. An optimised 63cm length of single-mode polarization-preserving fibre (from York Technology) provided the nonlinearity in the resonant external cavity which was twice the length of the main cavity. Figure 4 shows the experimental configuration. A stabilization scheme based on the design by Mitschke and Mollenauer [5] was incorporated to interferometrically match the main and external cavity lengths. For an average power of 200mW in the fibre, subpicosecond pulses as short as 770fs were obtained for the first

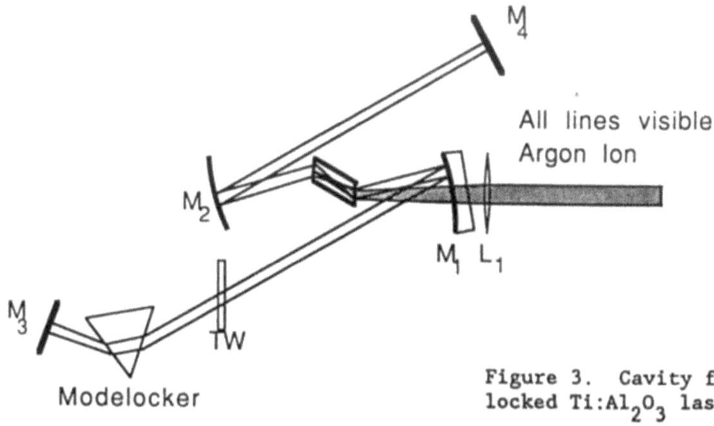

Figure 3. Cavity for the actively mode locked $Ti:Al_2O_3$ laser.

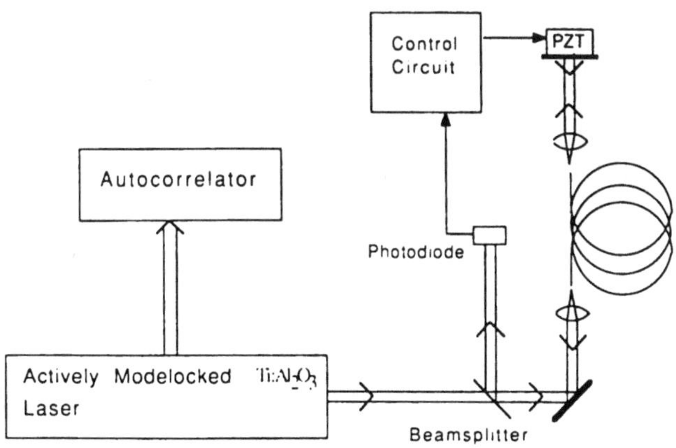

Figure 4. Experimental configuration for the actively mode locked $Ti:Al_2O_3$ laser with nonlinear external cavity feedback.

time [6]. Figure 5 shows the autocorrelation trace obtained at 788nm. The output was highly sensitive to the interferometric mismatch between the cavity lengths (i.e. less than ~0.5μm) but the external cavity length could be coarsely adjusted over several mm without broadening the autocorrelation trace. It should be understood that at 788nm the fibre displayed positive group velocity dispersion and so this laser is certainly not a "soliton laser". The pulse compression mechanism has yet to be elucidated but possibly involves (a) spectral broadening in the external cavity; (b) the intensity-dependent refractive index of the fibre resulting in more constructive interference between the peaks of the pulses in the main and external cavities than between the wings of the pulses (described as Additive Pulse Mode locking in [7]); (c) a beating between the (different) frequency modes of the main and external cavities, inducing a stronger coupling between the modes of the laser system.

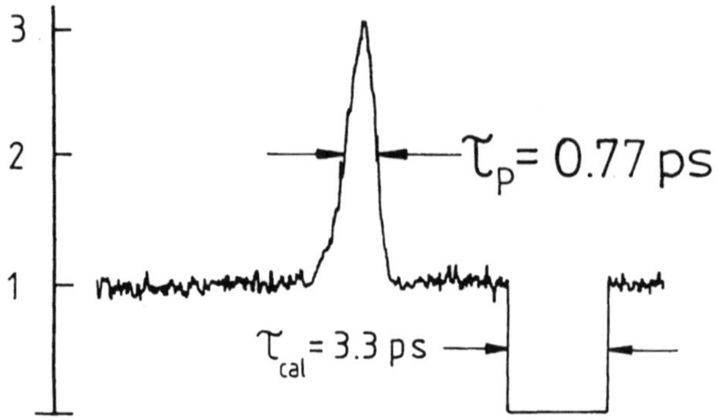

Figure 5. Autocorrelation trace of pulses obtained at 788nm.

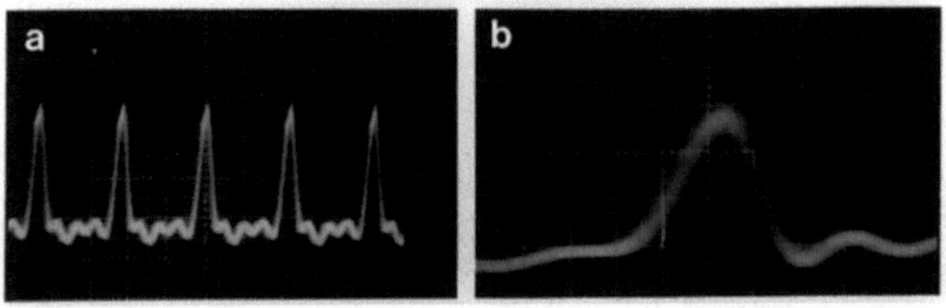

Figure 6. (a) mode locked pulse train (2ns/div), and (b) sampling oscilloscope trace (50ps/div) recorded for c.w. Ti:Al_2O_3 laser mode locked with a linear external cavity.

We have recently demonstrated that the laser system of Figure 4 yields mode locked pulses even when there is no active mode locking. This was first demonstrated by Goodberlet et al. [8] who measured chirped pulses of 2ps duration from the "passively mode locked" laser which were linearly compressed to 560fs. We have also shown that a linear (i.e. empty) external cavity coupled to a c.w. Ti:Al_2O_3 laser can result in mode locked pulse trains. This is observed as the external (or main) cavity length is slowly adjusted - either manually or by a mirror mounted on a shaker which moves it through ~2mm at a frequency of ~20Hz. Figure 6(a) shows the mode locked pulse train and Figure 6(b) shows the trace obtained from a sampling oscilloscope with a response of ~70ps. Synchroscan streak camera measurements have been made of pulses as short as 40ps. The physical mechanisms behind this very recent experimental observation are not yet understood. Kelly [9] has, however, numerically studied the influence of a linear external cavity on a synchronously mode locked colour centre laser, where he speculates that some pulse compression can arise as a consequence of the radiation in the linear and external cavities beating together.

References

1. P.F. Moulton, J.Opt.Soc.Am. B3, 125 (1986).
2. J.D. Kafka, A.J. Alfrey and T. Baer in Ultrafast Phenomena VI, T. Yajima, K. Yashihara, C.B. Harris and S. Shionoya, eds (Springr-Verlag, Berlin 1989) p64.
3. K.J. Blow and D. Wood, J.Opt.Soc.Am. B.5, 629 (1988).
4. K.J. Blow and B.P. Nelson, Opt.Lett.
5. F.M. Mitschke and L.F. Mollenauer, IEEE J.Quantum Electron. QE-22, 2242 (1986).
6. P.M.W. French, J.A.R. Williams and J.R. Taylor, Opt.Lett. 14, 686 (1989).
7. J. Mark, L.Y. Liu, K.L. Hall, H.A. Haus and E.P. Ippen, Opt.Lett. 14, 48 (1989).
8. J. Goodberlet, J. Wang, J.G. Fujimoto, P.A. Schulz and S. Henion, CLEO conference, paper FQ4, Baltimore 1989.
9. S.M.J. Kelly, Opt.Commun. 70, 495 (1989).

Colliding Pulse vs Mirror Contacted Dye Cell Mode-Locking Performances of a Nd:YAG Laser

G.P. Banfi [1], *G. Gabetta* [1], *P.G. Gobbi* [2], *and G.C. Reali* [1]

[1]Università di Pavia, Dipartimento di Elettronica, Via Abbiategrasso 209, I-27100 Pavia, Italy

[2]Laser Science, R & D, Viale Lombardia 5, I-20131 Milano, Italy

Abstract. We report the results of passive colliding pulse mode-locking (CPM) of a pulsed Nd:YAG laser oscillator, and compare them with those obtained with mirror contacted dye cell mode-locking (MCDCM). We find that CPM out-performs MCDCM by a factor of almost two as far as the pulse duration is concerned, maintaining a good amplitude stability and reliability in the short term, but exhibiting a poorer long term stability. Possible explanations of the pulse shortening mechanism and of the long term failure are discussed.

1. Introduction

In the last few years attempts have been made to generate, from solid state lasers, pulses as short as a few picoseconds or less. Several paths have been taken both in the direction of new materials, with broader gain bandwidth, and of new techniques able to use up all or a greater part of the available bandwidth.

Colliding pulse mode-locking (CPM), using an antiresonant ring interferometer as one of the cavity mirrors [1], has been proposed as a possible technique. Since there are many doubts about the advantages of such a configuration when comparing its performance with the mirror contacted dye cell mode-locking (MCDCM), we report here the CPM results obtained in our laboratory and discuss a possible interpretation in an attempt to reconcile the previous findings [2-5].

2. Experimental set-up and results

The arrangement of the CPM cavity was prepared by the replacement of the mirror-cell combination in the standard MCDCM cavity with the antiresonant ring plus the Brewster angled dye cell.

The cavity was close to a hemifocal configuration, which we tested to be very convenient for stable resonators. Transverse mode selection was enhanced using a pinhole just before the antiresonant ring . With the mirror-cell combination, a smaller diameter pinhole was needed to select the lowest order mode. In CPM the dye flowed in a 0.5 mm thick cell, set at the Brewster angle with respect to the direction of the counter-propagating beams to minimize the reflection losses. The cell was mounted on a micro-positioner to allow its fine tuning to the middle of the ring interferometer. The equivalent reflectivity of the interferometer, without the dye, was found to be always better than 96%, but always lower than the 99% expected on the basis of a purely theoretical estimate of the losses. As a matter of fact, we always detected a small radiation leakage at the exit arm of the

Springer Proceedings in Physics, Vol. 49 Ultrafast Phenomena in Spectroscopy
Editors: E. Klose · B. Wilhelmi

interferometer which should be silent (we often used to monitor the silent arm when the oscillator was free-running during the alignment procedures). The other parts of the cavity were standard except for a polarization output coupling, through a quarter wave plate-polarizer combination, which allowed a more careful output coupling control. The active medium was a 3"x1/4" Nd:YAG rod.

The pulse duration was measured by a background-free, type II, second harmonic autocorrelation technique. All the mode-locked fundamental train signals and second harmonic (SH) train signals from the correlator, set at different delays from shot to shot, were recorded by a transient digitizer, with an overall resolution better than 1 ns. In this way no pulse selection was necessary to recover the single pulse-width.

The laser was first operated in passive MCDCM, using EK 9860 dye dissolved in 1.2 dichloroethane, with different single-pass low-signal dye transmissions T_{OD} and with different output couplings. For T_{OD} = 26% and 75% output coupling, the FWHM pulse-width resulted in 37.1±0.9 ps assuming a gaussian shaped pulse, and 32.7±0.5 ps assuming a squared hyperbolic secant shaped pulse, while a negligible failure rate and a 15% standard deviation in peak intensity were observed. Both time duration and intensity fluctuation refer to the main pulse (1 mJ) in the train of 5 pulses, typically, within the FWHM of the envelope.

The laser was then operated in passive CPM, with the antiresonant ring in place of the mirror-cell combination. The results of a set of autocorrelation measurements are reported in fig.1.

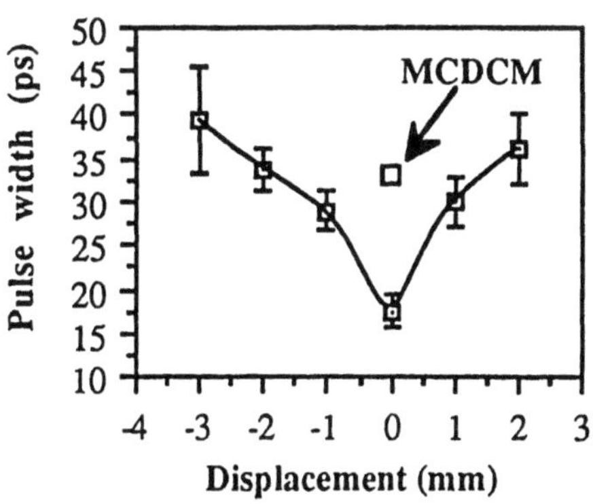

Fig.1. Pulse duration versus displacement of the dye cell from the midway of the optical path along the ring.

With the dye cell perfectly tuned to the midway of the interferometer path, the pulse duration is drastically reduced to 18 ps, shortening the pulse by a factor of almost 2. These operating conditions were the best obtained in terms of intensity stability, which turned out to be slightly better than MCDCM. By slightly displacing the dye cell from the middle of the ring, the counter-propagating pulses no longer collide symmetrically in the dye cell and we see a rather sharp change in pulse duration, a displacement of only 2 mm being sufficient to broaden the pulse from 18 to over 30 ps. Furthermore, moving the dye cell far from the perfect tune up, mode-locking quickly deteriorates: the intensity stability is decreased (much worse than 15%), the pulse energy falls, and the failure probability of the mode-locking process increases. This is also documented in fig.1 by the increased spread of the pulse duration for this case.

Our measurements revealed a low long term stability of CPM in the sense that optimal working conditions were not maintained over long periods of time, but longer and less reproducible pulses started to be produced within a few hours. The simultaneous recording of the fundamental train and its SH conversion allowed a quick glance at these pulse-width changes during the mode-locking train.

In the small signal regime of SH generation obtained in a KDP crystal, if all the pulses in the train have the same temporal shape, the SH energy of every pulse is expected to show a square dependence on the energy of the corresponding fundamental pulse. If this ratio is not kept, but deviations toward lower/higher SH are observed, a corresponding widening/ narrowing of the pulse can be inferred. Results of this monitoring show that two pulse trains, both recorded under apparently perfect tune up conditions of CPM, but with peak pulse durations found to be 18 ps and over 30 ps, respectively, follow the scaling law of the intensity ratios between the fundamental and the second harmonic reasonably well in the first case, while in the second one the strong deviation observed indicates that changes in the pulse duration are occurring even in the well-developed train. This finding was so regular that it has been used to predict the pulse-width just by monitoring the regularity of the SH train.

3. Discussion and conclusions

Our results indicate that CPM leads to shorter pulse durations than MCDCM, at least with the experimental parameters and dye we have adopted.

It has been pointed out [4-7] that the coherent interference between the two colliding pulses produces a standing-wave grating (involving both absorption and refractive index) in the saturable absorber that enhances the nonlinear reflectivity of the interferometer for the colliding pulses as compared to the two pulses passing through the absorber separately. These grating effects may thus play a significant role in a greater effectiveness of the mode-locking process. However, while this argument accounts well for the behavior of fig.1, we do agree with [4] and [5] that there is no reason in principle to distinguish between CPM and MCDCM in the way the grating effects occur in the two configurations, at least when the two beams in the arms of the interferometer are perfectly balanced.

Another argument which deserves consideration is that, if properly aligned, the antiresonant ring favors single lowest order transverse mode operation [4]. As is well known, single lowest order transverse mode operation leads to shorter pulses than transverse multimode operation of a mode-locked laser. This is because there is a more efficient discrimination at the onset of the second threshold for the mode-locking and a better redistribution of the energy in the oscillating (longitudinal) modes. Any small misalignment of the ring sets up a transverse substructure (even in a limited percentage, the same as we can consider is always present in MCDCM) of each longitudinal mode, which lowers this discrimination and deteriorates the mode-locking performances, even if it does not show any detectable increment of the lasing threshold. The low long term stability is thus connected with the very stringent angular acceptance of the antiresonant ring and its balancing, which turned out to be very sensitive in our experimental set-up to mechanical or thermal perturbations.

To this end we notice that the beam splitter is a main source of asymmetry in the CPM with anti-resonant ring configuration. Its behavior is still to be explored both in the linear case, where a detailed analysis of the routing of different order modes through the antiresonant ring for different beam splitters is still lacking, and in the nonlinear case, which occurs when the beam reaches high intensity levels. An imbalance of the intensity or phase along the two paths of the interferometer, in this case, can produce deep effects, since it causes an asymmetric nonlinear phase transfer among the beams inside the dye cell, changing the way they recombine at the beam splitter.

Acknowledgments. This work has been performed with the partial support of the Progetto Finalizzato Telecomunicazioni of the Italian Comitato Nazionale delle Ricerche.

References

[1] A.E.Siegman, Optics Lett., **6**, 334, 1981
[2] H.Vanherzeele, J.L.Van Eck, A.E.Siegman, Appl.Opt., **20**, 3484, 1981
[3] J.M.Buchert, D.K.Basa, C.Tzu, R.R.Alfano, J.Appl.Phys., **55**, 683, 1984
[4] V.I.Prokhorenko, E.A.Tikhonov, D.Ya.Yatskiv, E.N.Bushmakin, Sov.J.Quantum Electron., **17**, 505, 1987
[5] A.Daugvilla, A.Deringas, V.Kalbelka, A.V.Masalov, A.Miliuskas, Ultrafast Phenomena in Spectroscopy, Z.Rudzikas ed., 56, World Scientific, Singapore, 1988
[6] A.E.Siegman, Lasers, University Science Books, Mill Valley CA, 1986
[7] J. Hermann, F.Weidner, and B.Wilhelmi, Appl. Phys. B **26**, 197, 1981

Diode Pumped cw Mode-Locked Nd:Glass Laser: Performance and Limitations

T. Brabec, F. Krausz, E. Wintner, and A.J. Schmidt

IAEE, Abteilung Quantenelektronik und Lasertechnik, TU Wien,
Gußhausstr. 27, A-1040 Wien, Austria

Abstract. The performance of a continuous wave actively mode-locked Nd:phosphate glass laser longitudinally pumped by a multistripe diode laser is described. The use of glass as a host material yields a broader bandwidth and consequently results in shorter pulse durations. On the other hand glass has a bad heat conductivity, which can result in thermal effects limiting the performance. Thermal effects are discusssed together with other design considerations and it is shown that for pumping powers lower than 500 mW thermal problems can be avoided. Furthermore, a simple arrangement for focusing the multistripe laser diode is demonstrated. The laser operates at 1.054 μm. The pump threshold and the slope efficiency are found to be 120 mW and 11%. The shortest pulse duration is 7 ps, which is approximately twice as short as predicted by the theory of amplitude modulated mode locking. We explain the improved performance by additional frequency modulation due to the nonlinear index of the active material.

1. Introduction

Short pulse generation with solid state lasers traditionally has been realized by active mode-locking of cw Nd:YAG lasers and by passive mode-locking of pulsed Nd:glass lasers. The advantage of shorter pulses from glass lasers was significantly impaired by their low repetition rate and poor stability. The pulsing of the glass laser was necessary to avoid thermal problems. To combine the advantages of cw active mode-locking with the large bandwidth of a glass laser became possible by using an argon laser (514 nm) instead of lamps as a pump source [1-3]. The appearance of GaAlAs diode lasers offered the possibility of further improving the performance of laser pumped glass lasers.

Diode lasers are compact, efficient and reliable sources of coherent radiation. Their use as a pump source provides significant advantages over argon lasers: mainly higher efficiency, reduced thermal problems and improved stability. Recently, diode pumped mode-locked Nd lasers have been built using both crystals (YAG [4], YLF [5]) and glass [6] as a host material. As expected, the glass laser generates the shortest pulses. However, because of the low pump power delivered by the single stripe diode laser the pulse energy was too low to measure the pulse duration accurately [6]. Stimulated by this work we started to develop a multistripe diode pumped Nd:glass laser with tolerable threshold.

In section 2 the choice of the length of the laser medium and the concentration of the Nd doping is discussed. Effects limiting the laser performance, such as thermal damage, thermal lensing and birefringence are considered. Section 3 deals with the laser resonator and the design of the focusing lenses. In section 4 mode-locked pulse generation and frequency modulation due to the nonlinear index, which is an additional pulse shortening effect, are discussed.

Springer Proceedings in Physics, Vol. 49 Ultrafast Phenomena in Spectroscopy
Editors: E. Klose · B. Wilhelmi

2. Design Considerations and Performance Limitations of the Laser Material

For the design of a cw laser three important parameters have to be considered: The pump threshold, the slope efficiency and the highest possible pump level at which no performance limiting effects break in. The last parameter is especially important for nature. Heat dissipation in a solid laser material pumped by narrow line radiation arises from the radiationless transitions in the material as a result of a quantum efficiency $\eta_q = h\nu_l/h\nu_p$ less than unity. ν_l is the frequency of the resonator beam, ν_p is the frequency of the pump beam. Heating and cooling of the laser material result in a transverse temperature distribution, which leads to a distortion of the laser beam due to a temperature and stress dependent variation of the refractive index, and due to bending at the end surfaces. These effects have two possibly severe consequences [7] : thermal lensing (TL) and thermal birefringence (TB). Besides TL and TB there is a third thermal effect limiting the pump power: thermal damage (TD) of the laser material. TD occurs when the temperature along the cavity axis in the active medium reaches the softening point T_s of the host.

As the laser is operated at low pump powers, it is reasonable to employ phosphate glass, which has a higher stimulated emission cross section and a weaker concentration quenching than silicate glass, in order to obtain a lower pump threshold. For minimum threshold the dopant concentration (N) should maximize the product $N\tau(N)$. $\tau(N)$ is the lifetime of the upper laser level. The doping level determines the length L of the active medium by the requirement of a high absorption of the pump power. In our case these considerations led to a Nd:phosphate glass LG760 with a doping concentration of 8 weight percent. For this concentration the absorption coefficient $\alpha = 18\text{cm}^{-1}$ and L=1.2 mm yields a 90 % absorption of the pump power at a wavelength of 0.8μm.

In [8] it has been shown that the most severe effect for a phosphate glass with very high doping concentration is TD. TL and TB only become important at higher pump levels. Furthermore, no TD appears until a pump power of 500 mW. If the pump power is in a range, where no thermal effects appear [8], the average spot size of the pump beam should be chosen equal to or less than the average spot size of the resonator beam, in order to gain a high overlap of pump and resonator beam and consecutively a high slope efficiency. From the length of the laser medium the minimum average signal spot size can be calculated, when the pump beam is a Gaussian beam. From the requirement of a high overlap the average signal spot size of the pump beam can be determined. For a multistripe diode, which does not emit a Gaussian beam, it is very difficult to obtain high overlap of pump and signal beam.

3. Design of the Laser Resonator and of the Focusing Arrangement

Fig.1 shows a schematic diagram of our laser cavity. The 1mm thick plate of LG 760 Nd:glass with an 8% doping level was inserted at Brewster's angle into the resonator. The laser medium was placed at the beam waist of the cavity mode. To facilitate heat removal, the glass plate was sandwiched between two copper plates contacted to a heat sink. The astigmatism of the oblique-incidence mirror M2 (R = 20 cm) was compensated for by that of the active material at $\vartheta = 9^o$ [9]. The use of the plane mirror M3 for folding the long arm of the cavity made a compact setup possible.

The pump source was a 500 mW ten stripe GaAlAs diode array (SDL 2432) temperature tuned by a Peltier-cooler for operation at 798nm. The extensions of the laser beam at the diode surface are 100 μm parallel and 1 μm perpendicular to the diode array. The divergency of the diode is 10° parallel and 35° perpendicular to the array. The average spot size of the pump beam in the active medium was calculated to be approximately 45μm [10]. Due to the limited spatial coherence of a multistripe diode laser this value is about twice the minimum value for a diffraction limited Gaussian beam

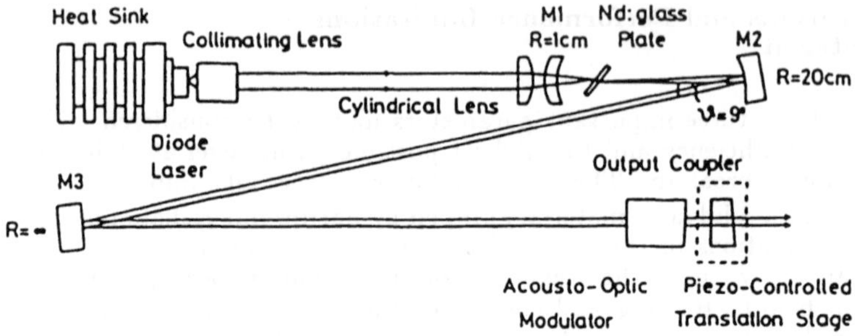

Fig.1 Schematic diagram of the diode-pumped, mode-locked Nd:glass laser.

at this wavelength. To achieve high slope efficiency the spot size of the pump beam must not greatly exceed the spot size of the cavity mode. We met this requirement by appropriate choice of the radius of curvature of the folding mirror M2. $R_2 = 20$cm and a cavity length of $L_c \approx 2$m yielded an average spot size of the cavity mode of about 40 μm. It must be mentioned that although the average spot sizes of pump and resonator beam are approximately equal no optimal overlap can be obtained, as the divergency of resonator and pump beam is different.

The focusing can be achieved by employing two lenses, a spherical and a cylindrical lens. For collecting the emitted radiation of the laser diode and focusing the direction parallel to the array a compound lens of 8 mm focal length and a numerical aperture of 0.5 (Melles Griot 06GLC002) was used. The imaging can be controlled by the ratio of object to image distance. For the focusing of the direction parallel to the array an additional cylindrical lens is necessary. The object for the calculation of the image and the object distance is not the surface of the laser diode, but the field distribution of the diode radiation at the rear focal point of the spherical lens. At this point the beam has its smallest diameter. It can be calculated that a cylindrical lens with a focal length of 2 cm, inserted in the beam approximately 2 cm from the focus will image the direction parallel to the array to the above demanded extension.

4. Active Mode-Locking

The active mode-locking of the thin plate Nd:glass laser was accomplished by an acousto-optic standing wave Bragg modulator (Intra Action ML-38J) inserted close to the output coupler in the cavity. Both windows of the modulator are AR coated at 1.05 μm. The modulator was driven at about $f_m = 38$MHz with 5 W of RF power. The ultrastable frequency synthesizer generating the RF signal was tuned near the cavity resonance and locked to avoid frequency drift. Fine adjustment was performed by positioning the output coupler with a piezo translator. With a 2.5 % output coupler the pump threshold of the mode-locked laser described above was measured to be 120 mW. Although the output was not exactly a linear function of the input power [7], we defined an effective slope efficiency, which was found to be $\approx 11\%$.

The mode-locked laser pulses were monitored by a picosecond diode and sampling oscilloscope. Simultanously, the pulse widths were measured by the standard collinear intensity autocorrelation method. The spectra of the pulses were investigated by a 1 m double grating spectrometer with a resolution of 0.1 Å. The shortest pulses were generated at pumping levels so low ($\leq 2P_{th}$) that stable pulses just developed. The autocorrelation trace at an average output power of ≈ 5mW is shown in Fig.2. The full width at half maximum (FWHM) of the trace is 10 ps. Assuming a Gaussian pulse shape we obtained the pulse duration $\tau_p = (10/\sqrt{2})$ps ≈ 7ps. The laser bandwidth was

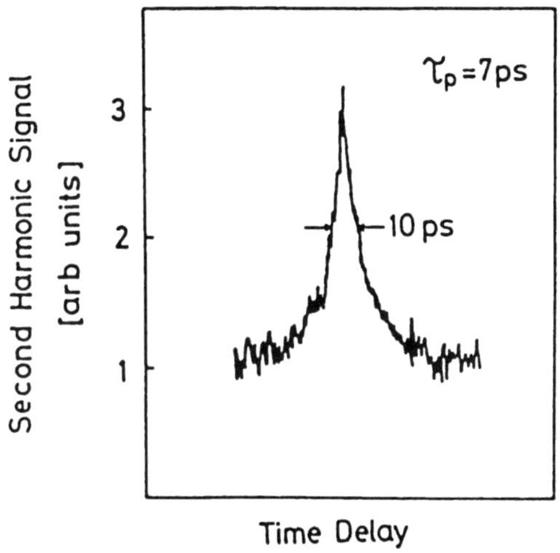

Fig.2 Autocorrelation trace of the laser pulses at an average output power of approximately 5 mW.

measured to be 3.8 Å, and revealed a modulated structure. The measured data yielded a time-bandwidth product of 0.71, a value which is 1.6 times that for a bandwidth-limited Gaussian pulse (0.44). The coherence-spike-free autocorrelation trace and the modulated spectrum of the laser suggest that the pulses are phase modulated. Let us compare these results with predictions of the self-consistent field theory of active mode-locking developed by Kuizenga and Siegman [11]. Inserting the values of our laser in this theory [12], one obtains $\tau_p = 13$ps. However, instead of bandwidth-limited pulses of 13 ps duration, the laser generates phase modulated pulses of 7 ps duration. Two possible reasons for this behaviour are group-velocity dispersion (GVD) and self-phase modulation (SPM) by intracavity elements. Whereas GVD is expected to be small in a picosecond laser, noticeable SPM may occur in the glass plate, where the cavity mode is tightly focused. Expanding the theory of Kuizenga and Siegmann for the effect of SPM and inserting again the values for our laser [12], we obtained $\tau_p \approx 5$ps and a time-bandwidth product of 0.62. As the approximations of our calculation somewhat overestimate the extent of pulse shaping by SPM the agreement with experimental results can be considered satisfactory. One would expect further pulse shortening at higher output levels due to an increased SPM. However, numerical calculations by Haus and Silberberg [13] have shown that this kind of pulse shaping becomes unstable when the pulse shortening with respect to the pulse width without SPM exceeds a factor of approximately 2. Although no signs of instabilities were observed in our laser at increased pump powers, significant pulse broadening occurred, resulting in pulse widths of 12 - 20 ps for 10 - 30 mW of output power ($P_{pump} \leq 400$mW).

5. Conclusion

In summary, we have developed a cw mode-locked Nd:glass laser pumped by a high-power multistripe diode laser. It was shown that thermal problems play no important role for our laser medium in the pump regime up to 500 mW. The complete utilization of the weak concentration quenching in phosphate glass allowed to achieve a low threshold inspite of the poor spatial coherence of the pump beam. The shortest pulses generated by the laser are twice as short as that predicted by the theory of active mode-locking. We found self-phase modulation in the active medium to be a possible source of additional pulse shortening. This effect might have played a role also in previous short pulse generation experiments [3-6]. The conclusion may be drawn that modulation frequencies around or higher than 100 MHz should result in pulses shorter than 5 ps from an actively mode-locked cw Nd:glass laser. Diode pumped regenerative amplification of this output will allow efficient, stable generation of high energy femtosecond pulses by the chirped pulse amplification technique [14].

Acknowledgements

This work was supported by the Austrian Science Research Foundation under project number P 7282.

References

[1] S.Kishida, K.Washio and S.Yoshikawa, Appl. Phys. Lett. 34, 273 (1979)

[2] S.A.Stobel, P.T.Ho, C.H.Lee and G.L.Burdge, Appl. Phys. Lett. 45, 1171 (1984)

[3] L.Yan, J.D.Ling, P.T.Ho and C.H.Lee, Opt. Lett. 11, 502 (1986)

[4] G.T.Maker, S.J.Keen and A.I.Ferguson, Appl. Phys. Lett. 53, 1675 (1988)

[5] G.T.Maker and A.I.Ferguson, Appl. Phys. Lett. 54, 403 (1989)

[6] S.Basu and R.L. Byer, Opt. Lett. 13, 458 (1988)

[7] W.Koechner, Solid State Laser Engineering (Springer, Berlin, Heidelberg 1976)

[8] F.Krausz, E.Wintner, A.J.Schmidt and A.Dienes, to be published in IEEE J. Quantum Electron.

[9] H.Kogelnik, E.P.Ippen, A.Dienes and C.V.Shank, IEEE J. Quantum Electron. QE-8, 373 (1972)

[10] T.Brabec, F.Krausz, E.Wintner, A.Schmidt to be published in Applied Optics

[11] D.J.Kuizenga and A.E.Siegman, IEEE J. Quantum Electron. QE-6, 694 (1970)

[12] F.Krausz, T.Brabec, E.Wintner, and A.J.Schmidt to be published in Appl. Phys. Lett.

[13] H.A.Haus and Y.Silberberg, IEEE J. Quantum Electron. QE-22, 325 (1986)

[14] P.Maine, D.Strickland, P.Bado, M.Pessot and G.Mourou, IEEE J. Quantum Electron. QE-24, 398 (1988)

Substructure in ps Light Pulses Emitted by Gain Switched Semiconductor Lasers

A.G.Weber, E.H. Böttcher, and D. Bimberg

Institut für Festkörperphysik 1 der Technischen Universität Berlin, Hardenbergstr. 36, D-1000 Berlin 12, Germany

Abstract. Short optical pulses obtained from gain-switched multimode semiconductor lasers show a periodic spikelike substructure varying from pulse to pulse. A theory is presented based on the interference of longitudinal laser modes having randomly distributed phases and a transient shift of the envelope of the mode spectrum during emission. Using this theory single shot streak camera records and autocorrelation traces are simulated and found to be in good agreement with experimental data.

1. Introduction

Semiconductor lasers are commonly used to produce optical pulses in the picosecond regime. There are two different methods generally applied to produce such short pulses: gain-switching by direct current modulation [1] and mode-locking [2]. Whereas mode-locking provides pulses with a duration below 1 ps, the shortest pulses obtained by electrical gain-switching having a reasonable peak power of 200 mW are smaller than 15 ps[1,3]. In addition, emission of gain-switched ultrashort optical spikes as narrow as 4 ps [4], 2 ps [1] and 1.8 ps [5] was reported which cannot be explained by simple rate equation models. Optical pulse production by gain-switching is commonly used in spite of the higher pulsewidth in comparison to mode-locking because of the experimental simplicity [1]. Only a laser diode and an electrical driving circuit are necessary whereas mode-locking requires in addition a stable aligned external cavity or a saturable absorber introduced into the internal or external cavity. An important factor in terms of application, which determines the performance of a picosecond optical pulse train, is the reproducibility of the optical waveform from pulse to pulse.

Single shot streak camera records at highest resolutions of optical pulses obtained from multimode gain-switched semiconductor lasers show a distinct substructure which varies randomly from shot to shot. A typical example for the substructure and its variation is presented in Fig. 1. It shows two single shot streak camera records of optical pulses emitted by a gain-switched GaAs V-groove laser taken at identical injection conditions at a time resolution of about 1.5 ps [7]. The driving voltage of the laser is adjusted to an amplitude which ensures the emission of the first relaxation oscillation only. Obviously the pulse consists of a sequence of more or less distinct pulses varying in amplitude from shot to shot. The time between two consecutive spikes is approximately equal to the cavity roundtrip time t_r with $t_r = 2nl/c$ where l

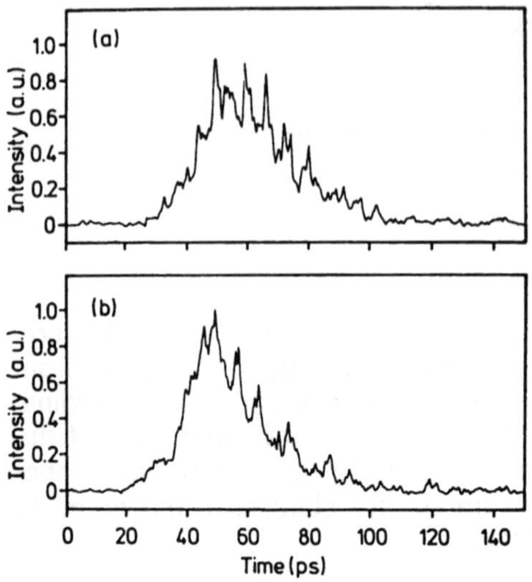

Fig. 1 : Two typical examples of single shot streak images obtained from a gain-switched multimode GaAs DH laser (length l = 280 μm, camera resolution τ = 1.5 ps), a and b show traces recorded under identical injection conditions .

is the laser cavity length and c/n is the speed of light in the cavity. The spikewidth is limited by the resolution of the streak camera. This randomly varying substructure in the optical single pulse waveform of gain-switched GaAs-based injection lasers has been observed by several groups using equivalent experimental equipment [4,5,8,9].

2 . Simulation

In order to model the observed finestructure we extend a description which has been applied to simulate SHG (second harmonic generation) correlation traces of short pulses from directly modulated AlGaAs lasers [10]. The essential extension of this treatment is the consideration of the time-dependent shift of the envelope of the longitudinal mode spectrum due to the time-varying carrier density in the laser cavity. The experimentally observed time periodicity of the substructure indicates that a consistent approach has to be based on interference effects. Clearly, the interference of two adjacent longitudinal laser modes results in a beat signal and the time period of the intensity is given by the cavity roundtrip time t_r. Consequently, we assume that the temporal evolution of the laser intensity I(t) can be expressed by the sum of the longitudinal modes [10]. Thus

$$I(t) = \left[\sum_{i=1}^{N} a(\omega_i) \sin(\omega_i t + \varphi_i) \right]^2 F(t) \tag{1}$$

where $a(\omega_i)$ is the amplitude, ω_i is the frequency, and φ_i is the phase of the i-th mode. F(t) describes the transient mode intensity. For the sake of simplicity, it is assumed to be equal for all modes. For the simulations, the following model parameters corresponding to typical GaAs-based lasers are used: a cavity length of 300 μm, a central wavelength of 800 nm, and a longitudinal mode spacing of 0.3 nm. The profiles F(t) and $a(\omega_i)$ are approximated by Gaussian functions with a full width at half maximum (FWHM) of 25 ps and 7 modes, respectively. For the simulation of each single optical pulse, a different set of the phase values φ_i randomly distributed within the interval $[0,2\pi]$ is used. This is reasonable because mode coupling does not occur. Computer simulations are carried out without and with the inclusion of a temporal linear shift of the envelope of the mode spectrum. A value of 0.012 nm ps^{-1} is used [11].

Three simulated pulse profiles are shown in Fig.2. The results depicted in Fig. 2a and Fig. 2b are calculated for the absence of mode shifting, whereas the simulated waveform in Fig.2c includes this effect. Regularly spaced substructure shows up. It is evident, that in the absence of envelope shifting, the substructure is strictly periodical in time and relative amplitude hight during the pulse emission with a time period given by the cavity roundtrip time. The transient mode shifting manifests itself by a violation of the strict periodicity, i. e., the substructure is no more regular during the entire pulse duration and the time period varies slightly. In order to compare these result with experimentally observed streak records we convolute the simulated pulse profiles with the temporal resolution of the streak camera. The convoluted waveforms are displayed in the insets of Fig.2. The convolution is performed assuming a temporal resolution of 1.5 ps. The features of the resulting pulse profile shown in the inset of Fig.2c are in excellent agreement with the presented streak camera traces and previous experimental observations [4,5,8,9].

In addition to the simulation of pulse profiles, the corresponding autocorrelation profiles are calculated by averaging over a number of 100 consecutive single pulses with different sets of random phases. A simulated autocorrelation trace which includes envelope shifting is shown in Fig.3. The background has a Gaussian shape which is superimposed by sharp peaks appearing at regular intervals and reflecting the time period t_r of the substructure similar to the observations [10,12]. Due to envelope shifting the ratio of the amplitudes of the spikes to the background decreases with increasing delay time because of the simultaneously reduced regularity of the substructure. This result is also in good agreement with experimental data [10,12,13].

In contrast to the model here proposed, an alternative theoretical view based on Monte Carlo simulations using spatially dependent single mode rate equations has been presented very recently in order to explain the fluctuating substructure [14]. Investigations on dynamical *single mode* lasers can serve as a test of the latter model. Complete absence of regularly spaced sharp peaks in the autocorrelation trace [12] was demonstrated which should be present under the assumptions of [14]. Therefore this model can also not be taken into

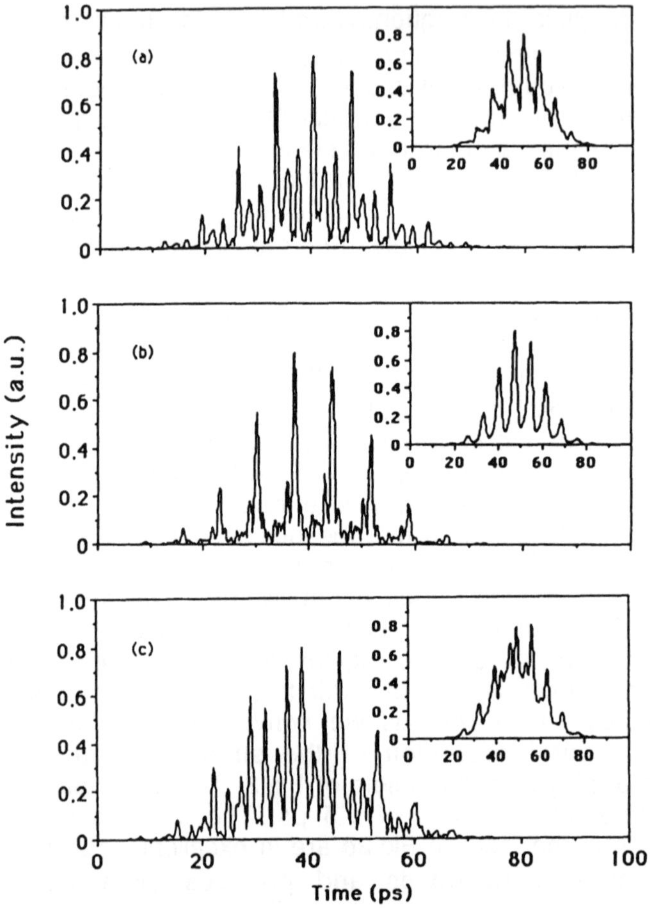

Fig. 2 : Simulated temporal profiles of single optical pulses emitted by a typical multimode injection laser under gain-switched operation. Simulations without (a,b) and with (c) the inclusion of the transient longitudinal mode shift are shown. The profiles are calculated using different sets of random phases for the longitudinal modes. Insets: Simulation of the corresponding streak records by convolution of the pulse profiles assuming a camera resolution of 1.5 ps.

consideration for the explanation of the substructure in multimode lasers. Of course, theoretically it cannot be excluded that the stochastic process of spontaneous emission affects the substructure. The importance and the validity of our approach is strongly supported by the good agreement between the simulations and experimental results from streak camera and autocorrelation measurements. In addition, the results on the energy variation of optical pulses from gain-switched AlGaAs diode lasers lying in the 2-4 % range [6] are consistent with the presented model, too.

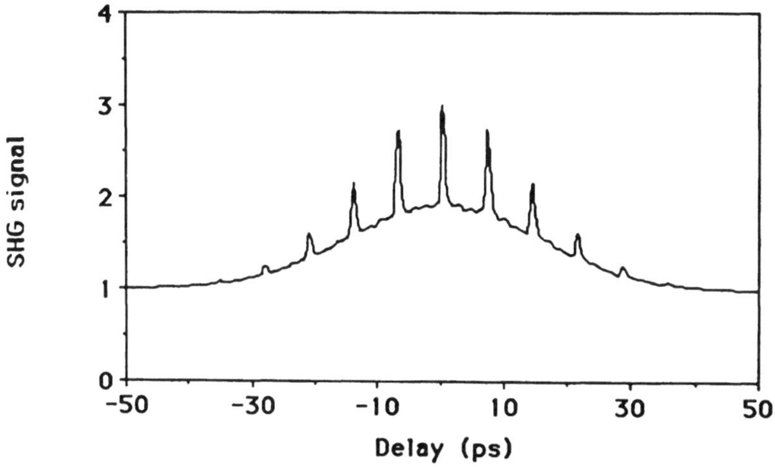

Fig. 3 : Simulated SHG autocorrelation trace including shift of the envelope of the mode spectrum obtained by averaging over 100 different pulses, using the same parameters as for Fig. 2 .

3 . Conclusion

In summary, the random substructure observed by streak camera records and autocorrelation traces in optical picosecond pulses from gain-switched multimode semiconductor lasers is consistently understood for the first time. A model based on the interference of the longitudinal laser modes is found to describe the substructure. Simulations of temporal pulse profiles and the corresponding autocorrelation traces are performed. The results demonstrate that the recently reported transient shift of the envelope of the longitudinal mode spectrum is of importance in order to achieve a good agreement between the simulations and the experimental data.

The authors appreciate funding of parts of this work by the BMFT.

REFERENCES

[1] for a review on direct current modulation of semiconductor injection lasers, see: K. Y. Lau and A. Yariv, in *Semiconductors and Semimetals*, edited by W. T. Tsang (Academic, Orlando, FL, 1985), Vol. 22,Part B, p.71, and D. Bimberg, K. Ketterer, E. H. Böttcher, and E. Schöll, Int. J. Electronics **60**, 23 (1986).

[2] for a review on mode-locking of semiconductor lasers, see: J. P. van der Ziel, in *Semiconductors and Semimetals*, edited by W. T. Tsang (Academic, Orlando, FL, 1985), Vol. 22, Part B, p.1.

[3] H.F.Liu, M.Fukazawa, Y. Kawai, and T. Kamiya, IEEE J. Quantum Electron. **QE-25**, 1417 (1989).

[4] D. Bimberg, K. Ketterer, E. Schöll, and H. P. Vollmer, Electron. Lett. **20**, 640 (1984).

[5] Y. Arakawa, T. Sogawa, M. Nishioka, M. Tanaka, and H. Sakaki, Appl. Phys. Lett. **51**, 1295 (1987).

[6] R. T. Hawkins, J. H. Goll, and N. C. Anheier, in *Picosecond Electronics and Optoelectronics II*, edited by F. J. Leonberger, C. H. Lee, F. Capasso, and H. Morkoc (Springer, Berlin, 1987) Springer Series in Electronics and Photonics Vol. 24, p. 198.

[7] for experimental details, see Ref. 4.

[8] Y. Tsuchiya, M. Miwa, and M. Koishi, J. Appl. Phys. **53**, 6671 (1982).

[9] R. A. Elliot, Huang DeXiu, R. K. DeFreez, J. M. Hunt, P. G. Rickman, Appl. Phys. Lett. **42**, 1012 (1983).

[10] H. Ito, H. Yokoyama, S. Murata, and H. Inaba, IEEE J. Quantum Electron. **QE-17**, 663 (1981).

[11] K. Ketterer, E. H. Böttcher, and D. Bimberg, Appl. Phys. Lett. **53**, 2263 (1988).

[12] N. Onodera, H. Ito, and H. Inaba, IEEE J. Quantum Electron. **QE-21**, 568 (1985).

[13] H. Ito, H. Yokoyama, S. Murata, and H. Inaba, Electron. Lett. **15**, 738 (1979).

[14] T. Sogawa, Y. Arakawa, M. Tanaka, and H. Sakaki, Appl. Phys. Lett. **53**, 1580 (1988).

Generation of Tunable Femtosecond Pulses by Selective Amplification of a Continuum

J. Hebling and J. Kuhl*

Max-Planck-Institut für Festkörperforschung, Heisenbergstr. 1,
D-7000 Stuttgart 80, Fed. Rep. of Germany
*On leave from Joszef Attila University, Szeged, Hungary

Abstract We describe generation of tunable femtosecond pulses with 3 MW peak power around 700nm at a repetition rate of 7kHz. The system is based on the output of a colliding pulse mode-locked dye laser operating at 625nm and amplified by a Cu vapor laser. Tunability is achieved by white light continuum excitation and subsequent amplification of selected spectral bands in a single pass amplifier pumped by part of the amplified femtosecond output at 625nm.

1 Introduction

Time-resolved spectroscopy of femtosecond phenomena in molecules and solids is usually accomplished by excite-and-probe experiments with femtosecond optical pulses. In many cases this technique necessitates a frequency tunable probe pulse which is highly synchronized to the excitation pulse. Most experiments described in the literature, so far, have utilized a white light continuum excited in an ethylen glycol jet by the amplified output of a colliding pulse mode-locked (CPM) dye laser as the probe. Since the spectral intensity in the white-light-continuum is relatively low, subsequent amplification of selected spectral bands of the broad continuum may be indispensable for many applications. [1,2]

The application of this technique for the generation of tunable optical pulses in the near-infrared regime between 800-860nm is described in Ref.[1]. In the experiment 50fs long pulses of a CPM laser were amplified in a 6-pass amplifier pumped by a first Cu-vapor laser and then focused into an ethylen glycol jet for white light continuum excitation. A spectrally filtered band of the continuum was finally amplified in a further six-pass amplifier pumped by a second Cu vapor laser. The severe drawback of this complex system, which provides tunable optical pulses with a duration comparable to the CPM output, is the large amount of expensive equipment.

2 Experimental Results and Discussion

In this paper we present a distinctly simpler system which needs only one Cu vapor laser and one multipass amplifier.

This twofold advantage compared to the device of Ref.1 is obtained by employing part of the amplified output of the six pass amplifier as a pump pulse for a single pass longitudinally or transversally pumped amplifier cuvette which is used for the amplification of the spectrally filtered white light continuum.

2.1 Longitudinally Pumped Amplifier

Figure 1 depicts the experimental arrangement for the case of the longitudinally pumped amplifier. The amplified output pulses of the CPM laser (pulse energy 10μJ, pulse duration 80fs, wavelength 625nm) are split into two equal parts. The first part excites the white light continuum, the second part pumps the flowing dye cell amplifier. A variable part of the continuum is selected by an interference filter before injection into the amplifier cuvette.

With Rh 700 (concentration 0.7g/l) this amplifier generates 120fs output pulses with a pulse energy of 0.3μJ at 720nm corresponding to an overall efficiency of 3% for the conversion of the 625nm light into the output at 720nm. The spatial distribution of the emission of this longitudinally pumped amplifier depends sensitively on the beam profile of the injected continuum and the tunability range is limited to about 800nm at the long wavelength edge owing to its relatively small gain and the rapid decrease of the spectral intensity in the continuum with increasing wavelength. (Variation of the wavelength from 700 to 800nm results in a decrease of the spectral intensity in the continuum by 2-3 orders of magnitude).

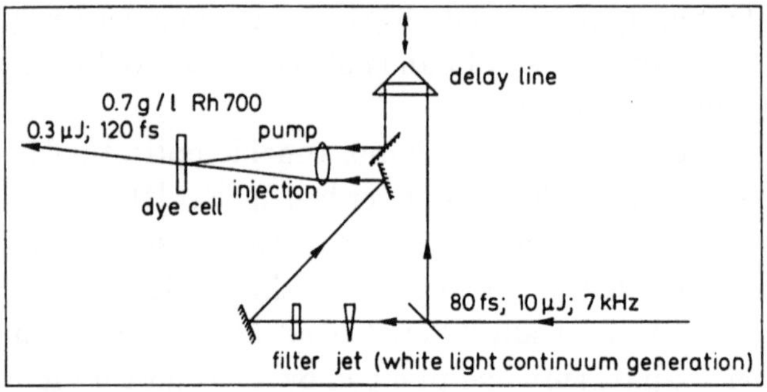

Fig. 1: Experimental set-up with the longitudinally pumped amplifier.

2.2 Transversally Pumped Traveling Wave Amplifier

These disadvantages can be eliminated if the amplifier for the continuum is pumped transversally and operated in the saturation regime. [3] For the present case of femtosecond pulse amplification we utilized a traveling wave pumping scheme, [4,5] in order to eliminate the competing process of amplified spontaneous emission which is observed in conventional transversally pumped amplifiers (see Fig.2).

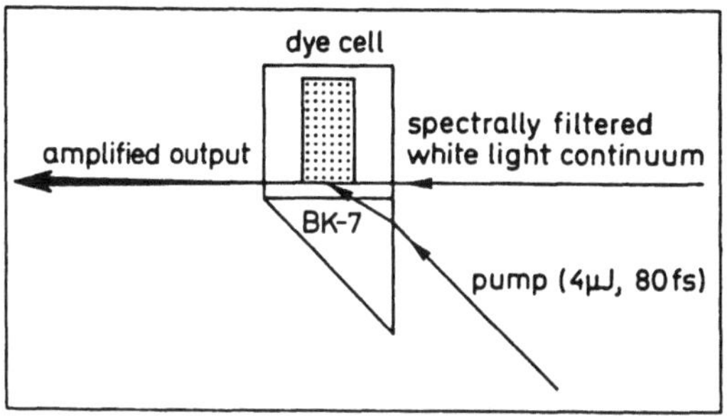

Fig. 2: Scheme of the transversally pumped traveling wave amplifier.

Traveling wave excitation of the dye solution is provided by the BK-7 glass prism glued to the surface of the dye cell. Refraction of the pump pulse at the prism surface as well as the large angle of incidence of the pump beam onto the dye solution sweep the pump pulse along the dye amplifier with a velocity which can be exactly matched to the velocity of the propagating continuum pulse by adjusting the direction of incidence. It should be noted that tuning of the output in the case of the transversally pumped amplifier requires tilting of the interference filter for tuning of the injected wavelength and a simultaneous variation of the time delay between the pump and the injected pulse owing to the distinct frequency chirp of the white light continuum. In contrast a small delay between pump and injection pulse has no influence on the operation of the longitudinally pumped amplifier since the lifetime of the population inversion is much larger because of the lower gain and the corresponding lower amplified spontaneous emission. A further advantage of the longitudinal pumping scheme is the considerably easier adjustment. On the other hand the transverse pumping provides a higher gain and consequently it is expected that the tuning range extends distinctly further to long wavelengths where the injected power of the continuum rapidly decreases. In addition the dependence of the beam quality on the spatial distribution of the white light continuum is less critical for the transverse configuration because of saturation effects associated with the higher gain.
The transversally pumped amplifier operated with Rh700 (concentration 0.6g/l) yields 110fs (see Fig.3a) output pulses with an energy of 0.4μJ and a bandwidth of 10nm at 692nm (see Fig.3b). Further pulse shortening is attainable by subsequent pulse compression as demonstrated below. The energy conversion efficiency of the system (5%) is slightly higher than that of the longitudinally pumped configuration. Spectral filtering of the continuum and wavelength tuning of the injection pulse has been accomplished by a 700nm interference filter with a bandwidth (FWHM) of 10nm. Tilting of the interference filter permits continuous tuning of the output from 700-680nm.

The long wavelength limit observed in the present experiments is determined by the available interference filter. A proper interference filter is expected to shift the long wavelength edge of the tuning curve for Rh 700 to at least 720-730nm.

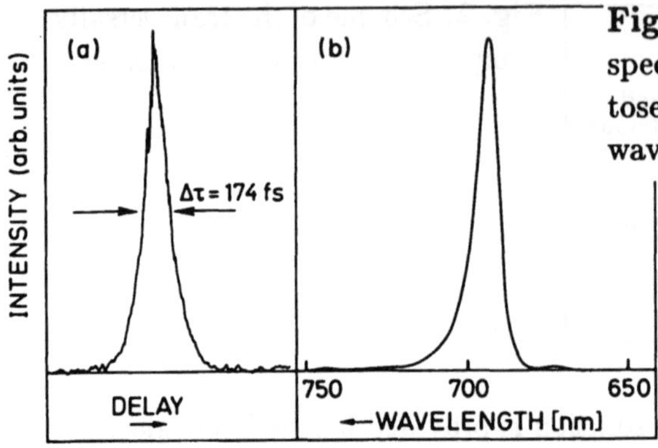

Fig. 3: Autocorrelation trace (a) and spectrum (b) of the amplified femtosecond output from the traveling wave amplifier operated with Rh 700.

Using Oxazine 170 (concentration 0.8g/l) as the gain medium we have generated in the traveling-wave pumping geometry 180fs output pulses with an energy of 140nJ at 670nm. The spectral bandwidth of these pulses amounts to 14nm and 17nm for injection of the continuum with and without an interference filter, respectively. In both cases the time-bandwidth product is considerably larger than the Fourier transform limit indicating the presence of significant frequency chirping. Considerable pulse shortening is attainable after compression by a grating pair with 300 lines/mm and a separation of 72mm. After compression the pulse duration is 60fs as demonstrated by the autocorrelation curve of Fig.4a.

The energy of the amplified pulses at 670nm is high enough to study nonlinear optical processes. The potentials of this tunable femtosecond source have been

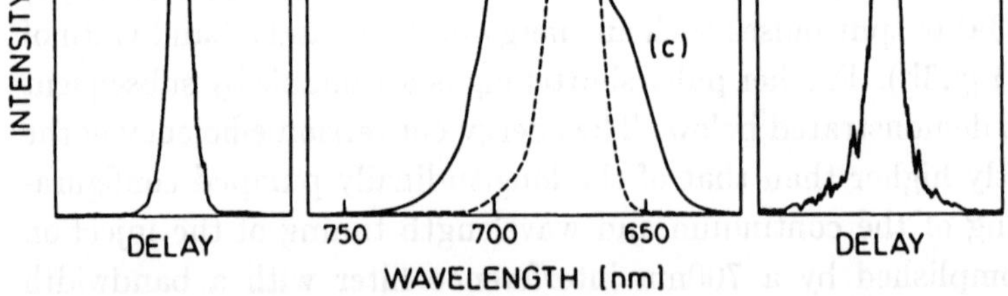

Fig.4 (a) : Autocorrelation curve of an amplified pulse for the Oxazine 170 amplifier after compression by a grating pair.
(b) Amplifier output spectrum and spectral broadening(c) via self-phase modulation in a 43mm long optical fiber.
(d) Autocorrelation trace of the fiber output after compression by a grating pair.

proven by performing a self-phase modulation experiment. In this experiment part of the amplified pulse has been sent through a 43mm long single mode optical fiber. The pulse spectrum on the fiber output, depicted as curve c in Fig. 4, reveals significant spectral broadening due to self-phase modulation. Subsequent compression of the fiber output by a grating pair results in a pulse duration of 33ps (see Fig.4c) which is still much larger than the transform limited value corresponding to the observed spectral width of 50nm. Still shorter pulses should be obtainable if the fiber length is optimized and if the 60fs precompressed pulses are injected into the fiber instead of the 180fs pulses provided by the amplifier.

3 Conclusion

In conclusion, we have described simple single pass amplifier configurations which allow easy amplification of the white light continuum excited by high power femtosecond optical pulses. The new approach to femtosecond continuum amplification does not require additional expensive equipment since the amplifier is pumped by part of the femtosecond pulse which is used for the continuum excitation.
Finally, we would like to point out that the ultimate potential of the traveling wave excitation technique is not fully utilized in the present experiment. As demonstrated earlier the output pulses of a traveling wave amplifier can be as short as the pump pulses independently on the length of the injection.[6] That means such an amplifier functions like an optical gate and can generate femtosecond output pulses even for the case of a cw input.[7]

References

[1] P.C. Becker, H.L. Fragnito, R.L. Fork, F.A. Beisser, and C.V. Shank, Appl. Phys. Lett. **54**, 411 (1988)

[2] J.P. Chambaret, A. Dos Santos,G. Hamoniaux, A. Migus, and A. Antonetti, Optics Comm. **69**, 401 (1989)

[3] R.L. Fork, C.V. Shank, C. Hirlimann, R. Yen, and W.J. Tomlinson, Optics Lett. **8**, 1 (1989)

[4] Z.S. Bor and B. Rácz, Optics Comm. **54**, 165 (1985)

[5] J. Hebling and J. Kuhl, Optics Lett. **14**, 278 (1989)

[6] J. Hebling and J. Kuhl, Optics Comm., accepted for publication

[7] J. Hebling and J. Kuhl, to be published.

Spectral Diagnostic Method for Temporal Study of Subpicosecond Distributed Feedback Dye Lasers

P. Simon[1], *H. Gerhardt*[1], *and S. Szatmári*[2]

[1]Laser-Laboratorium Göttingen e.V., Robert-Bosch-Breite 10, D-3400 Göttingen, Fed. Rep. of Germany

[2]Max-Planck-Institut für biophysikalische Chemie, Am Fassberg, D-3400 Göttingen, Fed. Rep. of Germany

Abstract. The spectral and temporal behaviour of subpicosecond DFDLs are studied. A simple and sensitive spectral diagnostic method is proposed to exhibit the presence and determine the temporal separation and relative amplitude of any following pulse.

1. Introduction

Distributed feedback dye lasers (DFDLs) are able to generate transform limited ultrashort pulses in a wide spectral range [1]. Due to their easy tunability and automatic synchronization to the following amplifiers DFDL based hybrid excimer-dye laser systems have proved to be the simplest tools for the generation of femtosecond pulses at most of the excimer wavelengths [2-4].

As it is shown in [1] the DFDL generates a single transform limited pulse when pumped just above its threshold. With stronger pumping multiple pulse generation is expected. The single pulse generation regime in picosecond DFDLs could easily be controlled by spectral measurements, since the spectra of the successive pulses were slightly shifted [5,6]. For subpicosecond DFDLs different spectral behaviour is found in [3], but no explanation is given there.

Since the exclusion of any trailing pulse is crucial in many spectroscopic applications, we have studied the correlation between the spectral and temporal behaviour of subpicosecond DFDLs. In this paper we show a method to control the temporal behaviour through spectral measurements.

2. Experimental

The scheme of our experimental setup is just the same as described in [3]. One tube of a Lambda Physik EMG 150 MSC double discharge excimer laser, filled with XeCl, pumps an intermediate dye laser whose output is used to pump the DFDL master oscillator [3]. The subpicosecond 496 nm pulses are amplified in a two stage dye amplifier pumped by the same XeCl laser.. After frequency doubling the 248 nm, fs pulses are amplified in the other discharge channel of the excimer laser, filled with KrF. In a double pass arrangement 9 mJ output energy is obtained.

3. Results

Simultaneous spectral and temporal study of the 496 nm pulses, just before frequency doubling, was carried out. With pump energies up to two times the DFDL threshold single pulse operation was observed. A typical pair of a streak-camera trace and spectrum is shown in

Springer Proceedings in Physics, Vol. 49 **Ultrafast Phenomena in Spectroscopy**
Editors: E. Klose · B. Wilhelmi

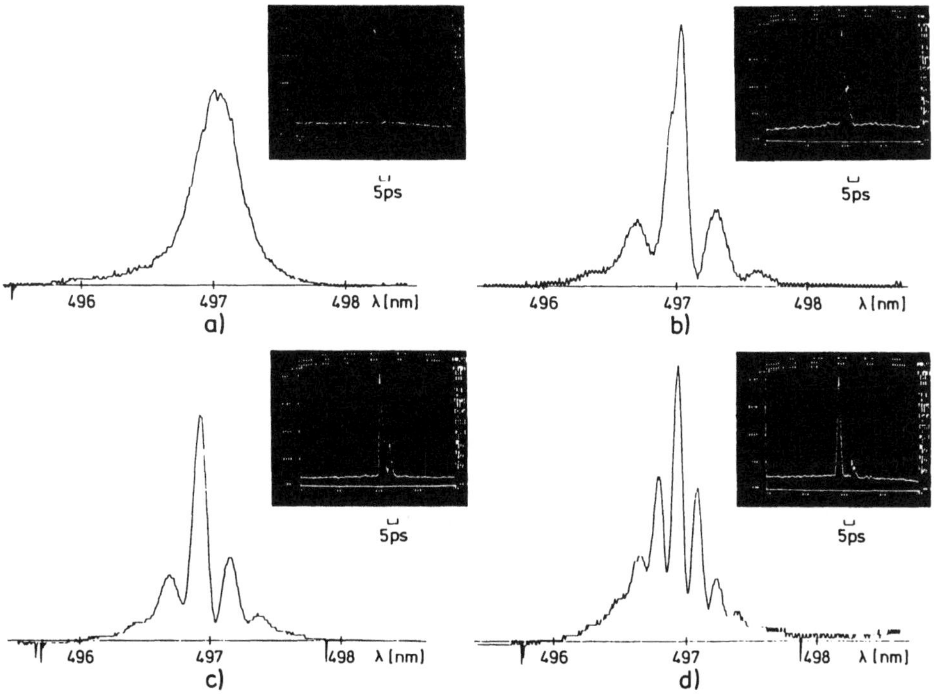

Fig. 1 Simultaneously measured spectra and streak camera traces of the amplified DFDL pulses. The DFDL operates a) ~1.5, b) ~4, c) ~3, d) ~2 times above threshold

Fig. 1a. The pulsewidth is limited by the resolution of the streak-camera (2 ps) as was checked by autocorrelation measurements. The typical pulse duration was measured to be 650 fs, close to the transform limit of the 4 Å wide spectrum, supposing $sech^2(t)$ shape for the pulses.

When the DFDL is pumped 2-5 times above its threshold, two or sometimes three pulses are generated, as shown on the streak-camera traces in Fig. 1b-d. It is seen on the corresponding spectra that multiple pulsing is always accompanied by a structured spectrum. Note that larger temporal separation of the subsequent pulses results in smaller modulation period in the spectrum, and a trailing pulse of smaller intensity causes smaller modulation depth.

Using the results of the simultaneous spectral and temporal measurements of 50 shots with the DFDL pumped well above threshold, the modulation period observed in the spectra ($\Delta\lambda$) versus the temporal separation of the main and trailing pulses (ΔT) is plotted in Fig. 2.

4. Discussion

One can fit the experimental results (Fig. 2) by

$$\Delta\lambda = (\lambda^2/c) * (1/\Delta T) \qquad (1)$$

where λ is the wavelength of radiation and c is the speed of light.

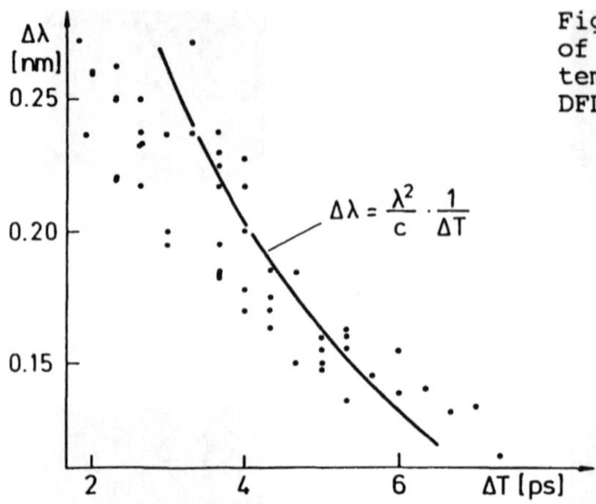

Fig. 2 Modulation period of the spectra versus temporal separation of the DFDL pulses

Equation (1) is set up with the assumption that the DFDL, when pumped well above threshold, emits two coherent pulses, and the spectrograph displays the interference of these pulses.

A very important feature of this spectral trailing pulse detection method is its extremely high sensitivity. For comparison, let us regard the most often used second order autocorrelation technique. If a trailing pulse is present, the autocorrelation function has side maxima on both sides of the main peak. The contrast ratio between the side maxima and background is

$$\text{Max/Min} = 1 + 4/(I_2/I_1 + I_1/I_2) \tag{2}$$

where I_1 and I_2 are the intensities of the two pulses.

With the use of our spectral method the contrast ratio between the maximum and the minimum of the modulation of the spectrum is given by

$$\text{Max/Min} = (1 + \sqrt{I_2/I_1})^2/(1 - \sqrt{I_2/I_1})^2. \tag{3}$$

Equations (2) and (3) are plotted in Fig. 3 for different ratios of I_1 and I_2. It is seen that the spectrograph gives much higher contrast.

It is worth noting that in contrast to picosecond DFDLs where the trailing pulses have slightly shifted spectra due to gain induced refractive index [5,6], the pulse train generated by a femtosecond DFDL is coherent. This behaviour has been observed with different dyes (Coumarin 47, Coumarin 102, Coumarin 307, Rhodamine B, DCM). The different behaviour we attribute to the fact that the bandwidth of fs DFDL pulses is much broader than the spectral shift.

When the DFDL is pumped well above threshold some structure appears even in the spectrum of the final UV output indicating the presence of trailing pulses. An important feature of the UV spectrum is its weak modulation which can be explained by the pulse selection effect of amplifiers [7]. In saturated amplifiers, because of strong gain depletion by the first pulse being amplified, the trailing pulses see much less gain. A maximum of 0.1 % relative energy content of the trailing pulse with respect to the main pulse is typically evaluated. In applications where even this trailing pulse is disturbing, stable single pulse generation can be achieved by keeping the energy of the XeCl pump laser within ± 7 %.

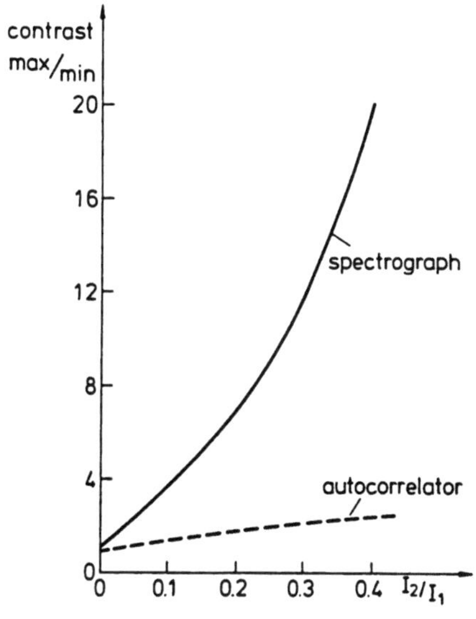

Fig.3 Calculated sensitivity of the spectral and autocorrelation methods for the detection of trailing pulses

In conclusion we have studied the spectral and temporal behaviour of subpicosecond DFDLs. A simple and very sensitive spectral method is proposed for the detection of trailing pulses, for determination of their relative amplitude and temporal separation.

The permanent address of P.S. and S.S. is JATE Univ., Szeged, Dóm tér 9., H-6720, HUNGARY

References

1. Zs. Bor: IEEE J. Quantum Electron. QE-16, 517 (1980)
2. S. Szatmàri, B. Rácz, F.P. Schäfer: Optics Comm. 62, 271 (1987)
3. S. Szatmàri, F.P. Schäfer: Optics Comm. 68, 196 (1988)
4. Q. Zhao, S. Szatmàri, F.P. Schäfer: Appl. Phys. B in press
5. Zs. Bor: Appl. Phys. 19, 39 (1979)
6. S. Szatmàri, Zs. Bor: Appl. Phys. B. 34, 29 (1984)
7. S. Szatmàri, F.P. Schäfer: Opt. and Quant. Electron 16, 277 (1984)

Intensity Dependent Transmission of Window Materials at 248 nm

P. Simon[1], *H. Gerhardt*[1], *and S. Szatmári*[2]

[1]Laser-Laboratorium Göttingen e.V., Robert-Bosch-Breite 10, D-3400 Göttingen, Fed. Rep. of Germany
[2]Max-Planck-Institut für biophysikalische Chemie, Am Fassberg, D-3400 Göttingen, Fed. Rep. of Germany

Abstract. Transmission of fused silica, CaF_2, LiF, and MgF_2 is measured using 450 fs, 248 nm pulses in the range 10 - 120 GW/cm². Different loss mechanisms such as scattering of transmitted radiation, color-center formation and multiphoton absorption were studied separately. For fused silica a two-photon absorption mechanism is found, while for CaF_2, LiF, and MgF_2 three-photon absorption and absorption due to color-center formation are found as dominant absorption mechanisms.

1. Introduction

With existing high-brightness KrF laser-systems, optical powers in the order of typically 10-100 GW are generated.[1-7] The power densities of the output beams of these systems typically range from 10 to 100 GW/cm². At those high power densities nonlinear-optical properties of window materials become important.

Recently, measurements of nonlinear absorption of window materials at 248 nm were reported in Refs. 6, 8 and 9. However, the relative contributions of the different loss mechanisms (multiphoton absorption, color-center formation, scattering) were not considered. In addition, the results reported so far are not in total agreement.

We made comparative measurements of the intensity-dependent transmission of UV windows of different materials. Here we report on the nonlinear loss mechanisms observed in fused silica, CaF_2, LiF, and MgF_2, using 450 fs pulses at 248 nm.

2. Experimental

For the transmission measurements 8 mJ, 450 fs pulses at 248 nm from a high-power KrF laser-system [6] were used. The amplified spontaneous emission (ASE) is ≈5% of the total energy.

The relatively homogeneous middle part of the beam is selected by a ≈3.5 mm diameter circular aperture. The plane of this aperture is imaged onto the sample by a 1 m focal length fused silica lens. The energy incident on the sample is varied by a variable attenuator put just before the aperture. The pulse duration seen by the sample was determined by an autocorrelation measurement at the position of the sample. The intensity on the sample ranged from 10 to 120 GW/cm².

The results of the transmission measurements showed that the overall transmission of the samples is given by the combined effect of multiphoton absorption, light scattering and color-center formation. In order to get information on the net effect of multiphoton absorption the sample was illuminated only for a limited number of shots, until absorption due to color-center formation became important. Care was taken to collect all the transmitted energy including the scattered part which was distributed in a solid angle of $\approx 10^{-2}$ sr. The transmission values showing the net effect

Springer Proceedings in Physics, Vol. 49 Ultrafast Phenomena in Spectroscopy
Editors: E. Klose · B. Wilhelmi

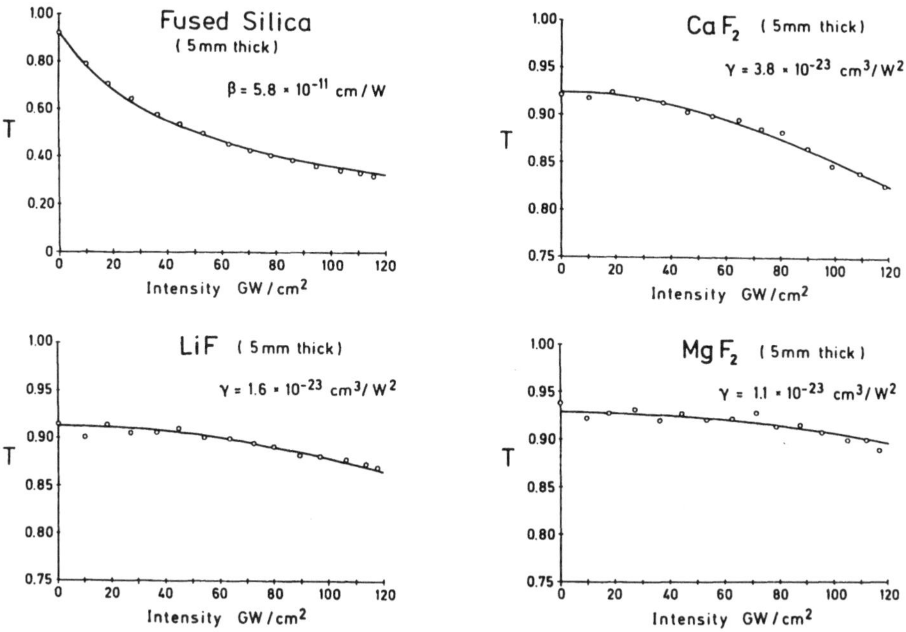

Fig. 1. Measured values of the transmission as a function of intensity. The solid lines are the best fits to the measured data assuming a two-photon absorption mechanism for fused silica, and three-photon absorption for CaF_2, LiF, and MgF_2.

of multiphoton absorption are indicated in Fig. 1 for different materials as a function of power density incident on the sample.

3. Discussion

For fused silica the measured data can be fitted assuming a two-photon-absorption process, while for CaF_2, LiF, and MgF_2 best fits are found assuming a three-photon-absorption mechanism. To determine the two-photon- and three-photon-absorption coefficients we assumed that only a single multiphoton process is involved, and that the input intensity $I_{in}(t)$ is spatially uniform and temporally $sech^2(t)$ shaped. The general expression describing the change of intensity as the pulse propagates through a sample is $dI/dz = -\alpha I^n$, where α is the absorption coefficient. n = 2 and 3 denotes two-photon and three-photon absorption, respectively. From this equation, considering a $sech^2(t)$ pulse shape and the reflection R at each surface of the sample of thickness ℓ, the transmission can be obtained as

$$T = \frac{(1-R)^2}{2} \int_{-\infty}^{\infty} \frac{dt}{[(n-1)\alpha \ell \{(1-R)I_1\}^{(n-1)} + \{\mathrm{sech}(t)\}^{-2(n-1)}]^{1/(n-1)}}. \quad (1)$$

Best fits of Eq. (1) to the measured data are displayed in Fig. 1 by solid curves. For the absorption coefficients the values listed in Table 1 are found. The relative uncertainty of the transmission measurements is ±10 %. (Note that each experimental point in Fig. 1

Table 1. Multiphoton absorption coefficients at 248 nm

Material	n-photon Absorption	Absorption coefficient
Fused silica	n = 2	$\beta = 5.8 \cdot 10^{-11}$ cm/W
CaF_2	n = 3	$\gamma = 3.8 \cdot 10^{-23}$ cm^3/W^2
LiF	n = 3	$\gamma = 1.6 \cdot 10^{-23}$ cm^3/W^2
MgF_2	n = 3	$\gamma = 1.1 \cdot 10^{-23}$ cm^3/W^2

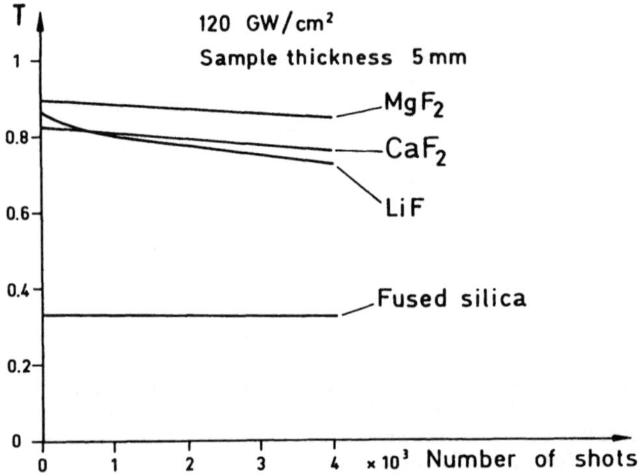

Fig. 2. Transmission of fused silica, CaF_2, LiF, and MgF_2 as a function of the number of shots at 120 GW/cm^2.

represents an average over 10 shots.) Besides the multi-photon processes the other loss mechanisms, excluded in the above measurements, were also studied in a separate measurement. One of these processes is absorption due to color-center formation. We tested this effect at 120 GW/cm^2. In Fig. 2 the change of transmission of the samples is displayed as the number of pulses incident on the same area of the sample increases.

An additional observation is strong scattering of transmitted radiation at powers exceeding ≈100 GW/cm^2. This effect is likely due to nonlinear refraction. Considering that the intensity distribution is not flat-topped but modulated, any kind of intensity dependent refractive index change results in significant scattering. At 120 GW/cm^2 we measured that portion of the transmitted energy which travels to the original direction defined as the direction of the beam at low (10 GW/cm^2) intensity. The remaining scattered portion we then related to the incident energy, as listed in Table 2. Although the values in Table 2 are also connected to our specific beam profile, their relative magnitudes are characteristic of the different materials.

In conclusion, intensity dependent loss mechanisms occurring in window materials at 248 nm were studied. It is found that the loss is partly due to light scattering and absorption. Absorption in fused silica is mainly two-photon absorption, while in CaF_2, LiF, and MgF_2 the combined effect of color-center formation and

Table 2. Ratio of the scattered to the incident energy for different materials at 120 GW/cm²

	Fused silica	CaF_2	LiF	MgF_2
$\frac{I_{scattered}}{I_{incident}}$ at 120 GW/cm²	0.06	0.33	0.05	0.11

three-photon absorption must be considered. The evolution of the absorption due to color-center formation and the values of the two- and three-photon-absorption coefficients for the different materials are also given.

The permanent adress of P.S. and S.S. is JATE Univ., Szeged, Dóm tér 9., H-6720, HUNGARY

References

1. A.P. Schwarzenbach, T.S. Luk, I.A. McIntyre, U. Johann, A. McPherson, K. Boyer, and C.K. Rhodes, Opt. Lett. 11, 499 (1986).
2. J.H. Glownia, J. Misewich, and P.P. Sorokin, J. Opt. Soc. Am. B 3, 1573 (1986).
3. S. Szatmári, F.P. Schäfer, E. Müller-Horsche, and W. Mückenheim, Opt. Commun. 63, 305 (1987).
4. J.P. Roberts, A.J. Taylor, P.H.Y. Lee, and R.B. Gibson, Opt. Lett. 13, 734 (1988).
5. M. Watanabe, A. Endoh, N. Sarukura, and S. Watanabe, Appl. Phys. B 48, 417 (1989).
6. S. Szatmári and F.P. Schäfer, Opt. Commun. 68, 196 (1988).
7. J.R.M. Barr, N.J. Everall, C.J. Hooker, I.N. Ross, M.J. Shaw, W.T. Toner, Opt. Commun. 66, 127 (1988).
8. A.J. Taylor, R.B. Gibson, and J.P. Roberts, Opt. Lett. 13, 814 (1988).
9. A. Endoh, private communication.

On the Extension of the Matrix Formalism for the Calculation of Group Delay and Group Velocity Dispersion in Linear Optical Elements

E. Döpel

Sektion Physik, Friedrich-Schiller-Universität Jena, Max-Wien-Platz 1, DDR-6900 Jena, German Dem. Rep.

Abstract. In the frame of geometrical optics a formalism for the calculation of group delay and group velocity dispersion in linear optical elements from their angular dispersion and the wavelength dependence of their ABCD-parameters will be demonstrated.

1. Introduction

The parameters of ultrashort light pulses can be considerably influenced in linear optical elements by a frequency dependent phase shift $\phi(\omega)$ leading to group delay (GD) and group velocity dispersion (GVD). This phase shift can originate (i) from a frequency dependent reflection or transmission especially at dielectric coated elements or (ii) from a frequency dependent optical path length in systems with angular dispersive elements. For the first type of elements there exist well known general formalisms for the calculation of $\phi(\omega)$. But elements of the second type have been treated in numerous publications only for special arrangements. However, it would be desirable to have a general formalism for such systems too. In the paraxial approximation there is the well known ABCD-matrix formalism for the calculation of the imaging properties of such elements. Recently Martinez /1/ has shown how this formalism can be extended to the description of the GVD of Gaussian beams in arbitrary elements. In /2/ it is shown in the frame of geometrical optics how the optical path length within linear optical elements (as an equivalent of the phase delay) can be concluded from their propagation matrix and their angular dispersion.

2. The optical path length in systems of linear optical elements

We consider an arbitrary system consisting of a sequence of N elements. In the paraxial approximation we have to assume for each element (i=1..N) a linear dependence of the output beam parameters (x_i, x'_i) on the input beam parameters (x_{i-1}, x'_{i-1}) which leads to the well known ABCD-matrix formalism /3/:

$$\begin{bmatrix} x_i \\ x'_i \\ 1 \end{bmatrix} = \begin{bmatrix} A_i & B_i & a_i \\ C_i & D_i & a'_i \\ 0 & 0 & 1 \end{bmatrix} \begin{bmatrix} x_{i-1} \\ x'_{i-1} \\ 1 \end{bmatrix} . \tag{1}$$

The matrix elements A_i, B_i, C_i and D_i may exhibit a wavelength dependence. The parameters a_i and a'_i describe an additional lateral shift and angular deviation of a beam of arbitrary wavelength from that of the reference wavelength ω_0 respectively.

The frequency dependent optical path length originates from the material dispersion and different paths of the spectral content of the input beams due to angular dispersion at the interfaces.

Springer Proceedings in Physics, Vol. 49 Ultrafast Phenomena in Spectroscopy
Editors: E. Klose · B. Wilhelmi

We assume a reference beam with reference frequency ω_o to be the optical axis of the arbitrary system. This beam traverses the N elements of this system, which consist of material with refractive index $n_i(\omega)$, with a geometrical path length l_i.

We decompose the system into elements with
(a): $a_i=a'_i=0$, but $l_i\neq0$ (for instance centered lens systems) or
(b): $a_i=l_i=0$, but $a'_i\neq0$ (for instance gratings). In /2/ the optical pathlength of an arbitrary sequence of N elements is shown to be

$$P = \sum_{i=1}^{N} (n_i l_i + \frac{1}{2} x_i a'_i) + \frac{1}{2} (x_N x'_N - x_o x'_o). \tag{2}$$

According to our decomposition there is possible only the first term in the sum for elements of type (a) or the second term for elements of type (b). The beam parameters x_i, x_N and x'_N have to be calculated by successive application of equation (1) from x_o and x'_o. Due to the quadratic dependence of P on the beam parameters equation (2) is valid only in the paraxial approximation. The influence of imaging errors on P will occur in higher orders of the beam parameters. Owing to a possible λ-dependence of these beam parameters the general validity of the λ-derivatives of P is consequently also restricted up to the second order.

From the frequency dependent phase shift $\phi(\omega)=\omega P(\omega)/c$ and $\omega=2\pi c/\lambda$ the GD τ and GVD $d\tau/d\omega$

$$\tau = \frac{d\phi}{d\omega} = \frac{1}{c} (P - \lambda \frac{dP}{d\lambda}) \tag{3}$$

$$\frac{d\tau}{d\omega} = \frac{d^2\phi}{d\omega^2} = \frac{\lambda^3}{2\pi c^2} \frac{d^2P}{d\lambda^2} \tag{4}$$

can be calculated.

3. Examples

In /2/ different examples of the application of this formalism are presented.

With the help of equations (2) and (3) the differences of the GD ($\sim c\tau$) and phase delay ($\sim P$) in dependence on the output coordinate x_N can be calculated. In the case of elements with angular dispersion ($da'/d\lambda\neq0$) these differences lead to a small tilting angle α of the pulse and phase fronts respectively (/4/) which is described by equation (5)

$$\alpha = - \lambda \frac{da'}{d\lambda}. \tag{5}$$

The application to focusing elements ($A=D=1$, $C\neq0$, $B=l/n$) with chromatical aberrations ($dC/d\lambda\neq0$) allows us to calculate the dependence of the GD on x_N. For a plane input wave we get

$$c\tau(x_N) = (n - \frac{dn}{d\lambda}) l + \frac{1}{2} (C - \lambda\frac{dC}{d\lambda}) x_N^2. \tag{6}$$

From equation (6) it can be concluded that the curvature of the pulse front ($C-\lambda*dC/d\lambda$) is different from that of the pulse front C (/5/).

The analysis of equations (2) and (4) enables the discussion of GVD effects in any arrangement with dispersive elements. In /2/ it was shown that the application of this formalism to the two-prism compressor leads to the result of /4/.

Another interesting example is the application to a laser cavity with dispersive elements. For instance, in femtosecond lasers prism sequences for intracavity chirp control are necessary. In the case of a two-prism compressor there is a remaining lateral shift for different wavelengths. In general the action of these elements can be understood as a wavelength dependent misalignment which causes different optical axes and different cavity perimeters for the spectral content of the pulses. Therefore the dispersion characteristics of these systems inside and outside the cavity may differ considerably.

From the condition of the reproduction of the optical axis every round trip, its wavelength dependent beam parameters at an arbitrary starting point $(\bar{x}_0(\lambda), \bar{x}'_0(\lambda))$ can be found from equation (7):

$$\begin{bmatrix} \bar{x}_0 \\ \bar{x}'_0 \\ 1 \end{bmatrix} = \prod_{i=1}^{N} \begin{bmatrix} A_i & B_i & 0 \\ C_i & D_i & a'_i \\ 0 & 0 & 1 \end{bmatrix} \begin{bmatrix} \bar{x}_0 \\ \bar{x}'_0 \\ 1 \end{bmatrix}, \tag{7}$$

which has a nontrivial solution if a'_i is not vanishing for all elements (i=1..N). With equation (2) we achieve for the cavity perimeter

$$P(\lambda) = \sum_{i=1}^{N} \left(n_i(\lambda) l_i + \frac{1}{2} \bar{x}_i(\lambda) a'_i(\lambda) \right) . \tag{8}$$

From equations (4) and (8) the resulting GVD per cavity round trip can be calculated. Examples for special cavities are given in /6/ and /7/.

4. Conclusion

In conclusion we have demonstrated that the wavelength dependence of the optical path length of arbitrary beams in systems of linear optical elements in the paraxial approximation can be determined from their angular dispersion, the wavelength dependence of the ABCD-parameters and the material dispersion of the reference beam. This general formalism is suitable for the calculation of group delay and group velocity dispersion for any system either analytically or numerically. Especially this has been shown for laser cavities with angular dispersive elements.

References.

1. O.E. Martinez: IEEE Journ.Quant.Electr. **QE24**,2530 (1988)
2. E. Döpel: Journ. of Mod. Optics, in press
3. A. Gerrard, J.M. Burch: Introduction to matrix methods in optics (Wiley, London 1975)
4. Z. Bor, B. Rasz: Opt. Commun. **54**, 165 (1985)
5. Z. Bor: Opt. Lett. **14**, 119 (1989)
6. O.E. Martinez: IEEE Journ. Quant. Electr. **QE25**, 296 (1989)
7. E. Döpel: Journ. of Mod. Optics **36**, 1329 (1989)

Limits of Pulse Generation in cw Mode-Locked Dye Lasers

V. Petrov[1], *W. Rudolph*[2], *U. Stamm*[2], *and B. Wilhelmi*[3]

[1]Department of Physics, Sophia University, B-1126 Sophia, Bulgaria
[2]Sektion Physik, Friedrich-Schiller-Universität Jena, Max-Wien-Platz 1, DDR-6900 Jena, German Dem. Rep.
[3]Zentralinstitut für Optik und Spektroskopie, Akademie der Wissenschaften, Rudower Chaussee 6, DDR-1199 Berlin, German Dem. Rep.

1. Introduction

For the study of very fast phenomena, the periodic trains of light pulses generated by cw mode-locked lasers are very attractive since powerful signal averaging and processing techniques are available. Various methods for pulse generation have been developed to satisfy different requirements for the pulse parameters.

Here we discuss results of a numerical model with respect to essential mechanisms limiting the achievable pulse duration in synchronously pumped, hybridly mode-locked and passively mode-locked cw subpicosecond dye lasers. The action of spontaneously emitted light is shown to be the detrimental factor in synchronous mode-locking. The influence of this factor can be suppressed by a saturable absorber in the hybridly mode-locked lasers. In this case, as well as in the case of passive mode-locking, the combined action of group velocity dispersion and Kerr-effect-type nonlinearity, though serving as an additional shortening mechanism, causes fluctuations of the pulse parameters and determines the limit of the steady-state regime.

2. Round trip model

The round trip model rests on a general resonator scheme which is shown above. A steady-state for this system can be looked for in the sense of a slowly varying complex amplitude of the electric field of

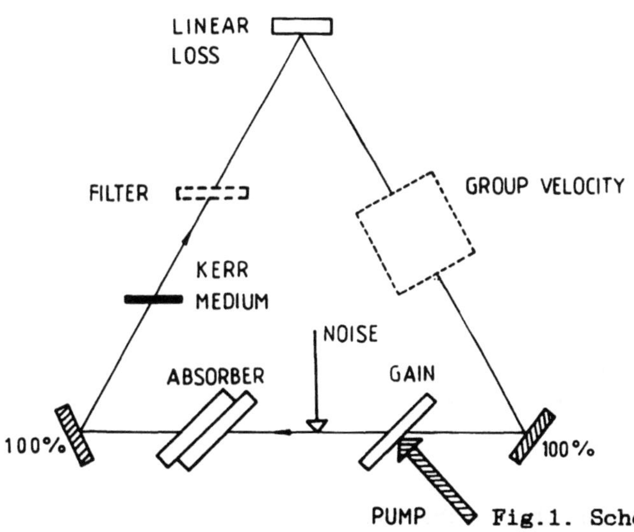

Fig.1. Schematic of the cavity model

Springer Proceedings in Physics, Vol. 49 Ultrafast Phenomena in Spectroscopy
Editors: E. Klose · B. Wilhelmi

the intracavity pulse $\bar{E}(\eta) = |\bar{E}(\eta)| \cdot \exp(i\varphi(\eta))$ that reproduces itself after one cavity round trip. The quantities characterizing the steady-state pulse are the pulse intensity and the pulse chirp $\bar{\dot{\omega}} \propto \frac{d}{d\eta}\varphi(\eta)$.

3. Synchronous Mode-Locking

From the ring cavity model we have taken into account the active dye, the nonsaturable loss and the filter element. The pulse evolution is simulated from noise by calculating the change of the electric field round trip after round trip /1/ according to

$$\bar{E}^{(i+1)}(\eta+h) = G(\eta)\,\bar{E}^{(i)}(\eta) + \bar{E}_S(\eta) \tag{1}$$

with $G(\eta)$ as resonator transfer function

and $\bar{E}_S(\eta)$ as spontaneously emitted light.

The steady-state is defined to be reached if the intensity reproduction is better than 0.1 % after one round trip.

The results of the computer simulation show that spontaneous emission leads to a considerable pulse broadening and modifies the range of the geometrical cavity mismatch in which single pulses can evolve. In Fig. 2 the normalized pulse intensity is plotted for different cavity length mismatch h/τ_F between pump laser and dye laser. The substructure of the pulses generated for cavity mismatch h/τ_F of 1.25 is caused by the stochastic background of spontaneous emission. Therefore all pump and laser parameters as pump energy and intracavity filter bandwidth have to be adjusted so that the cavity mismatch necessary for single pulse operation is less than 1. This in fact limits the minimum pulse width achievable by pure synchronously pumped lasers.

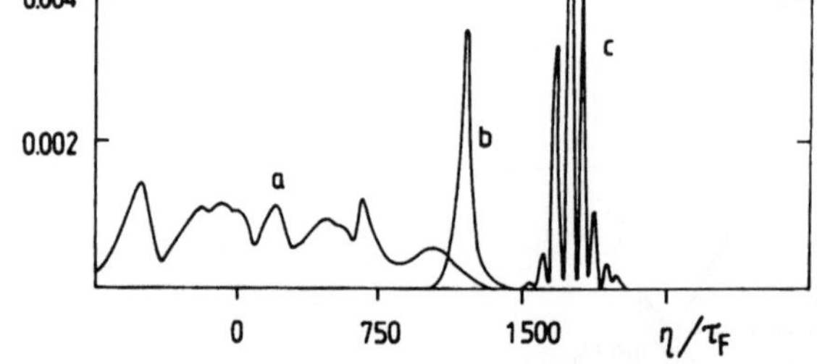

Fig.2. Normalized pulse intensity in the case of synchronous mode-locking for different cavity length mismatch h/τ_F =-1.25 (a), 0 (b), and 1.25 (c).

4. Hybrid Mode-Locking

In addition to the synchronous pumping scheme from the last section a saturable absorber is placed in the cavity and $\bar{G}(\eta)$ in (1) is supplemented by the corresponding term. This leads to a suppression of spontaneously emitted light at the leading edge of the laser pulse where the absorber is not saturated. In Fig. 3 the cavity mismatch behaviour of the hybridly mode-locked laser is considered. The parameters are chosen in such a way that a direct comparison with Fig.2 is possible. In particular, the losses have equal magnitude, but now they consist of two equal parts - a saturable and a non-saturable contribution. We find that hybrid mode-locking provides shorter pulse duration and increased peak intensity and the steady-state covers a broader h -range.

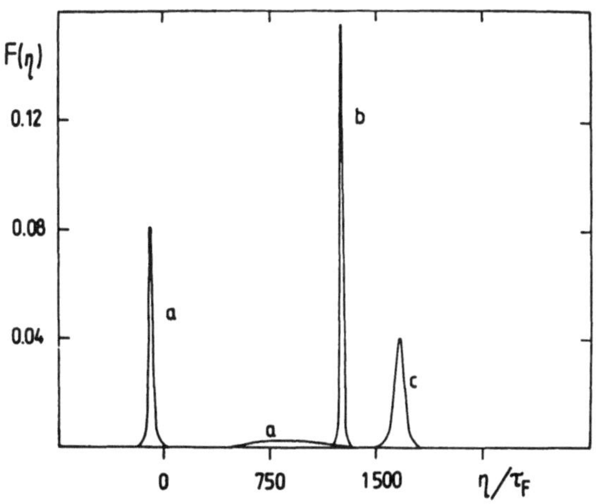

Fig.3. Normalized pulse intensity in the case of hybrid mode-locking for different cavity length mismatch (same values as in Fig.2).

5. Passive Mode-Locking

The round trip model includes all elements presented in Fig. 1 except the passive frequency filter. The only bandwidth limiting effect arises then from the active dyes and the saturable absorber through action of their finite transition profiles described by the complete set of Maxwell and Bloch equations.
The electric field changes in one round trip according to

$$\bar{E}'(\eta+h) = \bar{E}(\eta) + \sum_i \delta_i \bar{E}(\eta) \qquad (2)$$

where $\delta_i \bar{E}(\eta)$ describes the pulse modification in the respective element (i) and for a steady-state

$$\sum \delta_i \bar{E} = 0$$

is required.

We find that the combination of intracavity dispersion with nonlinearity of the opposite sign provides, however, an additional shortening mechanism. The shortest pulses occur at maximum possible nonlinearity just before instabilities set in and they are slightly chirped (see Fig.4).

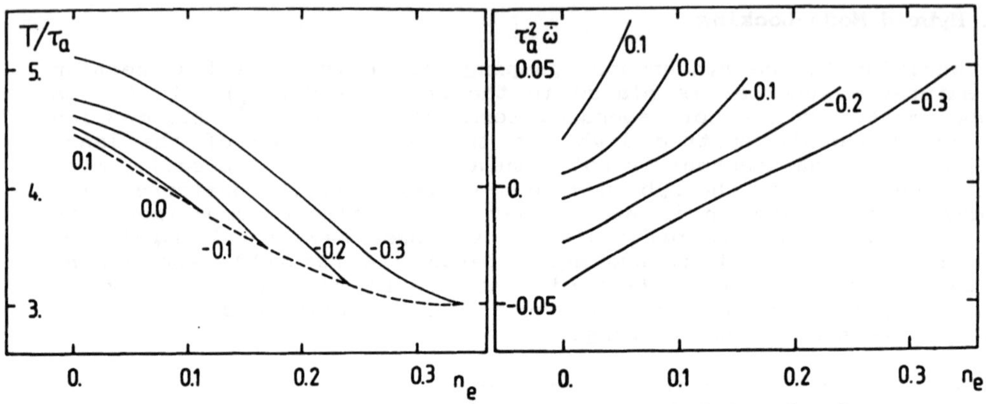

Fig.4. Normalized steady-state pulse duration (left) and pulse mean chirp parameter vs nonlinearity n_e and dispersion r (values indicated on the curves).

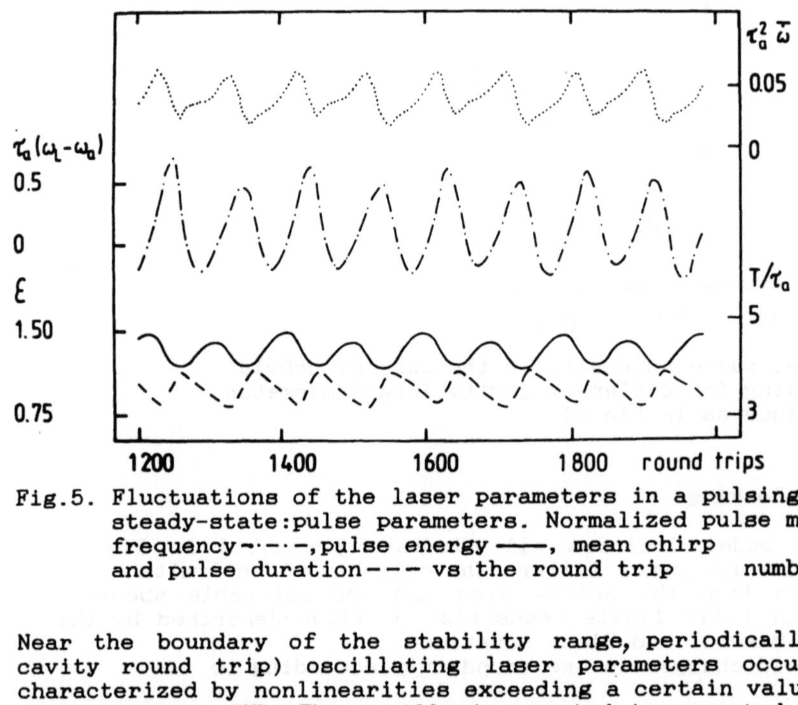

Fig.5. Fluctuations of the laser parameters in a pulsing steady-state:pulse parameters. Normalized pulse mid frequency -·-·-, pulse energy ——, mean chirp and pulse duration - - - vs the round trip number.

Near the boundary of the stability range, periodically (with the cavity round trip) oscillating laser parameters occur. They are characterized by nonlinearities exceeding a certain value and by too small negative GVD. The oscillation period is expected to be in the µs range which agrees with experimental findings /2,3/.

References

1. V. Petrov, W. Rudolph, U. Stamm, B. Wilhelmi, Phys. Rev. A 40, 1474 (1989).
2. H. Avramopoulos, P. M. W. French, J. A. R. Williams, G. H. C. New, J. R. Taylor, IEEE J. QE-24, 1884 (1988).
3. F. Salin, P. Grangier, G. Roger, A. Brun, Phys. Rev. Lett. 56, 1132 (1986).

A Dye Laser Synchronously Pumped by Intense Subpicosecond Pulses from a Negative Feedback Stabilized Nd:Glass Laser

T. Damm[1], *U. Stamm*[1], *K. Vogler*[1], *Š. Burdulis*[2], *R. Grigonis*[2], *A. Piskarskas*[2], *G. Sinkevičius*[2], *and V. Sirutkaitis*

[1]Sektion Physik, Friedrich-Schiller-Universität Jena, Max-Wien-Platz 1, DDR-6900 Jena, German Dem. Rep.

[2]Laser Research Centre, Vilnius University, Sauletekio Ave. 10, 232054 Vilnius, Lithuania

Due to their large amplification bandwidth dye lasers are broadly used to generate very short light pulses. There are several reasons to improve the parameters of these lasers. One among other goals is to have a very simple and reliable laser system which provides a stable source of subpicosecond laser pulses. In this paper we report on a jet stream dye laser which is mode locked by synchronously pumping with the pulse train of a solid state laser. The schematic of the experimental equipment is shown in Fig.1. The pump source for the dye laser is a passively mode-locked Nd:glass laser which is stabilized by an actively controlled negative feedback loop. This loop is introduced in order to obtain an increased duration of the emitted pulse train in connection with a reduced width of the pulse itself.

A part of the intracavity radiation is directed onto the photodetector PD (response time about 6 ns) and the electrical signal is applied to the Pockels´ cell PC after amplification to 500V with a time delay of about 15 ns. This arrangement enables us to keep the amplitude of the circulating light pulse on a certain level, which could be chosen by means of adjusting the amplification of the PD signal. As a result, the nonlinear stage of the pulse build-up period could be extended to more than 10^3 cavity round trips. Provided the intensity of the laser pulse is kept approximately on the level of saturation intensity of the saturable absorber SA, such an extension results in the reduction of the pulse duration to less than 1 ps. In addition, the instability of both the duration and the energy of the pulses is reduced. A more detailed description of the pump laser system is given in /1/. The output of the laser is a flat pulse train of 10-12μs duration. It passes the two stage amplifier

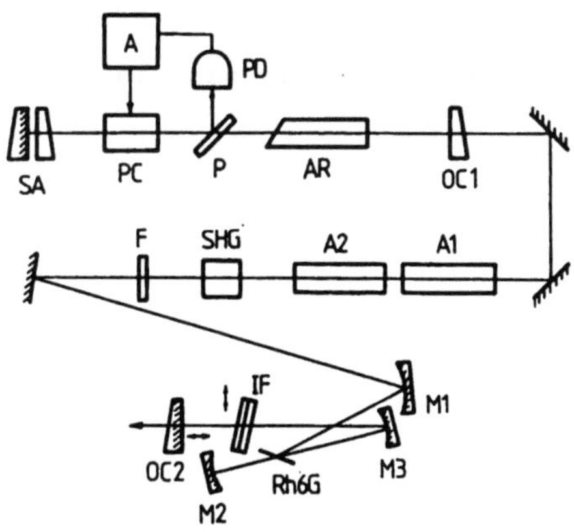

Fig.1. Scheme of the experimental equipment

system A1 and A2 and is frequency doubled in a KDP crystal. A maximum energy of a single second harmonic pulse of 10μJ is obtained. The pump beam is focused into the jet stream (Rh6G, 0.3mm thickness) by the mirror M1 of 30cm radius curvature. The linear cavity of the dye laser is formed by two concave mirrors M2 and M3 (radii of curvature 10cm and 7.5cm, both 100% reflectivity) and a flat 70% transmission outcoupling mirror OC2 mounted on a translation stage. The dye laser pulse width is measured by noncollinear second harmonic generation in a 0.3mm thick KDP crystal, whereas the spectrum was monitored by a grating spectrograph.

The oscillation threshold of the dye laser without an intracavity filter was reached at a single pump pulse energy of 1μJ. At a pump energy of about 3μJ, the shape of a dye laser train is still similar to that of a pump laser, whereas at a higher pump energies the trailing part of the train is suppressed. All the subsequent investigations were carried out under a moderate pump power, corresponding to a threshold exceeding factor of 3 to 4.

The average energy of a single dye laser pulse is 100nJ and it did not change considerably within a cavity length detuning of several millimeters. A more accurate cavity length adjustment is obtained by monitoring the intensity of the second harmonic pulse of the dye laser. This reveals a several tens of microns wide range of the cavity detuning where the shortest pulses were expected. Fig.2 shows typical autocorrelation functions. The relative cavity detuning value is considered to be 0 where the shortest pulses have been recorded. The autocorrelation function of the shortest pulse is depicted in Fig.2b. Obviously a noise burst of about 300fs width at quarter-maximum /2/ is emitted, indicated by the coherence spike of the autocorrelation trace.

The spectral width of these pulses is up to 15nm, centered at 572nm. Detuning the cavity length for a few microns results in the generation of backround-free pulses, however, their duration is increased (Fig.2 a, c) while the spectral width is decreased.
In order to obtain a perfect mode-locking, the amplification bandwidth of the dye laser is limited by introducing an intracavity filter IF (Fig.1).

In a first experiment a Fabry-Perot interferometer (FPI) was used as intracavity filter. The FPI had a base of 6μm and the reflectivity of its mirrors was 77% resulting in a single-pass transmission bandwidth of 2.5 nm. The obtained autocorrelation functions revealed a good mode-locking, however, the pulse width was found to be not less than 900fs FWHM, while the spectral bandwidth was about 1 nm. It should be mentioned that a tuning of the laser wavelength in a range from 560 nm up to 590 nm has been achieved by tilting the interferometer.

Secondly, an interference filter of 10 nm transmission bandwidth was used. Fig.3 shows the autocorrelation functions obtained in this case. The pulses are practically background-free and represent a

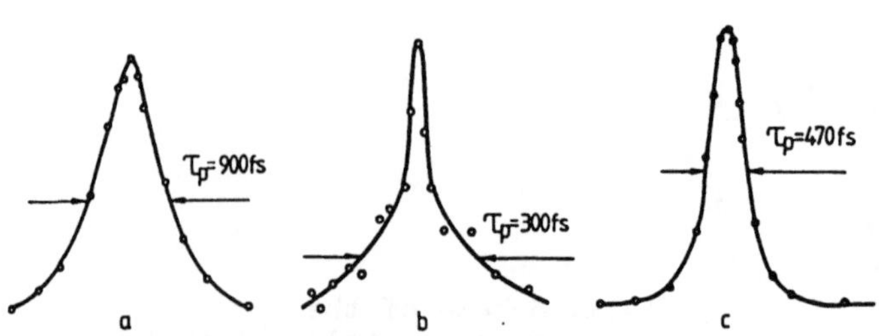

Fig.2. Autocorrelation functions of the dye laser pulses at:
(a):-3μm, (b):0μm, (c):4μm cavity detuning (no filter used)

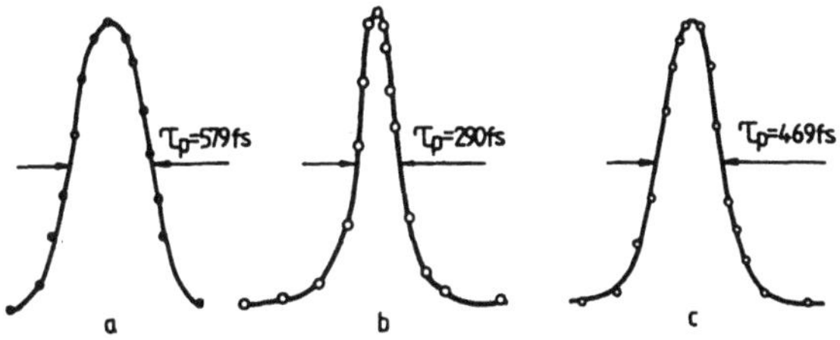

Fig.3. Autocorrelation functions of the dye laser pulse with an interference filter of 10 nm bandwidth at: (a):-22μm, (b):0μm, (c):28μm cavity length mismatch

fairly good mode-locking. The minimum pulse width of 290 fs FWHM is obtained from the autocorrelation function in Fig.3b by deconvolution. The spectral bandwidth of this pulse was found to be 4nm, which results in a time-bandwidth product of nearly 1 (a transform-limited $sech^2$ shaped pulse would yield 0.315). The energy conversion efficiency of the dye laser with an intracavity filter was found to be approximately 3% which leads to an energy of a single dye laser pulse of up to 60nJ.

References

/1/ K.Burneika, R.Grigonis, A.Piskarskas, G.Sinkevicius, V.Sirutkaitis: Kvantovaja Elektronika 15, 1658 (1988)
/2/ A.M.Johnson, W.M.Simpson: J.Opt.Soc.Am.B 2, 619 (1985)

Beta-Barium Borate Widely Tunable Picosecond Optical Parametric Oscillators

Š. Burdulis [1], *R. Grigonis* [1], *A. Piskarskas* [1], *G. Sinkevičius* [1], *V. Sirutkaitis* [1], *J. Nolting* [2], *and R. Wallenstein* [2]

[1] Laser Research Centre, Vilnius University, Sauletekio Ave. 10, 232054 Vilnius, Lithuania

[2] Institut für Quantenoptik, Universität Hannover, D-3000 Hannover, Fed. Rep. of Germany

Beta-barium borate or BBO was reported as a new type of nonlinear medium [1,2] and soon its efficiency in frequency upconversion to UV [3,4] as well as in optical parametric oscillation [5,6] was investigated. More recently, experiments on a widely tunable optical parametric amplification pumped by the third harmonic of a Q-switched YAG:Nd laser [7] and an optical parametric oscillation in the BBO synchronously pumped at 532nm by picosecond pulses [8] have been carried out. The BBO was shown to be an attractive medium for the nonlinear optical frequency conversion especially in visible and UV regions due to being transparent in 200-2600 nm range together with both the optical damage threshold and the second order nonlinearity several times higher than those of KDP. In addition, a strong birefringence of the BBO provides phase-matching conditions for a wide range of wavelengths.

Here we report on the operation of BBO optical parametric oscillators (OPO) synchronously pumped by the second and third harmonics of a picosecond YAP:Nd laser. In contrast to previous works, widely tunable oscillators were run and the investigation of the output spectrum narrowing has been carried out.

The master oscillator was a passively mode-locked YAP:Nd laser providing trains of ultrashort pulses (USP). The average train consisted of 10 pulses of 25ps duration. The oscillator was separated from the amplifiers by the Pockels' cell shutter in order to suppress the positive feedback during the initial stage of the USP build-up.

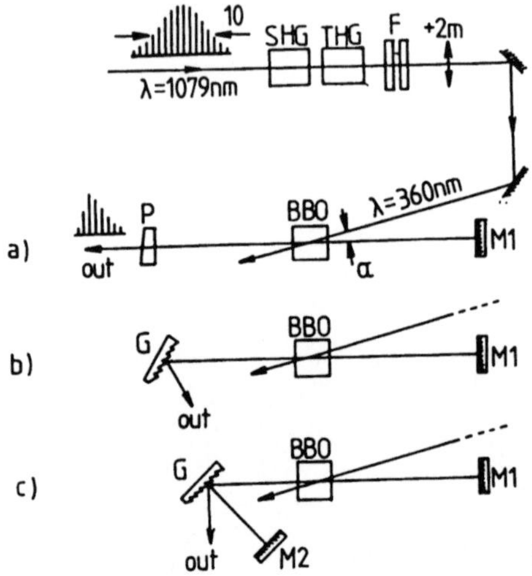

Fig. 1. Experimental set-up.

Springer Proceedings in Physics, Vol. 49 **Ultrafast Phenomena in Spectroscopy**
Editors: E. Klose · B. Wilhelmi

At the output of the amplification system the total train energy was 50mJ. The rest of the experimental system is shown in Fig.1. A KDP crystal of 25mm length was used for the type-I phase-matched frequency doubling (SHG) and another KDP crystal of 30mm length was added when the OPO was being pumped by the third harmonic for the type-II phase-matched sum frequency generation (THG). The filter F was used to suppress the residual fundamental and second harmonic radiation. The maximum total pump train energy was 10mJ at 360nm. The pump beam was focused by a 2 meter spherical lens and the spot diameter was 1.5mm in the BBO sample (dimensions 6x6x12 mm) placed 30cm ahead of the beam focus. The values of the group velocity mismatch and the beam walk-off were calculated to be 7cm and 2cm respectively and in our case they could not be efficiency-limiting.

The cavity of the OPO was formed by a full reflector (M1) and the output coupler. The output coupler was either a glass plate (P) or a 1200 lines/mm diffraction grating (G). The M1 reflector was mounted on a translation stage to match the optical lengths of the OPO and the pump oscillators. The OPO was pumped noncolinearly to enable the use of an Al-coated full reflector for wide-range wavelength tuning.

The results in the cases of the second harmonic (SH) and the third harmonic (TH) pump are similar and here we shall concentrate on the latter case, though the main data for the former will be given too. At first the pump noncolinearity was approximately 15mrad (as low as allowed by the pump beam and M1 mirror spacing). A glass plate was used as the output coupler (Fig. 1a). The threshold of optical parametric oscillation was reached at 0.5GW/cm^2 pump intensity, while under the most intensive pump (2.2GW/cm^2) the total OPO energy conversion efficiency was 10%. The tuning range was 406-3170nm. (In the case of the SH pump 0.2GW/cm^2, 14%, 680-2570nm respectively).

In the case of the TH pump the parametric luminescence is visible, and its distinct pattern was observed several degrees aside of the OPO output. When the OPO cavity was arranged in the direction of the luminescence (roughly 50mrad with respect to the pump beam), the OPO efficiency increased significantly. This is demonstrated in Fig. 2. where pump intensity dependencies of the OPO efficiency are given for the cases of 15mrad (1) and 50mrad (2) noncolinearity. All the subsequent investigations of the OPO pumped by the TH were performed under the latter conditions. On the contrary, in the case of the SH pump the parametric luminescence was invisible and no such rearrangement was made.

The output spectrum bandwidth was found to vary from 50nm at degeneracy to a few nanometers at the edge of a tuning range. This is far above the transform limit assuming the 20ps pulse duration. Moreover, the microdensitograms of signal wave spectra (shown in

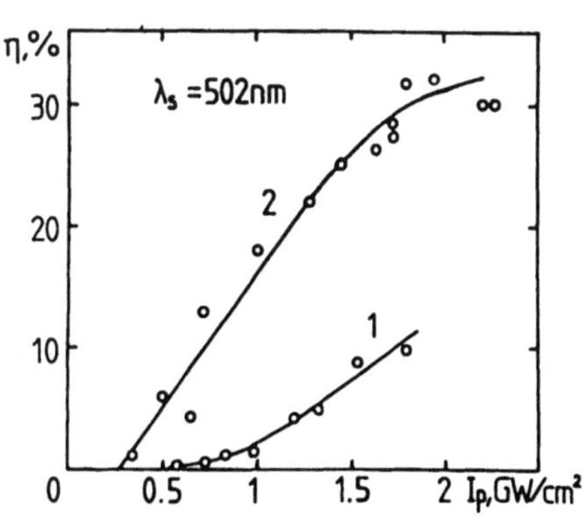

Fig. 2. Energy conversion efficiency versus pump intensity.

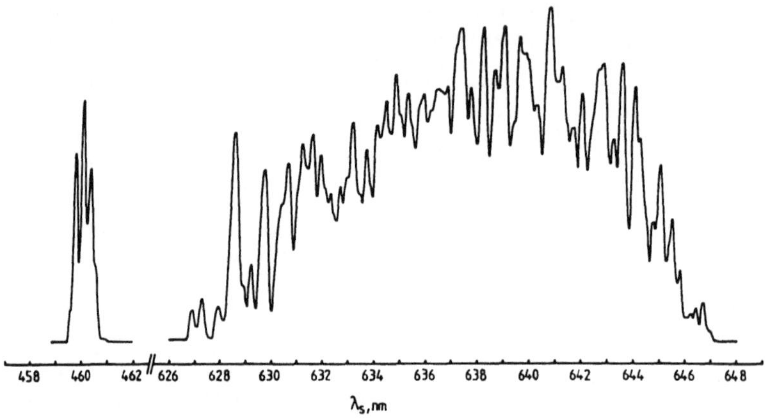

Fig. 3. Microdensitograms of the output spectra of the OPO without bandwidth limitation.

Fig. 3.) revealed a considerable modulation of the envelope which persists over the entire tuning range. The same feature was also found in the case of the SH pump.

In order to narrow the output spectrum a bandwidth limitation has been introduced into the OPO cavity. Fig. 1b. shows the OPO employing diffraction grating in an autocolimation mode as an output coupler. The zero-order of diffraction served as an output of the OPO. The resulting output spectrum bandwidth has been found to be 0.5nm (at a signal wavelength of 502nm). The total OPO efficiency has also dropped to 3% which, besides the spectral selection, may in part be explained by a change of output coupling rate from 4% to 70%.

Further narrowing of the output spectrum has been achieved by employing an additional full reflector M2 (Fig. 1c.). The diffraction grating was turned out of the autocolimation and the I-st order of diffraction was directed to the M2. The latter was adjusted to return the I-st order beam back to the grating and to the cavity. Hence, the angular dispersion has been increased. As a result, the output spectrum bandwidth has been reduced to 0.24nm (at 502nm). The energy conversion efficiency has increased to 8.5%, which may in part be due to a larger incidence angle on the grating, leading to a stronger reflection into zero-order used as an output.

Fig. 4. shows the sequence of the output spectrum envelopes recorded at the same 502nm wavelength and corresponding to the sequence of the optical parametric oscillators discussed above and shown in Fig. 1a-c. The last envelope (Fig. 4c.) has no evident modulation and gives a time-bandwidth product of 5, which is approximately 10 times higher than that for a transform-limited pulse, but, on the other hand, the time-bandwidth product of the pump has been measured to be approx. 6. Provided the pump pulse is transform-limited, the OPO output spectrum may be even more narrow.

In the case of the SH pump, a similar spectral behaviour was observed. However, energy conversion efficiency in the case of the OPO with the grating was hardly measurable and has been roughly evaluated to be 1%.

In conclusion, the picosecond widely tunable optical parametric oscillation in a type-I phase-matched BBO crystal synchronously pumped by either the second or the third harmonic of a YAP:Nd laser has been demonstrated. A summary of the experimental data is offered in Table 1.

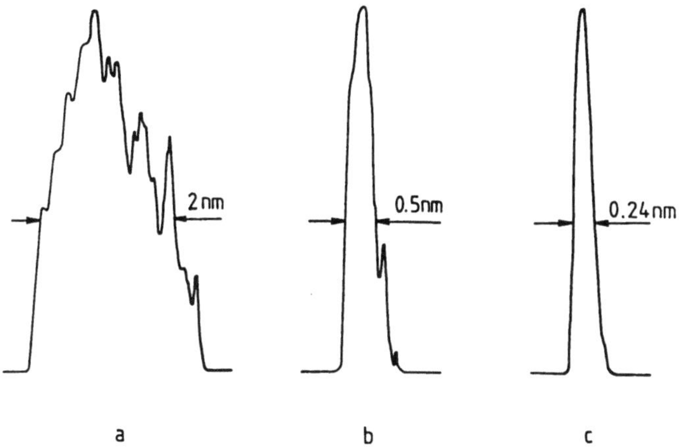

Fig. 4. Comparison of the output spectra of the OPO employing various cavity configurations.

Table 1. Summary of the experimental data.

Type of OPO as shown. in ...	SH pumped		TH pumped	
	energy conversion efficiency, %	spectrum width in nm at wavelength 690nm	energy conversion efficiency, %	spectrum width in nm at wavelength 502nm
Fig.1a	14	20	30	2
Fig.1b	1	0.6	3	0.5
Fig.1c	1	0.3	8.5	0.24

References

1. C.Chen, B.Wu, G.You, A.Jiang, Y.Huang: Dig.Tech.Papers XIII Int. Quant.Electr.Conf. (1984)
2. C.Chen, B.Wu, A.Jiang, G.You: Sci.Sinica(Ser.B) 28, 235 (1985)
3. K.Kato: IEEE J.QE-22, 1013 (1986)
4. C.Chen, Y.X.Fan, R.C.Eckardt, R.L.Byer: Digest CLEO'86, paper THQ4
5. Y.X.Fan, R.C.Eckardt, R.L.Byer, C.Chen, A.D.Jiang: Digest CLEO'86, postdeadline paper Tht4
6. C.Chen, Y.X.Fan, R.C.Eckardt, R.L.Byer: Proc.Soc.Photo-Opt. Instrum.Eng. 681, 12 (1986)
7. H.Vanherzeele, C.Chen: Appl. Opt. 27, 2534 (1988)
8. L.J.Bromley, A.Guy, D.C.Hanna: Opt. Commun. 67, 316 (1988)

Generation and Amplification of Picosecond Pulses in the 10 μm Region

V.M. Gordienko, Z.A. Biglov, and V.A. Slobodyanyuk

Moscow State University, Physical Department, 119899 Moscow, USSR

Abstract. Picosecond pulses, generated by parametric amplification of low-power cw 10 μm radiation, have been regeneratively amplified in a multiatmosphere CO_2 laser to an energy ≈25 mJ circulating in the cavity. The characteristics of the injected and amplified pulses have been investigated. Computer simulations of the parametric amplification process have been performed to investigate the potential possibilities of this method.

1. Introduction

The development of high-power picosecond sources in the midinfrared spectral region can open new areas of physics, chemistry and biology to experimental investigation. The potential applications may be connected not only with studying ultrafast processes, but also with obtaining the highest intensities of radiation.

This paper describes the generation of short 10 μm pulses using parametric amplification of weak cw CO_2 laser light in nonlinear crystals [1]. The low-power short pulses are regeneratively amplified in a multiatmosphere TE-CO_2 slave laser [2]. Computer simulation of the parametric amplification process was performed, and its brief description is also presented.

2. Computer simulation

Some discrepancies in the expected values of the parametric amplification efficiency and the experimental results have forced us to study the process numerically [3]. In our experiments, the proustite (Ag_3AsS_3) crystal was used as a medium for parametric amplification of the cw 10 μm radiation in the field of ultrashort pulse of $YAlO_3:Nd^{3+}$ laser operated on the 1.08 μm wavelength. As a semiconductor with the energy gap E_g=2.04 eV [4], the proustite crystal exhibits two-photon absorption (TPA) of 1.08 μm pulses, the TPA coefficient being equal to 0.15 cm/GW [5]. We considered the

Springer Proceedings in Physics, Vol. 49 **Ultrafast Phenomena in Spectroscopy**
Editors: E. Klose · B. Wilhelmi

influence of free charge carriers created in the medium by TPA of the travelling pump pulse on the parametric amplification process. This influence was accounted for by including to the right side of the well known truncated equations, describing the nonlinear interaction of three waves, a member

$$P_e = \frac{cn}{2\pi\omega}\eta E ,$$

which presents the free-carrier contribution to the polarisation [6]. Here $\eta=2\pi e^2 N_e(1+i\gamma)/cnm^* \omega(1+\gamma^2)$, with e the electron charge, m^* the effective mass, N_e the free-carrier density; $\gamma=(\tau_e\omega)^{-1}$, with τ_e the free-carrier momentum relaxation time. Ignoring the diffusion and the recombination of free carriers on picosecond time scale, the population density N_e is given by

$$\frac{\partial N}{\partial t} = \frac{\beta}{2\hbar\omega} I^2 ,$$

where I and ω denote the intensity and angular frequency of the pump pulse respectively.

The results of our numerical investigation agreed well with experimental data. The value of the effective amplification of seed cw CO_2 radiation $\ae = \varepsilon/\tau_p I_0$, with ε the energy density of 10 μm pulse, τ_p the pump radiation pulsewidth and I_0 the seed radiation

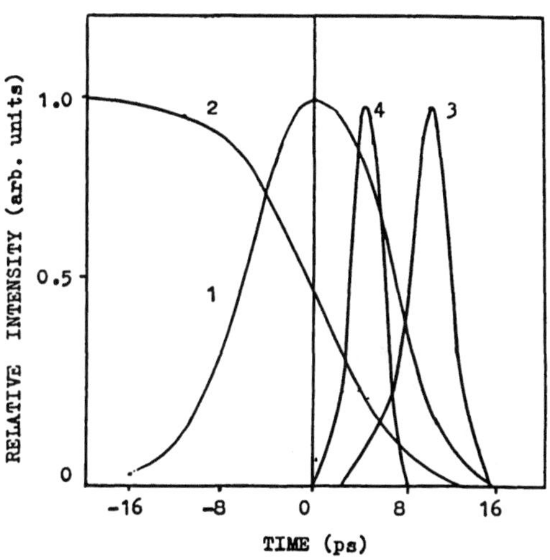

Fig.1. Pulse shapes for the pump pulse width 12 ps: 1 - pump pulse (λ=1.08 μm), 2 - electron density wave, 3 - signal wave (λ=10.6 μm), 4 - signal wave in the case of equal group velocities. The nonlinear crystal length is 1.7 cm.

intensity, is essentially (by 2 - 3 orders of magnitude) decreased by the influence of TPA.The other result was the shortening of the output 10 μm ps pulse without its chirping, which was expected to arise due to phase modulation by the electron-density wave. We suggest that these facts are connected, because the phase modulation leads to the phase mismatch of the nonlinear process and hence to the truncation of parametric amplification. It appears as if the 10 μm pulse were being pulled to the leading edge of pump pulse, where the carrier density is lower (fig.1).

3. Experimental

A schematic of the experimental setup is illustrated in fig.2. It consists of a low power 10 μm picosecond pulse generator and a slave multiatmosphere TE-CO_2 laser.

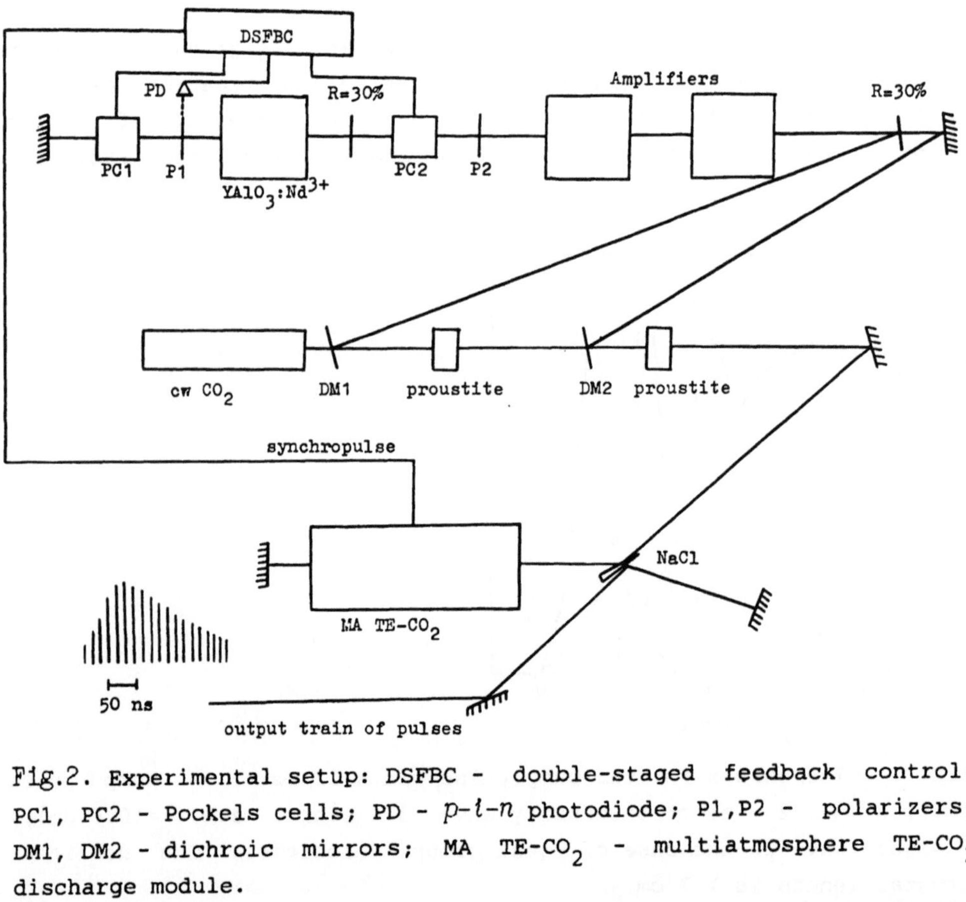

Fig.2. Experimental setup: DSFBC - double-staged feedback control; PC1, PC2 - Pockels cells; PD - *p-i-n* photodiode; P1,P2 - polarizers; DM1, DM2 - dichroic mirrors; MA TE-CO_2 - multiatmosphere TE-CO_2 discharge module.

Picosecond 10 μm pulses are generated by the two-stage parametric amplification of cw-CO_2 laser radiation (≈1 W/cm^2 power density) in proustite crystals. The cw CO_2 laser was operated on the 10 μm P(18) transition in our experiments. After the first stage 10 μm and 1.08 μm radiations are cut out and only the idler 1.18 μm pulse travels to the second stage, where it is amplified and in its turn gives rise to the 10 μm output pulse. The two stages are pumped separately by a single 17 ps (FWHM) ultrashort pulse of the $YAlO_3:Nd^{3+}$ passively mode-locked laser.This pulse is divided into two parts with energies of ≈1 mJ for the first stage and ≈2 mJ for the second. Separate pumping of the stages was necessary because of the high losses of pump radiation in the proustite crystal. On the other hand, this scheme allows us to change the arrival time of the pump pulse with respect to the idler 1.18 μm pulse in the second stage.

The solid-state picosecond source is a home-built one with negative feedback control [7]. It allows us to elaborate an electrical synchropulse which comes earlier than the optical pulse for 0.2-20 μs with a time jitter less than 10 ns, thus solving the problem of synchronization with the TE-CO_2 laser.The negative feedback control also improves the stability of the energy by as much as 1% and the pulsewidth by as much as 3%.

To amplify regeneratively the low-power picosecond 10 μm pulse, it is coupled into the slave cavity by reflection from one surface of the NaCl wedged beam-splitter which has been placed inside the laser resonator. The slave laser cavity consists of two plane copper mirrors separated by 220 cm. The experiments were performed on a home-built TE-CO_2 discharge module operated at a gas pressure of 10 atm and with 3:3:4:90::D_2:CO_2:N_2:He gas mix. The active medium is 50 cm long and has an aperture of 0.6×0.6 cm^2.

The NaCl beam-splitter is used to couple ≈3% of the intracavity flux per surface on each transit of amplified pulse in the resonator, so the slave laser output consists of a train of pulses. The total energy coupled from one NaCl surface in one direction is ≈8 mJ.

4 Results

In order to study the parameters of our installation, we measured durations of the pulses obtained. Since the application of the usual autocorrelation technique to the injected pulse duration measurement met with some difficulties due to its low energy (≈1 μJ), we applied a semiconductor transmission switch [8]. Silicon was used as the

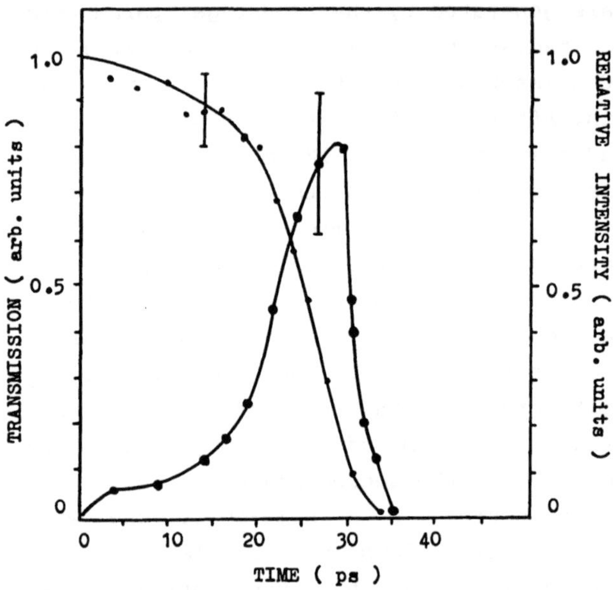

Fig.3. 10 μm pulse energy transmitted through the semiconductor switch plotted as a function of visible pulse arrival time and the 10 μm pulse temporal shape obtained by this curve differentiation.

material for switching the transmission. The switch was controlled by a 0.54 μm pulse with 11 ps pulsewidth, obtained by second harmonic generation of 1.08 μm pulse which was split from the pulse pumping the parametric amplifier. By monitoring the transmitted 10 μm energy as a function of the time the visible pulse reaches the transmission switch relative to the time of arrival of the 10 μm pulse and differentiating the curve obtained, the temporal profile of the 10 μm pulse can be determined. The transmission curve and temporal profile of 10 μm pulse are plotted in fig.3, showing the pulse duration 9 ps (FWHM). We believe that distortions introduced due to non-zero switching time can be neglected because the switch was controlled by the pulse with energy 10 times greater than that sufficient for switching.

To measure the temporal shape of the slave laser output, an infrared autocorrelator of standard design was constructed. The output radiation was split at a NaCl beam-splitter, mounted at proper angle of incidence. The two beams were delayed independently and combined at an angle of $\approx 3^{o}$ on a proustite crystal that was oriented for noncollinear SHG. The SHG signal was monitored by a 1 MHz Au:Ge detector, so we could only see the structure of an amplified pulse

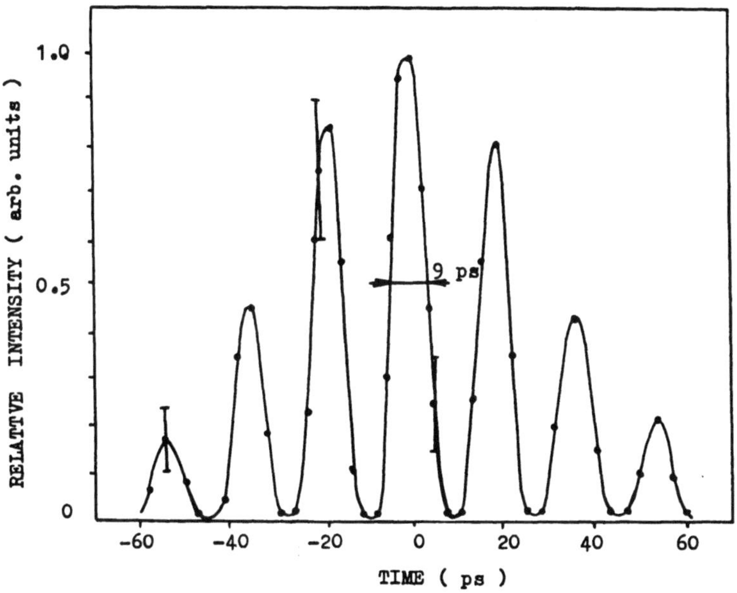

Fig.4. Autocorrelation trace of the regenerative amplifier output train.

averaged over the whole train. The data corresponding to each delay value were normalized on the independently monitored collinear SHG signal in each shot and then averaged over 20 shots. Fig.4 shows the autocorrelation trace obtained. One can see that the pulse in the train consists of a series of pulses, separated by 18 ps. This structure is the result of residual spectral modulation of the gain, as it was shown by P.B.Corkum [2].The individual feature has the pulsewidth ≈6 ps (FWHM), which is roughly consistent with the injected pulse duration. The shortening of the pulse during amplification is probably connected with the gain saturation, which leads to the pulse slope shortening.

We have performed the experiments on SHG in proustite crystal using output trains of pulses. The crystal was oriented for the collinear *eo-e* phase-matching. The energies of both 10 μm and 5 μm radiations were measured by a pyroelectric detector.The conversion efficiency of 13 % for train of 10 μm pulses was obtained. We also recorded both 10 μm and 5 μm trains on the fast 200 MHz HgCdTe detector, which serves as a sensitive energy meter on the picosecond time scale. Using oscilloscope traces of both trains, we have obtained the effective duration of an individual pulse in the train, given by expression $\tau_{eff}=\varepsilon_{1j}^2/\varepsilon_{2j}$, where ε_{1j} and ε_{2j} are the relative energies of the j-th 10 μm and 5 μm pulses respectively.Fig.5 shows

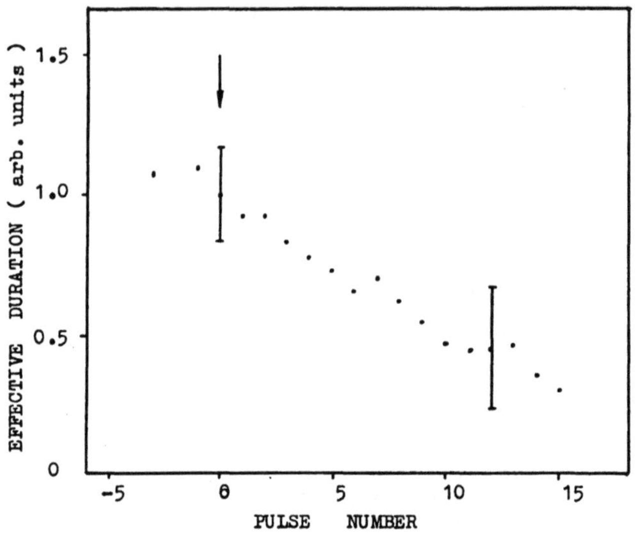

Fig.5. The dependence of the effective duration of output pulse on its position in the train. The arrow marks the peak energy pulse position.

how τ_{eff} behaves across the train. We believe the change of τ_{eff} value to be due to the change of relative magnitude of subpulses comparing with the first pulse, rather than to pulse compression. This change can be caused by the gain saturation. The pulse compression process observed in [2] isn't likely to appear in this case because of the wide beam aperture in the active medium and hence the pulse energy density which is too low to produce the plasma.

Using the train shapes of 10 μm and 5 μm radiations we deduced the SHG conversion efficiency for the peak energy pulse to be 18%. This is consistent with the value 25% reported in [2] for the single pulse selected from the train and is lower probably due to non-optimal type of phase matching.

5. Conclusions

The system we described here is capable of generating picosecond 10 μm pulses with energies of ≈1 mJ. The duration of the output pulse is determined by the pump pulse duration, which in its turn depends mainly on the active medium used in the solid state mode-locked laser. This system seems to be simpler compared with that described in [2], because we start forming 10 μm ps pulse with low power cw CO_2 laser instead of pulsed CO_2 laser, therefore avoiding at least the problem of synchronization at this point. However, ultimately short pulses cannot be obtained by our scheme due to restrictions caused by group velocity dispersion in nonlinear crystals. We believe that some new results in this direction can be achieved by compressing the 10 μm pulse chirped by cross phase-modulation in a semiconductor.

References

1. Z.A.Biglov, V.M.Gordienko, I.A.Kudinov, V.T.Platonenko, V.A.Slobodyanyuk, V.D.Taranukhin: Proc. XII Int. Conf. on Coherent and Nonlinear Optics, Minsk, USSR (1988), vol.3, p.119
2. P.B.Corkum: IEEE J. QE-21, 216 (1985)
3. I.M.Bayanov, Z.A.Biglov,V.Ya.Gayvoronsky, V.M.Gordienko:Kvantovaya elektronika 16, 1545 (1989)
4. Ya.O.Dovgii, N.I.Butsko,V.N.Korolyshin, E.T.Moroz: Sov.Phys. Solid State 13, 995 (1971)
5. B.Bareika, G.Dikcius, A.Piskarskas, V.Sirutkaitis: Proc. 2nd Int. Symp. "Ultrafast Phenomena in Spectroscopy", Jena, DDR (1980),p.14
6. V.T.Platonenko, V.D.Taranukhin: Izv. Acad. Nauk SSSR ser. fiz. 50, 786 (1986)
7. A.M.Valshin, V.M.Gordienko, S.V.Krayushkin, V.T.Platonenko, V.K.Popov: Kvantovaya elektronika 13, 1713 (1986)
8. A.J.Alcock, P.B.Corkum: Can. J. Phys. 57, 1280 (1979)

Part II

Measurement, Devices and Methods

Lateral High Speed Photodiodes on Semi-insulating InGaAs:Fe

D. Kuhl[1], *C.J. Wei*[1;*], *E.H. Böttcher*[1], *D. Bimberg*[1], *and E. Kuphal*[2]

[1]Technische Universität Berlin, Institut für Festkörperphysik, Hardenbergstr. 36, D-1000 Berlin 12, Germany
[2]Forschungsinstitut der DBP, Am Kavalleriesand 3, D-6100 Darmstadt, Fed. Rep. of Germany
* On leave from the Institute of Semiconductors of the Chinese Academy of Sciences, Beijing, People's Rep. of China

Abstract. The fabrication and characterization of laterally structured MSIM and PSIN photodiodes on high resistivity InGaAs:Fe acting as the photoabsorbing semiconductor layer is reported. Schottky contacts for the MSIM detectors were formed by the evaporation of Au. In the case of the PSIN diodes, the contacts were evaporated layers of Ni-AuGe-Ni-Au for the n-type and Ni-Au-Zn-Au for the p-type electrode, which were alloyed to achieve a diodelike behaviour. The interdigitated contact structure with finger distances of 2 μm - 10 μm was obtained by using a lift-off technique. The finger width is 2 - 4 μm. A mesa etch was formed around the contact structure and the bonding pads. The present MSIM and PSIN photodiodes exhibit a low total capacitance (< 70 fF), a very fast response (> 14 GHz), and an external quantum efficiency of about 40 % at 1.3 μm. The MSIM detectors show a dark current of about 250 nA and the PSIN diodes of about 1.5 μA at a bias voltage of 10 V.

1. Introduction

High speed photodiodes are required for numerous applications. They serve as convenient benchtop detectors for the characterization of ultrafast laser pulses and physics experiments like time resolved spectroscopy. Their major field of application is the optoelectronic conversion of high data rates in modern optic fiber systems. In order to meet future system requirements photodetectors for the 1.3 μm - 1.5 μm spectral range which can be monolithically integrated with InGaAs or InP FET's have to be developed. Due to their structural simplicity and compatibility with FET's, InGaAs photodiodes with a lateral planar structure are potential candidates for integrated photoreceivers. Metal-semiconductor-metal (MSM) as well as P-N photodiodes with a lateral structure on InGaAs have to be taken into consideration. Due to the low Schottky barrier height on InGaAs [1], however, the MSM concept is difficult to realize. A few attempts, based on various surface layers on n-type InGaAs, have been made most recently [2-4] to improve the Schottky-diode characteristic. Even less work has been done to realize laterally structured P-N photodiodes on InGaAs [5].

In this investigation, we present first results on the fabrication and characterization of MSIM and PSIN photodetectors on high resistivity

Springer Proceedings in Physics, Vol. 49 Ultrafast Phenomena in Spectroscopy
Editors: E. Klose · B. Wilhelmi

InGaAs:Fe. The acronyms MSIM and PSIN refer to the fact that the active layer of the photodiodes consists of a semi-insulating (SI) photoabsorbing semiconducting layer. The high performance of these devices is demonstrated by their large bandwidth (> 14 GHz), low capacitance (< 70 fF) and high quantum efficiency (> 40 %).

2. Detector requirements

Photodetector performance is determined by a low dark current, a high quantum efficiency and a large bandwidth. The dark current in an ideal PIN diode is determined by the generation recombination currents in the depletion layer. This current is given by [6]

$$I_{gr}=\frac{qn_i wA}{\tau} \tag{1}$$

where q is the elementary charge, n_i the intrinsic carrier concentration, A the junction area, w the depletion width, and t the carrier lifetime. Therefore a large carrier lifetime will decrease the dark current.

The carrier lifetime also influences the quantum efficiency. If the carrier lifetime is less than or comparable to the transit time of the photogenerated carriers, the quantum efficiency becomes small. These photogenerated carriers contribute to the photocurrent only during their lifetime. Consequently the carrier lifetime has to be larger than the transit time.

By incorporating Fe in InGaAs as a deep acceptor to achieve semi-insulating behaviour, the carrier lifetime will be reduced. In spite of that iron doping, dark currents of several nA for lateral PIN photodiodes have already been achieved [5]. From time resolved photoluminescence measurements, we obtained carrier lifetimes in the range of ns for semi-insulating InGaAs. We expect dark currents in the range of nA for our devices using equation 1. Also, the quantum efficiency should not be degraded by iron doping, because the expected transit time is about two orders of magnitude smaller then the carrier lifetime.

The major advantage of Fe-doped InGaAs is an expected higher speed of response and a larger barrier height for Schottky contacts than for undoped material. The higher speed of response is due to a lower capacitance. Therefore the bond pads of the device can be formed on the InGaAs surface without significantly increasing the stray capacitance.

3. Fabrication technique

Figure 1 shows a SEM photograph of a fabricated lateral photodetector with interdigitated contact structure. The InGaAs:Fe layer is grown on semi-insulating InP S.I. substrate by liquid phase epitaxy. In spite of the iron doping, mobilities up to 10000 cm^2/Vs were achieved [7]. Our Schottky

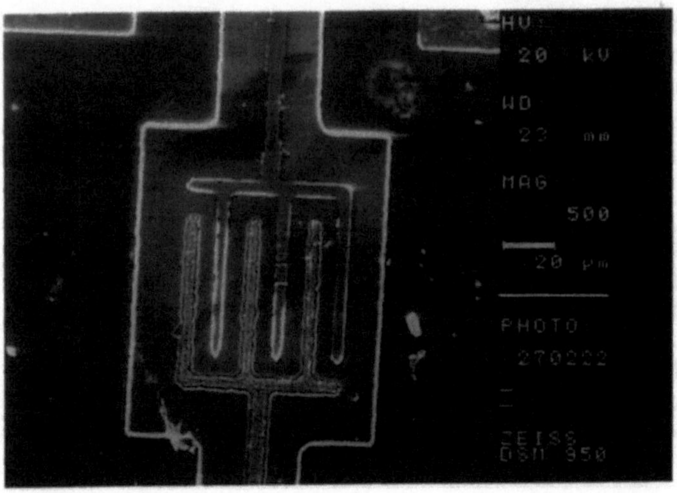

Figure 1: SEM photograph of a mesa etched PSIN structure with a finger spacing of 5 µm.

photodiodes M-SI-M (metal-semi-insulating-metal) are fabricated on InGaAs:Fe with a resistivity of 184 Ωcm, a carrier concentration of 1.5 x 10^{13} cm^{-3}, and a mobility of 2200 cm^2/Vs. The epitaxial layer thickness is about 2 µm. Au is used as the contact material.

The lateral PIN diodes, which we refer to as PSIN diodes (p-contact - semi-insulating - n-contact) are fabricated on InGaAs:Fe with a resistivity of 1600 Ωcm, a carrier concentration of 7 x 10^{11} cm^{-3}, and a mobility of 5700 cm^2/Vs. The layer thickness is about 2.7 µm. The material parameters were taken from Hall measurements.

In the first fabrication step of the PSIN diodes the n-contact is realized by the Ni-AuGe-Ni-Au metal system. The p-contact consisting of a Ni-Au-Zn-Au system is patterned in the following step. A mesa etch around the whole structure is introduced. The contacts are alloyed to form n- and p-doped regions spaced by the semi-insulating regions.

The contact patterning for the MSIM and PSIN structures is made by optical contact lithography and a lift-off technique. The contact finger distances vary from 2 µm to 10 µm, the finger width varies from 2 µm to 4 µm.

Finally the detectors are mounted on microwave SMA connectors. These connectors allow an operation up to 18 GHz.

4. Experimental results

Figure 2 shows the current voltage characteristics of a MSIM and PSIN photodetector with a finger spacing of 5 µm. The characteristic of the MSIM detector exhibits an almost symmetric behaviour with dark currents equal to 250 nA at 10 V bias. These dark currents are a promising result for MSIM diodes compared to MSM devices using lattice mismatched surface layers,

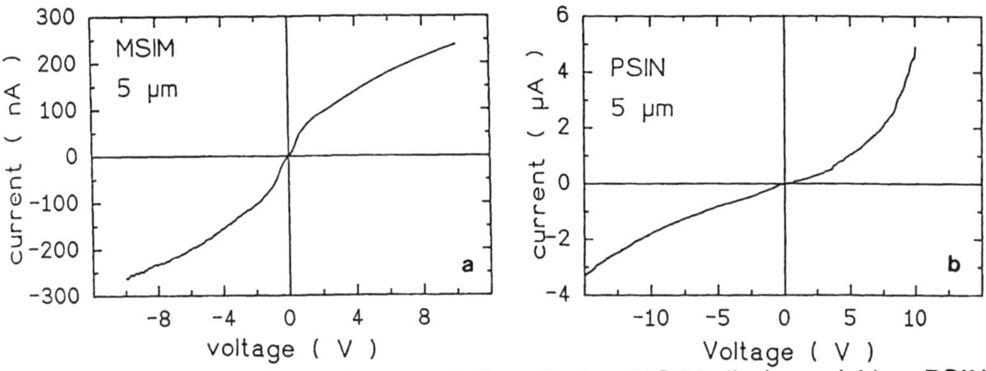

Figure 2: current voltage characteristics of a) a MSIM diode and b) a PSIN diode with 5 µm finger spacing.

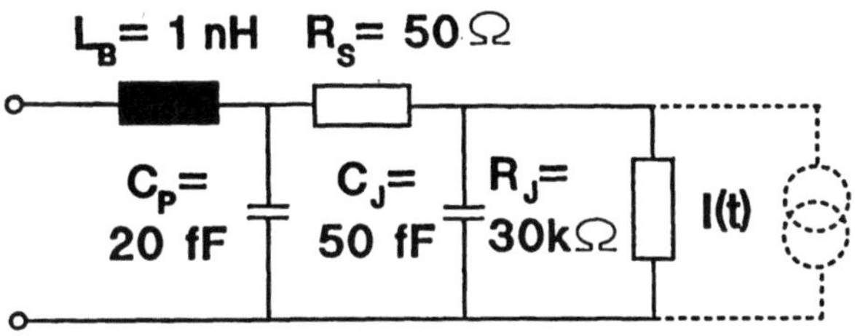

Figure 3: equivalent circuit of a photodetector with bond inductance Lb, stray capacitance Cp, series resistance Rs, junction capacitance CJ, and junction resistance RJ. The data were obtained from TDR and S-parameter measurements. An ideal current source (dashed lines) is introduced for a simulation of the time response.

which show a dark current of about several µA [3,4]. The PSIN structure shows a diodelike characteristic with a reverse and a forward regime. The dark current equals 1.5 µA at 10 V reverse bias. This relatively large dark current is caused by the existence of surface states and a not ideally formed junction. Therefore a reduction of surface states and contact optimization should decrease the dark current into the nA range.

The external quantum efficiency at 1.3 µm and 10 V reverse bias is about 40 %. The contact finger shadowing is taken into account. This result is comparable to other InGaAs detectors. Antireflection coatings will increase this quantum efficiency.

In order to investigate the influence of the packaging and the junction capacitance on the high frequency operation, we carried out network analysis in the time and frequency domain, namely time domain reflectometry (TDR) and S-parameter measurements up to 2.5 GHz. The equivalent circuit shown in figure 3 was assumed for the detectors. It consists of a bond inductance L_B, a stray capacitance C_P, a series resistance R_S, a junction resistance R_J, and a junction capacitance C_J. By varying these parameters we fit the measured data. The results of both measurement techniques are in good agreement. They

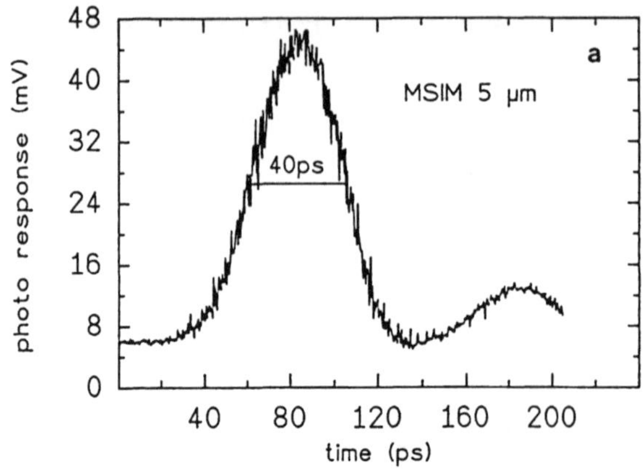

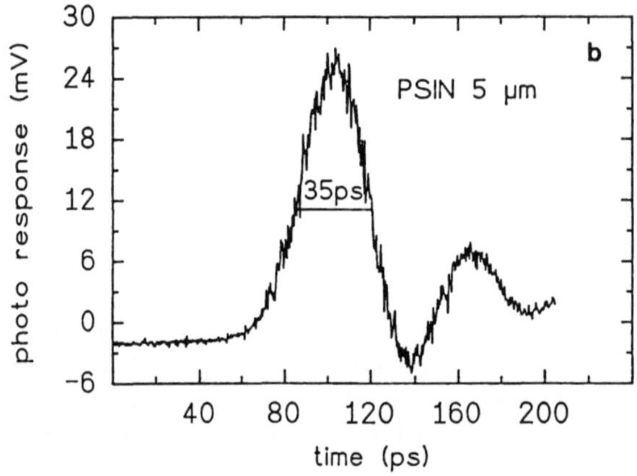

Figure 4: photoresponse to a 30 ps FWHM semiconductor laser pulse of the a) MSIM and b) PSIN diodes with 5 μm finger spacing.

are given in figure 3 for the PSIN diode. There is no essential difference to the MSIM diode. Note that the PSIN structure is depleted even at zero bias. This depletion allows operation of these devices at a low bias.

The time resolved measurements were performed by using a gain switched 1.3 μm semiconductor laser. This laser was driven by an avalanche generator. The FWHM of the emitted optical laser pulses is < 30 ps. The photocurrents of the detctors are monitored by a 20 GHz bandwidth Tektronix SD-24 sampling head. The response of the 5 μm MSIM and PSIN diodes is shown in figure 4. The FWHM of the MSIM and PSIN detctors is 40 ps and 35 ps, respectively. This corresponds to a bandwidth of 14 GHz. The measured signal is a convolution of the laser pulse width, the response of the sampling head, and the detector under test.

Therefore, the inherent bandwidth of the detectors is larger than 14 GHz.

The ringing of the photoresponse is caused by the bond inductance of about 1 nH. Using the data of the detectors´ equivalent circuit, the optical impulse response is simulated numerically. The "intrinsic diode" is represented by a current source (shown by dashed lines in Fig. 3). Neglecting carrier transit

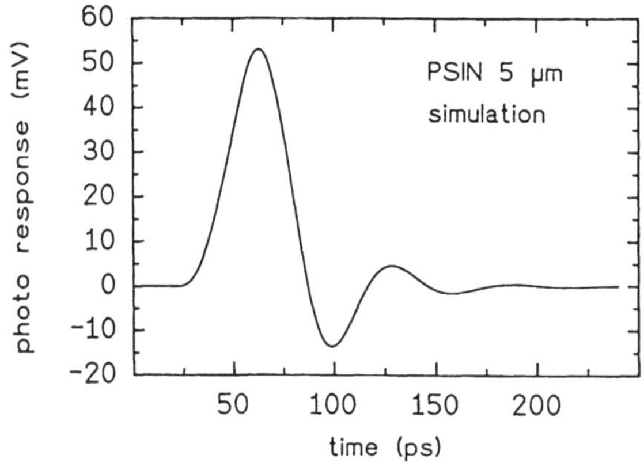

Figure 5: simulation of the photoresponse of the PSIN diode to an idealized 30 ps FWHM triangular current pulse.

time effects the response is calculated for a triangular laser pulse of 30 ps FWHM.

The result of that simulation for the 5 μm PSIN diode is shown in figure 5. The simulation agrees with the measured data. A reduction of the bond inductance results in disappearance of the ringing.

5. Summary

In conclusion, we have demonstrated the fabrication of MSM and PSIN photodetectors on high resistivity InGaAs. Laterally structured photodiodes are formed with interdigitated contact fingers. The devices show a high response bandwidth and are well suited for high-speed applications. Due to their ease of fabrication on high resistivity material, the photodetectors are potententially useful for monolithic optoelectronic integration.

6. Acknowledgements

The authors appreciate funding of this work by the Bundesministerium für Forschung und Technologie (TK 03077). One of the authors (C. J. Wei) was partly supported by the Alexander von Humboldt Stiftung and by the Stiftung Volkswagenwerk.

7. Literature

[1] C. Y. Chen, A. Y. Cho, K. Y. Cheng, P. A. Garbinski, Appl. Phys. Lett., 40, 401 - 403 (1982)

[2] O. Wada, H. Nobuhara, H. Hamaguchi, T. Mikawa, A. Tackeuchi, and T. Fuji, Appl. Phys. Lett., 54, 16 - 17 (1989)

[3] H. Schuhmacher, H. P. Le Blanc, J. Soole, and R. Bhat, 15th Symp. on GaAs and Related Compounds, Atlanta, GA (1988)

[4] W. P. Hong, G. K. Chang, and R. Bhat, IEEE Trans. Electron Dev., 36, 659 - 661 (1989)

[5] V. Diaduk, and S. H. Groves, Appl. Phys. Lett., 46, 157- 158 (1985)

[6] S. R. Forrest, IEEE J. of Quant. Elect., 17, 217 - 226 (1981)

[7] Z. Chen, T. Wolf, W. Korb, and D. Bimberg, J. of Appl. Phys., 64, 4574 - 4579 (1988)

High Resolution Optical Time Domain Reflectometry for the Characterization of Integrated Optical Devices

P. Beaud, J. Schütz, W. Hodel, and H.P. Weber

Institute of Applied Physics, University of Berne, Sidlerstr. 5, CH-3012 Berne, Switzerland

Abstract. An Optical Time Domain Reflectometry (OTDR) system designed for measurements on semiconductor optical waveguides is described. The OTDR system uses an infrared subpicosecond pulse source in conjunction with balanced heterodyne detection. Experimental results show a sensitivity of more than -110 dB and a resolution of 60 µm in air. We demonstrate the applicability of this system for the diagnostics of integrated optical devices.

1. Introduction

In the last few years great efforts have been made to develop an Optical Time Domain Reflectometry (OTDR) system for the characterization of integrated optical waveguides. The investigation of integrated devices requires a spatial resolution in the micrometer range and consequently a very high dynamic range of the system. For these purposes two different methods have been proposed. A first one uses ultrashort light pulses and intensity correlation techniques for the detection [1,2]. The second technique employs a cw short coherence light source (e.g. an edge emitting LED) and interferometric detection [3,4].

In a recent work we have described an OTDR system combining both methods mentioned above [5]. The light source is an infrared dye laser at 1.3 µm delivering pulses of 300 fsec duration. Detection of the backscattered light is performed in a Mach-Zehnder-type interferometer using a balanced heterodyne technique. It is important to note that in order to take advantage of the full capacities of the OTDR technique it is indispensable to detect a final level of backscattered light. The detection of Rayleigh backscattered light with fiber OTDR systems allows one to determine directly the fiber attenuation and to evaluate the loss of splices, connectors and defects. The aim of our work was to investigate whether similar results can be obtained from OTDR measurements on integrated optical devices.

2. Experimental set-up

Fig. 1 shows the experimental arrangement. The light source is an infrared dye laser synchronously pumped by the compressed pulses of a Nd:YAG laser [6]. The dye laser is tunable in the wavelength range between 1.25 and 1.35 µm, which is of particular interest for optical communication. The pulse duration is 300 fs (assuming a Gaussian pulse shape) and the maximum output peak power is 3 kW.

The laser beam is split into the Local Oscillator (LO) and the Probe Beam (PB). The LO passes an acousto-optic modulator (AOM) which operates in the deflection mode and imposes a frequency shift

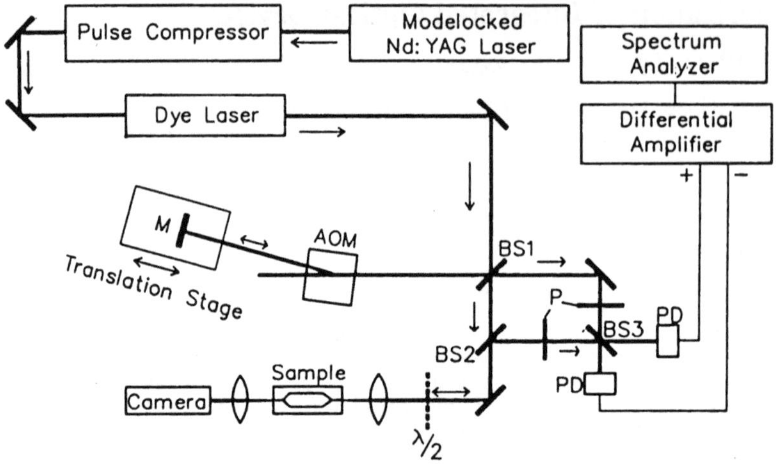

Fig. 1. Experimental set-up. AOM: acousto-optic modulator, BS: beamsplitter, $\lambda/2$: half-wave plate, P: polarizer, PD: photodiode , M: mirror.

$d\nu$ on the LO. The LO is then reflected back by a 100 % mirror (M) mounted on a precision translation stage. The PB is coupled into the structure under investigation. A fraction of the backreflected light is separated from the probe beam (BS2) and recombined with the LO in a Mach-Zehnder-type interferometer on a symmetric beamsplitter (BS3). Detecting the two resulting output beams with photodiodes (PD), beat signals at the difference frequency between the LO and the PB are observed. Due to the phase shifts introduced upon reflection at the beamsplitter these two beat signals have opposite signs. Therefore subtracting the two photodiode signals from each other allows one to suppress the noise contribution of the laser source. Ideally, exact symmetry of the beamsplitter would be required. The fact that the splitting ratio of the beamsplitter is slightly polarization dependent allows us to optically tune the balancing of the LO and PB with two polarizers (P) placed in each arm of the interferometer. This 'balanced heterodyne detection' technique [7] leads to a substantial reduction of the noise level (typically 45 dB in our experiment) compared to single receiver heterodyne detection where only one output arm of the beamsplitter is used for detection.

Because of the high acoustic carrier frequency of a commercial AOM (tens of MHz), one usually needs two of them operating at slightly different frequencies to create a conveniently low beat frequency. For a modelocked laser source, however, a single AOM can be used to create a frequency shift $d\nu$ close to the repetition frequency $\delta\nu$ of the laser system (82 MHz). A low beat frequency $\nu = \delta\nu - d\nu$ (= 20 kHz in the experiment) results in this case from the beating of every longitudinal mode in the probe beam with its frequency shifted neighbour in the local oscillator. Note that the generation of a beat frequency in this way depends on the fact that there exists a definite phase relationship between adjacent laser modes (as is the case for a modelocked laser).

In the experiment the two photocurrents are led to a differential amplifier and the resulting signal is detected with a spectrum analyzer. We use a PC for signal sampling and controlling the translation stage. Instead of continuous signal sampling we use a step-by-step measurement technique in order to avoid any Doppler shift introduced by movements of the translation stage.

3. System performance

The spatial resolution of the OTDR system is determined by the inverse spectral bandwidth ($\Delta\nu$ = 2.1 THz) of the laser emission and is 60 µm in air [5]. Figure 2 illustrates the sensitivity and linearity of the system. The sample is a BK7 mirror substrate. The trace is shown here on a logarithmic scale. The two highest peaks (at -14 dB) correspond to the 4 % reflections of the front (F) and back faces (B), respectively. Three orders of internal reflection echos (E1-E3) can be identified resulting in a sensitivity of more than -110 dB. Note that this sensitivity allows us to detect a relative index difference in the order of $5 \cdot 10^{-6}$. All peaks lie on a straight line, demonstrating the linearity of the system over the whole dynamic range. The exact drop in signal for two 4 percent reflections is 28 dB, which is in good agreement with our results (Fig. 2). The false signals (f) in Fig. 2 result from multiple reflections inside different optical components used in the experiment (mirrors, polarizers, attenuator) and could be reduced by using exclusively antireflection coated components.

One of the most important parameters of an OTDR system, apart from the spatial resolution and dynamic range, is the signal trace for a single reflection. This is because the wings of each large signal (as for example the strong entrance face reflection of semiconductor devices) can hide considerable parts of the waveguide under investigation (blindness of the apparatus). Assuming Gaussian pulses it can be shown that the signal shape is a Gaussian function of the time delay between the LO and the signal pulse [5]. Figure 3 shows a comparison between the measured and calculated signal trace on a logarithmic scale. Although the dye laser does not deliver

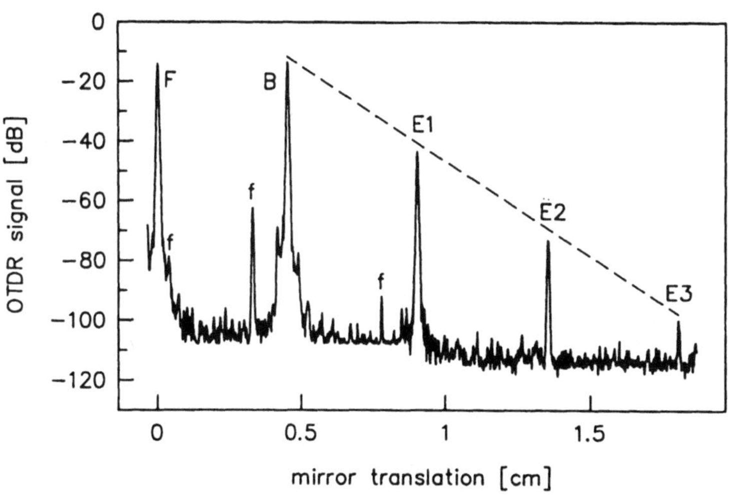

Fig. 2. OTDR measurement of a 3 mm thick glass plate (logarithmic scale). F,B: reflection from the front and back faces, respectively, E1-E3: internal echos, f: false signals.

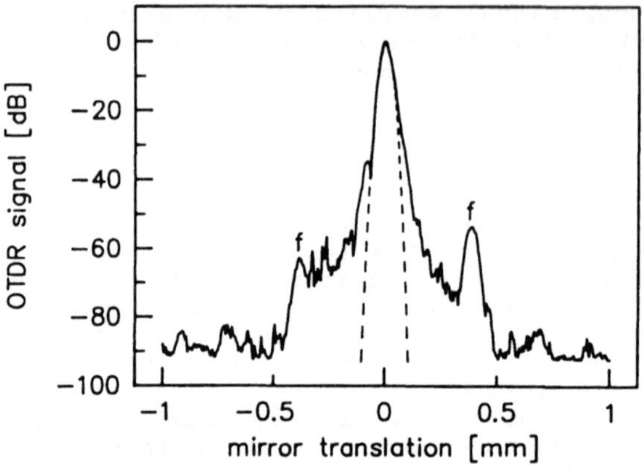

Fig. 3. Signal trace for a single reflection (logarithmic scale) Solid line: experiment; dashed line: calculation, f: false signals.

bandwidth limited Gaussian pulses we found good agreement between theory and experiment down to an intensity level of -50 dB. The Gaussian signal shape guarantees a minimal blindness of the OTDR system.

4. Waveguide measurement

An example of a measurement on an integrated optical waveguide is shown in Figure 4. The structure is a straight rib-waveguide made by dry etching of a n^-GaAs layer on a n^+GaAs substrate. The transverse dimensions of the waveguide are 6 µm x 3 µm, and the length is 7.85 mm. The front (F) and back faces (B) are coated and exhibit reflections of 2.5 %. Strong signals are recorded from these two facets as shown in Figure 4. From the optical distance of 28.5 mm between them we determine a group index of 3.63 at a wavelength of 1.27 µm.

In addition we observe a large number of scattering centers randomly distributed along the waveguide at a power level between -80 and -100 dB. The backscattered light is very inhomogeneous and therefore cannot be explained by Rayleigh scattering. It may be explained by Mie scattering from scattering centers located within the waveguide and from irregularities of the waveguide boundaries which dominate the loss of integrated waveguides [8].

Because of the large inhomogeneity the backscattered light cannot be used to determine the spatial distribution of the waveguide loss (as for example the loss of a localized defect). Nevertheless, an exponential drop of the backscattered power is clearly recognizable in Fig. 4. Assuming the loss to be distributed evenly along the waveguide we can evaluate an average loss coefficient α of 1.6 cm^{-1} from the mean slope of the backscattered light.

The average loss of the waveguide can also be determined directly by measuring the echo pulse which corresponds to the first internal reflection from the waveguide (the measured echo E is also shown in Fig. 4):

$$\alpha = +{}^{1}/_{2L} \cdot \ln\,(C \cdot r^2 \cdot P_B/P_E), \qquad (1)$$

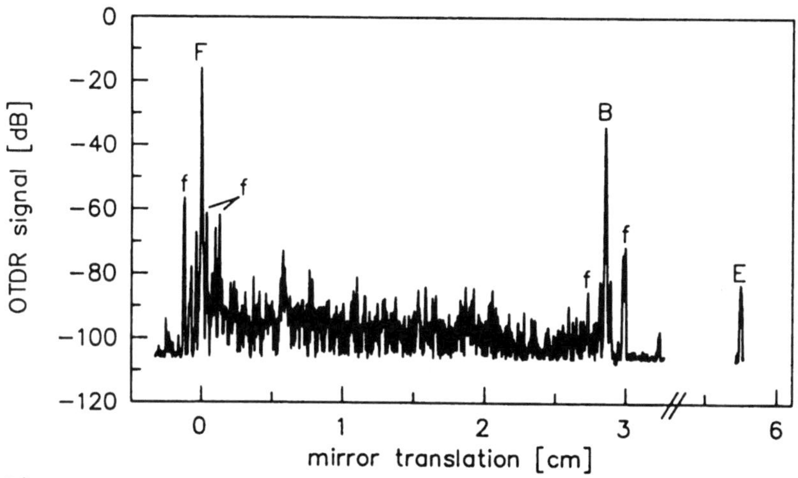

Fig. 4. OTDR measurement of a straight rib-waveguide. F, B: reflection from the front and back faces, respectively, E: internal echo, f: false signals.

where L is the length of the waveguide and r is the reflectivity of the end faces. The power ratio P_B/P_E of the echo (E) and the back reflex (B) can be taken from the figure. The factor C has been incorporated in Eq. 1 to take into account the influence of group velocity dispersion (GVD) which tends to broaden the duration of the pulse propagating in the waveguide:

$$C = \{ [1 + (k''\cdot L\cdot \Delta\omega^2/4\ln 2)^2] / [1+(k''\cdot L\cdot \Delta\omega 2/2\ln 2)^2] \}^{\frac{1}{2}}, \quad (2)$$

where $k''= \partial^2 k/\partial\omega^2$ is the group velocity dispersion parameter. From the measured wavelength dependence of the group index the group velocity dispersion can easily be evaluated. We obtain for k'' a value of 2.1 $\mathrm{psec}^2/\mathrm{m}$ at a wavelength of 1.27 µm. Eqs. 1 and 2 then give an average loss coefficient α of 2.3 cm^{-1} which is considerably larger than the value evaluated above. The difference may be due to the uncertainty of the values of r and C used in Eq. 1. On the other hand the assumption of an evenly distributed loss may not be justified. A clarification of these points will need some further investigations.

5. Conclusions

We have developed an OTDR system for the characterization of integrated optical devices using a subpicosecond pulse source and balanced heterodyne detection. The spatial resolution of 60 µm in air is determined by the spectral bandwidth of the laser emission. The high output power allows a high dynamic range of 110 dB. An advantage in using a modelocked laser source is that the nearly Gaussian shape of the laser spectrum leads to a fast drop of the wings of each strong signal and therefore minimizes the blindness of the OTDR system.

We have demonstrated the applicability of the system to performing OTDR measurements on integrated optical waveguides. In particular, we are able to determine an average loss coefficient of a waveguide and localize reflection sites inside the structure. Furthermore, it is possible to evaluate waveguide parameters such as the group index and group velocity dispersion. The measured value for k'' of 2.1 $psec^2/m$ at 1.27 µm indicates that pulse broadening due to GVD cannot be neglected when investigating semiconductor waveguides with our OTDR system.

Finally we have reported for the first time the detection of a final level of backscattered light due to Mie scattering from irregularities of the waveguide boundaries. The backscattering is inhomogeneously distributed along the waveguide and is observed at a very low power level (less then - 80 dB).

We would like to thank Drs. H.H. Gilgen, R.P. Salathé and K. Affolter for helpful discussions, M. Hösle for technical assistance and the Swiss PTT for financial support.

References

[1] J.J.Fontaine,J.-C.Diels, C.-Y.Wang, H.Sallaba, Opt. Lett. **6**, 405 (1981).
[2] H.H.Gilgen, R.P.Novàk, R.P.Salathé, B.Zysset, P.Beaud, W.Hodel, H.P.Weber, Proc. ECOC'87, Helsinki, VTT Finland, 365 (1987).
[3] R.C.Youngquist, S.Carr, D.E.N.Davies, Opt.Lett.**12**, 158 (1987).
[4] K.Takada, I.Yokohama, K.Chida, and J.Noda, Appl.Opt. **26**, 1603 (1987).
[5] P.Beaud, J.Schütz, W.Hodel, H.P.Weber, H.H.Gilgen, R.P. Salathé, J. Quantum Electron. **QE-25**, 755 (1989).
[6] P.Beaud, B.Zysset, A.P.Schwarzenbach, H.P.Weber, Opt. Lett. **11**, 24 (1986).
[7] G.L.Abbas, V.W.S.Chan and T.K.Yee, IEEE J. Lightwave Technol. **LT-3**, 1110 (1985).
[8] P.K. Tien , Appl. Opt. **10**, 2395 (1971).

Pulsed Excitation of a GaAlAs Semiconductor Laser by Picosecond Optoelectronic Switches

P.P. Vasil'ev[1], *D. Grosenick*[2], *E. Klose*[2], *A.V. Konyashchenko*[1], *A.B. Sergeev*[1], *and M. Lenzner*[3]

[1]P.N. Lebedev Physical Institute, USSR Academy of Sciences, 117924 Moscow, USSR
[2]Zentralinstitut für Optik und Spektroskopie, Akademie der Wissenschaften, Rudower Chaussee 6, DDR-1199 Berlin, German Dem. Rep.
[3]Sektion Physik, Friedrich-Schiller-Universität Jena, Max-Wien-Platz 1, DDR-6900 Jena, German Dem. Rep.

Abstract. We report the excitation of a GaAlAs semiconductor laser by powerful electrical pulses using optoelectronic switches. Relaxation oscillation behaviour of the optical signal is shown and single optical pulses of 40 ps duration are generated.

In order to use semiconductor lasers in high-speed communication systems, the short time parameters of these light sources have to be known. Of special interest in this connection is the gain switching behaviour, which has the advantage that no external cavity is required. Using short electrical pump pulses, several authors reported in this way the generation of optical picosecond pulses (see e. g. [1, 2]).

Not only pure electric but also optoelectronic generation of the electrical driving pulses has been realized employing semiconductor optoelectronic switches [3, 4].

We used such switches to excite a GaAlAs DH planar stripe laser by powerful electrical pulses to study its large signal behaviour without additional bias.

Figure 1 presents the experimental setup. The optoelectronic switches were made of semiinsulating GaAs:Cr with a 50 ohm microstripe structure having a gap width of 1 mm. The switch was incorporated in a 50 ohm coaxial line. The output side was connected in series with a 47 ohm resistor for impedance matching to the semiconductor laser.

The GaAlAs DH planar stripe laser used has a threshold current of 95 mA and a resonator length of 200 μm. The optoelectronic switch was excited by the second harmonic radiation of a Nd-YAlO$_3$ solid state laser system (540 nm, 10 ps FWHM, pulse energies 10 to 50 μJ). The emission of the semiconductor laser was studied by the Soviet-type streak camera "Agat-SF" (minimum time resolution 2 ps).

To get high output voltages at the optoelectronic switch we used material with fairly high carrier mobility causing a long carrier relaxation time. The generated electrical pulses show an asymmetric temporal behaviour with a short rising edge due to the short duration of the exciting laser pulse and a longer nearly exponential decay of 600 ps determined by the carrier relaxation time. This was measured by a real time oscilloscope.

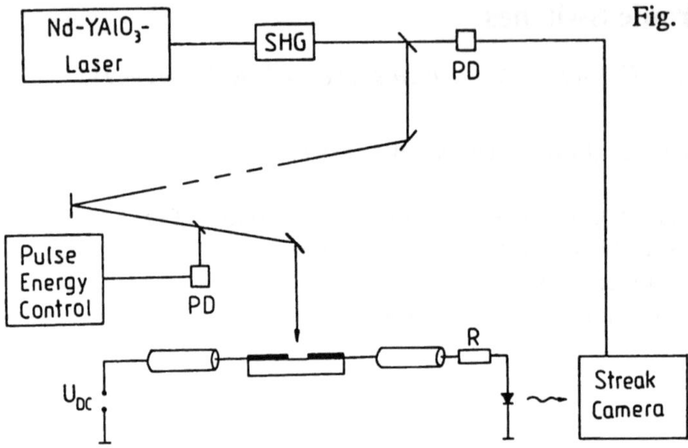

Fig. 1 Experimental setup

To get essentially shorter electrical pulses we covered a part of the gap with a material impermeable to light. Then there is a big resistance preventing carrier flow from the voltage bias into the gap area during the short time of switch activation. The electrical pulses of this switch were measured by a correlation method [5] using a switch made of amorphous semiconductor material ($GeSe_2$) with a relaxation time of less than 10 ps as a sampler. Exciting both switches with a passive mode-locked Nd-YAG laser (second harmonic 532 nm, 35 ps) we obtained a symmetric correlation function with 55 ps FWHM corresponding to an electrical pulse duration of 40 ps, giving us an upper limit for the case of the 10 ps exciting pulses used in the setup of figure 1.

A rather simple approach to theoretically describing the semiconductor laser output is a single mode rate equation system for the electron density n and the photon density s. Taking into account a band-to-band recombination rate B and a p-doping of the laser active material (acceptor concentration N_A) one gets

$$\frac{dn}{dt} = \frac{I}{eV} - A\,(n - n_0)\,s - Bn\,(n + N_A) \tag{1}$$

$$\frac{ds}{dt} = A\,(n - n_0)\,s + \beta\,Bn\,(n + N_A) - \frac{s}{\tau_R} \tag{2}$$

where I is the time-dependent injection current, V the active volume, e the electron charge, n_0 a parameter determining the laser threshold, $A(n - n_0)$ the optical gain, τ_R the overall photon lifetime and β the part of the spontaneous emission into the laser mode.

Solving this system numerically with the electrical driving pulses discussed above we get the photon emission illustrated in figure 2. The used parameter values are $A = 1.5{\cdot}10^{-7}\ cm^{-3}$, $B = 1.5{\cdot}10^{-10}\ cm^{-3}$, $n_0 = 1.3{\cdot}10^{18}\ cm^{-3}$, $N_A = 5{\cdot}10^{17}\ cm^{-3}$, $\beta = 10^{-3}$, $\tau_R = 1.25$ ps. Taking into account a deformation, especially for the very short driving pulse, due to dispersion, damping and capacity effects, the qualitative behaviour of the photon density (figure 2b) does not change.

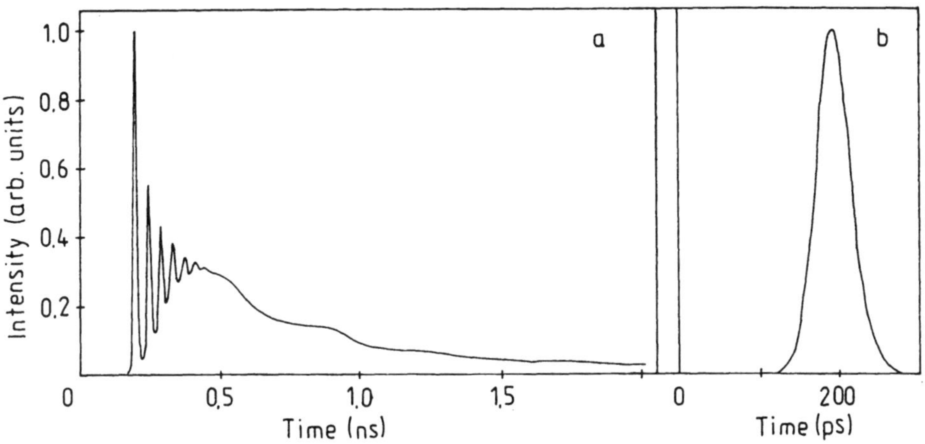

Fig. 2 Calculated semiconductor laser emission (peak current 25 I_{th})
a) driving pulse with 600 ps decay
b) near-Gaussian driving pulse with 40 ps duration

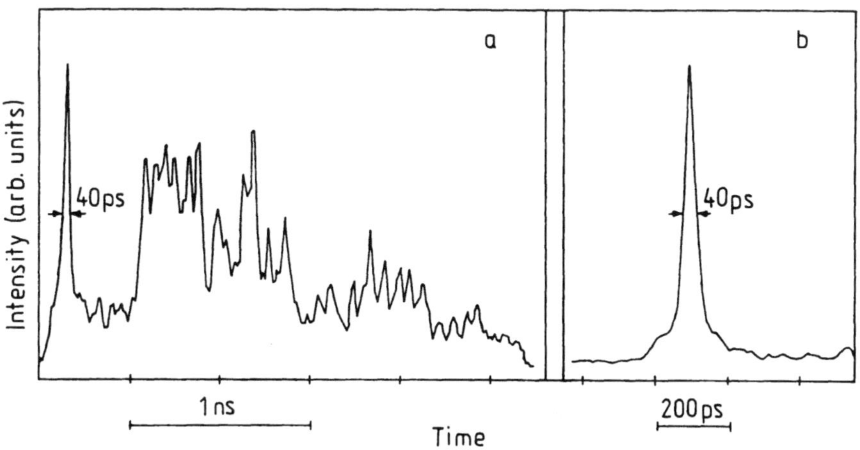

Fig. 3 Measured semiconductor laser emission (peak current 25 I_{th})
a) driving pulse with 600 ps decay
b) driving pulse with 40 ps duration

The experimental results for the two driving variants are shown in figure 3. With long driving pulses the semiconductor laser shows relaxation oscillation behaviour. After the first peak there is only small output for about 400 ps before the signal increases again. The very long time of the emission may be due to reflections of the electrical pulse.

The short pump pulse used allows one to obtain only the first peak of the semiconductor laser emission. The minimum pulse duration of 40 ps in both cases may be caused by the properties of the semiconductor laser or by nonperfect electrical connections and lines.

In conclusion, we investigated a simple scheme for the generation of picosecond light pulses with semiconductor lasers which were driven by ultrafast electrical pulses of optoelectronic switches. In comparison with [3] our system permits the generation of very powerful electrical pulses, giving us the possibility of unbiased excitation of all kinds of semiconductor lasers.

References

[1] D. Bimberg, K. Ketterer, E.-H. Böttcher, and E. Schöll, Int. J. Electron. 60, 23 (1986)

[2] H.-F. Liu, M. Fukazawa, Y. Kawai, and T. Kamiya, IEEE J. Quantum Electron. QE-25, 1417 (1989)

[3] E. O. Göbel, G. Veith, J. Kuhl, H.-U. Habermeier, K. Lübke, and A. Perger, Appl. Phys. Lett. 42, 25 (1983)

[4] J. M. Wiesenfeld, R. S. Tucker, P. M. Downey, and J. E. Bowers, Electron. Lett. 22, 396 (1986)

[5] V. Brückner, F. Kerstan, Electron. Lett. 20, 738 (1984)

Investigation of Internal Signal Propagation of an Integrated Circuit with High Temporal Resolution

H. Bergner, T. Damm, U. Stamm, M. Müller, K.-P. Stolberg, and H. Seidel

Sektion Physik, Friedrich-Schiller-Universität Jena, Max-Wien-Platz 1, DDR-6900 Jena, German Dem. Rep.

The development of high-speed semiconductor devices and highly integrated circuits (IC's) requires these devices to be tested with ultrafast time and microscopic spatial resolution. Several methods meeting these requirements have been developed, as time-resolved electron beam induced current /1/, picosecond scanning electron microscopy /2/, photoemission sampling /3/, electro-optic sampling /4/ and photoconductive switching /5/. All these methods work either under vacuum conditions or with special preparation of the sample.

Here we describe a novel method for the investigation of IC's with ultrafast time resolution and microscopic spatial resolution, which we will call the time-resolved optical beam induced current method (TOBIC). A ps laser pulse is focussed onto the surface of an IC near a pn-junction. Due to the light absorption, photocarriers are generated and they are separated by the internal field if the pn-junction is closed. In the outer circuit of the IC a photocurrent can be measured in addition to the normal operation current. If the illuminated pn-junction is shorted the generated photocarriers will not be separated efficiently and no additional current flows. Scanning the laser beam across the whole IC one gets a map of the logical states of all elements on the IC at the time of the pulse incidence. The input of the IC is connected with a source of short electrical pulses synchronized with the ps laser pulses. By varying the time delay between the electrical and optical pulses, changes in the logical states caused by the propagation of the electrical pulses inside the IC can be detected and the switching speed of certain elements, propagation delays and transition times in complex integrated circuits can be measured.

For the demonstration of the method we have combined a laser scanning microscope (LSM) /6/ with a cw mode-locked argon ion laser (514 nm wavelength, pulse width 100 ps, repetition rate 123 MHz) and achieved a mapping of the instantaneous optical beam induced current in an IC with a spatial resolution of better than 1 μm and a time resolution of 100 ps (Fig. 1).

The device under test is an 8 to 1 multiplexer type K500 ID 164 in ECL technology, which has a maximum clock frequency of 200 MHz.

The main part of the laser beam (excitation pulses, average power 50 mW) is directed via a delay line onto a fast photodiode (WF type SP102) to generate ultrashort electrical pulses. These pulses are amplified to U = 0.8 V (peak-peak, rise time 300 ps) and sent to the input terminal of the device under test. Thus the logical state of the input is synchronously switched with the laser pulses. An attenuated part of the laser beam (probe pulses, average power <5mW) is modulated by an acousto-optical modulator (modulation frequency 1 MHz, modulation depth 0.9), launched into the LSM and scanned across the IC. Only the current at the chopper frequency (1 MHz) arising from this modulated beam at the power sup-

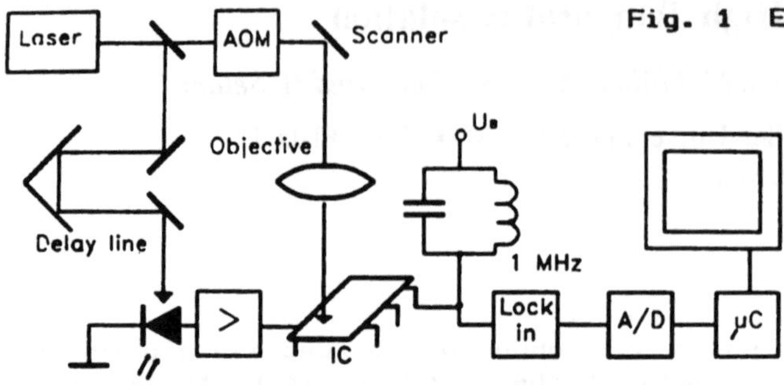

Fig. 1 Experimental setup

Fig. 2 Layout of the IC

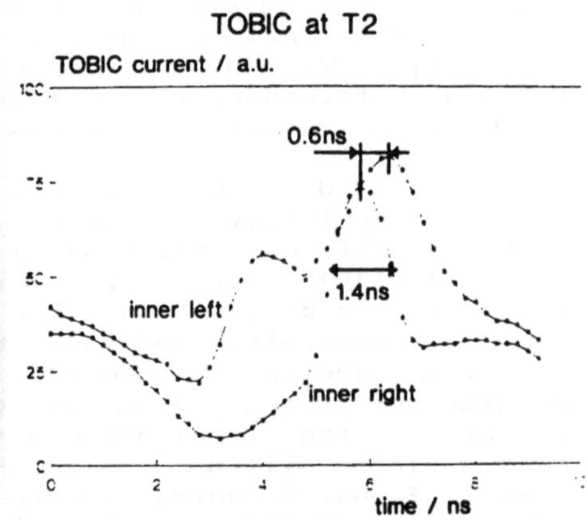

Fig. 3 Time dependence of the transistor current

ply pin is detected with a lock-in amplifier and used to build up the OBIC-image. One obtains an instantaneous OBIC-image for the given delay time between excitation and probe pulses. Due to the limited rise time of the photodiode and the amplifier a time resolution of approximately 300 ps is achieved. The delay between the optical and electrical pulses is adjusted by an optical delay line (0 to 9 ns time delay) which allows the recording of more than 1 period of the pulse repetition frequency.

All images are recorded with a 25x/0.5 long working distance objective providing a spatial resolution of less than 1 μm. Fig.2 shows the layout of a part of the IC. Visible are three npn-transistors (T1 upper left, T2 upper right, T3 lower left) recognizable by the 3 electrodes (emitter-collector-base) and two resistors (R1 lower middle, R2 lower right) formed by (vertical) stripes of doped semiconductor material.

In Fig.3 we have depicted the time dependence of the TOBIC signal at T2 in the region of the base-emitter-pn-junction (inner right) corresponding to the electrical input pulse and the TOBIC signal at the base-collector-pn-junction which represents the electrical output pulse of T2 (inner left). One can see that already the input pulse at T2 is longer than the electrical input pulse caused by the input stage. The time delay of input and output pulses was measured to be 0.6 ns. The deformation of the electrical output pulse of T2 is obvious.

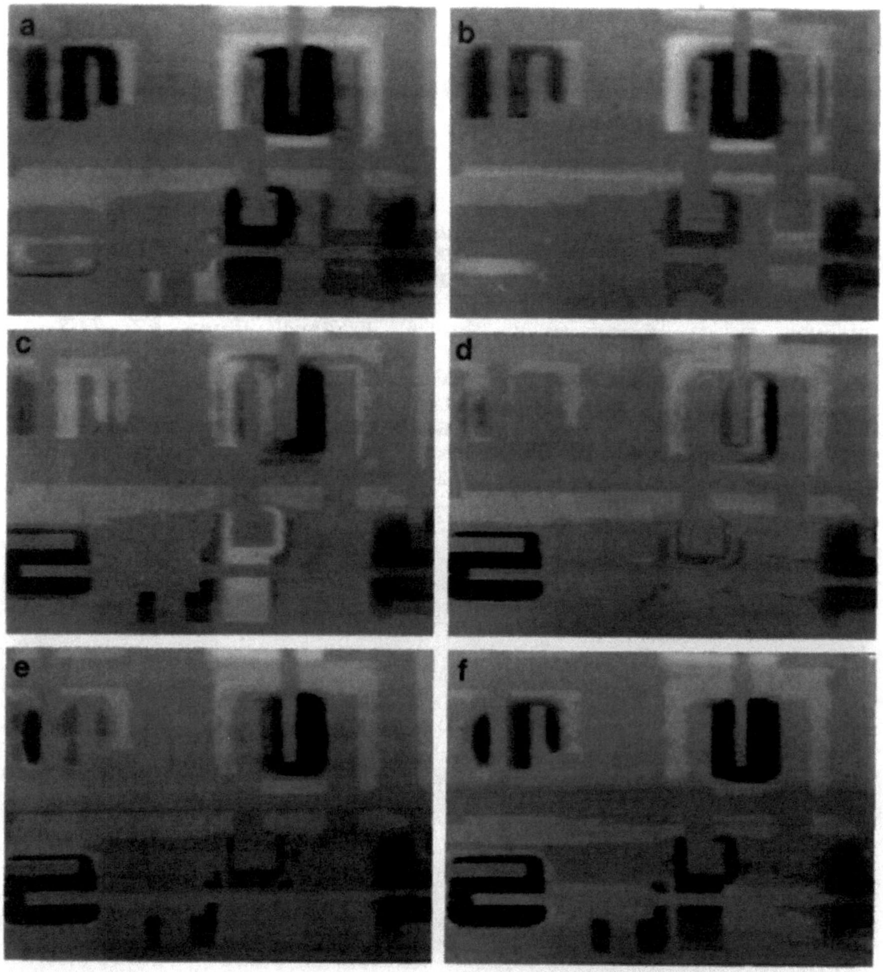

Figure 4
TOBIC-images corresponding to Fig.2 at delay times of
0 (a), 1.2 (b), 2.4 (c), 3.6 (d), 4.8 (e) and 6.0 (f) ns.

A set of 6 TOBIC images corresponding to the area recorded in Fig.2 which are obtained at a difference in the delay time of 1.2 ns is given in Fig.4a - 4f. The time dependence of the TOBIC signals indicates the transition of the transistors between conducting and nonconducting states and the changes of the voltage at the resistances, respectively. Thus the method is capable of giving detailed information as to where the slowest elements are located. In addition it is possible to follow the switching front along the propagation of the electrical input pulses and so propagation delays can be measured. Obviously the running time of the electrical front inside a single transistor (for example T2 at 3.6 to 6.0 ns) or at a resistor (compare R1 at 4.8 to 6.0 ns) can be recorded by this method too. This information is of interest especially for the development of very fast devices.

The time resolution of our method is obtained in principle by optical means, e.g. it is not limited by the electrical recording system. Processes on a time scale down to some tens of femtoseconds should be observable by using femtosecond laser pulses.

References

/1/ E. Wolfgang, R. Lindner, P. Fazekas, H. P. Feuerbaum, IEEE Trans.Electr.Dev. ED 26, 549 (1979)

/2/ J. M. Halbout, P. May, G. Chin, J. Modern Optics 35, 1995 (1988)

/3/ A. Blacha, R. Clauberg, H. K. Seitz, J. Appl. Phys. 62, 713 (1987)

/4/ J. A. Valdmanis, S. S. Pei, in Picosecond Electronics and Optoelectronics II, ed. by F.J. Leonberger, C.H. Lee, F. Capasso, H. Morkoc, Springer Ser. Electron. Photon., Vol. 24 (Springer, Berlin, Heidelberg, 1987)

/5/ D. H. Auston, Appl. Phys. Lett. 26, 101 (1975)

/6/ H. Bergner, T. Damm,; phys. stat. sol. (a), 115, 607 (1989)

Two-Photon Luminescence in CsI:Na and Its Application for the Measurement of UV Femtosecond Light Pulses

P. Heist[1], *R. Deich*[2], *F. Noack*[1], *W.E. Postovalov*[3], *W. Rudolph*[1], *and T. Schröder*[1]

[1]Sektion Physik, Friedrich-Schiller-Universität Jena, Max-Wien-Platz 1, DDR-6900 Jena, German Dem. Rep.
[2]Institute of Physics, Latvian Academy of Science, 229021 Riga, USSR
[3]Institute of General Physics IOFAN, 117924 Moscow W-333, USSR

Abstract: Using femtosecond light pulses at 308nm two-photon absorption (TPA) and subsequent two-photon luminescence (TPL) of CsI doped with Na is investigated. The TPL is shown to be a promising process for recording pulse autocorrelations.

Introduction

At present, high-power subpicosecond light pulses in the UV can easily be obtained by amplification of frequency-doubled femtosecond light pulses in excimer gain modules [1] , [2]. The handling of such pulses requires an exact knowledge of their parameters, in particular of the pulse duration. Common techniques, such as the measurement of pulse autocorrelations through second harmonic generation (SHG) in nonlinear crystals, fail due to the high one-photon absorption below 200nm. To overcome this difficulty surface SHG [3] was applied which, however, requires rather high intensities and a sensitive alignment for the sum frequency signal to be measurable. (Note, usually the repetition rate of the UV pulses is below several 10Hz.) Similar difficulties occur if two-photon ionization in vapor cells is utilized for the nonlinear process [4] . We propose the measurement of luminescence excited by two-photon absorption (TPL) in doped alkali halides. This technique distinguishes itself by its simplicity and sensitivity. Moreover, it provides the possibility of single-shot measurements similar to the two-photon fluorescence in organic dyes which has been applied for the measurement of ps pulses in the VIS and NIR for a long time, e.g. [5].

Firstly, we will describe measurements of two-photon absorption (TPA) coefficients in the alkali halide CsI with UV fs light pulses. The exact characterization of TPA in alkali halides is expected to gain increasing importance because high-power UV optics are often made from this class of materials and because of the fact that fundamental limits of material damage are mainly due to TPA [6] . Secondly, we describe the measurement of femtosecond pulse autocorrelations using TPL in CsI doped with Na, which is a promising candidate for this purpose because of its relatively high TPA coefficient and the fact that the TPA is not followed by the formation of color centers [7].

Measurement of TPA Coefficients in CsI Doped with Na

TPA coefficients can be determined from the measured dependence of the optical transmission on the incident light intensity.
The reciprocal energy transmittance for low input energy densities in a TPA process is given by

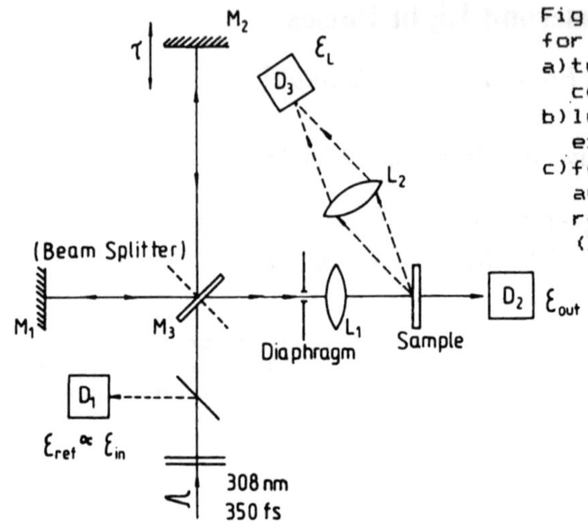

Fig.1: Experimental set-up for measuring
a) two-photon absorption coefficients
b) luminescence versus excitation energy
c) femtosecond pulse autocorrelation (M_3 is replaced by a beam splitter)
($D_{1,2,3}$; detector)

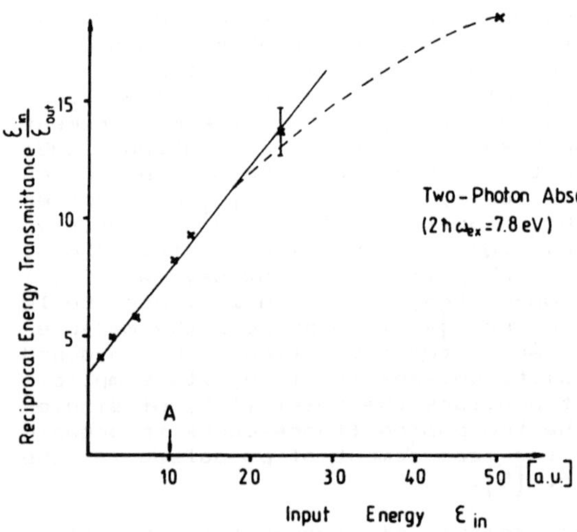

Fig.2: Reciprocal energy transmission as a function of the input energy for TPA in CsI (L 4mm, A 2mJ/cm²) (L is the sample length)

$$E_{in}/E_{out} = C_1+C_2 \quad E_{in}$$

where β is the TPA coefficient and C_1 and C_2 are numerical factors, determined by the experimental conditions (e.g. focusing pulse duration, pulse shape). The measurements were performed with the amplified pulses of a CPM laser [8] which after frequency doubling were reamplified in a XeCl module at 308nm. The pulse intensity was varied by means of color glass filters calibrated at 308nm, where care was taken to avoid contributions of nonlinear absorption. The corresponding experimental set-up is shown in Fig.1.
Fig.2 shows reciprocal energy transmittance versus input energy from which the TPA coefficient is estimated to be

$$\beta \approx (4,5\pm 2)\ 10^{-9}\ \text{cm/W for CsI}\ .$$

The error in this absolute value results mainly from uncertainities in the determination of the pulse and beam profile.

Two-photon Luminescence for Measuring Autocorrelation Functions of UV Femtosecond Light Pulses

The two-photon absorption in CsI doped with Na ($\approx 10^{18} cm^{-3}$) results in the creation of (e^-Na^+) centers within about 10ns which in turn decay under the emission of luminescence with a decay time of several μs [7]. The broad luminescence is centered at about 440nm and can easily be detected by ordinary photodiodes. To prove that the measured TPL obeys the quadratic law we measured the luminescence as function of the excitation energy, see Fig.3.
Obviously for input energies $\epsilon_{in}<A$, i.e. $\beta LI_o \ll 1$ (I_o is the peak input intensity) is satisfied, the linear dependence in the double-logarithmic plot with an ascent of 2 indicates an ideal quadratic dependence of ϵ_L on ϵ_{in}. For higher input energies the quadratic behavior is disturbed. Saturation of the luminescence could not be observed before the onset of material damage.
In autocorrelation measurements an ideal contrast ratio of 3 (defined as the ratio of the signals at zero pulse delay and infinite pulse delay) requires an exact quadratic dependence of the luminescent energy on the input pulse energy. Deviations from this ideal dependence which occur for $\beta LI_o \ll 1$ not being satisfied lead to a decrease of the contrast ratio, see Fig.4.
However, if β and ϵ_{in} are known, βLI_o can be estimated and the correction factor by which the measured autocorrelation width has to be multiplied can be determined. Also, if one can exclude other effects influencing the contrast ratio, the measured contrast ratio can serve for estimating the correction factor from Fig.4. In this manner autocorrelation widths with an accuracy better than 10% can be obtained for a broad range of input intensities.
Fig.5 shows a typical TPL pulse autocorrelation measured with the experimental arrangement sketched in Fig.1. The contrast ratio was about 2.1 resulting in a corrected correlation width of about 490fs. For Gaussian pulses this corresponds to a pulse duration of 350fs. The satellites are typical for femtosecond pulse amplification in XeCl in the weak saturation regime. This, as well as their constant separation of about 1.3ps, is due to the structured gain profile [9].

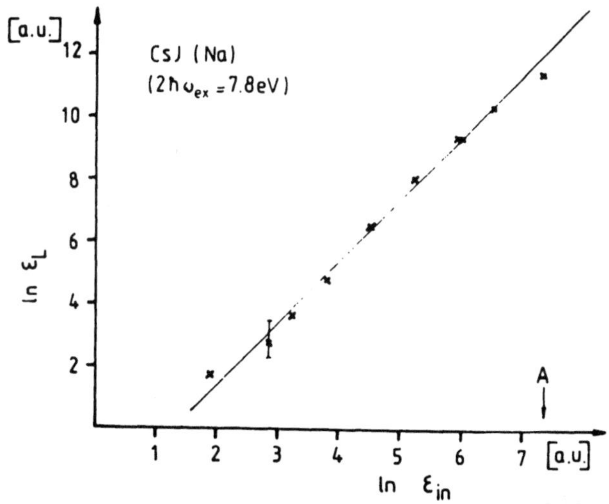

Fig.3: Two-photon luminescence of CsI:Na as a function of the input pulse energy ($A \approx 0.64 mJ/cm^2$, $T_L \approx 350fs$)

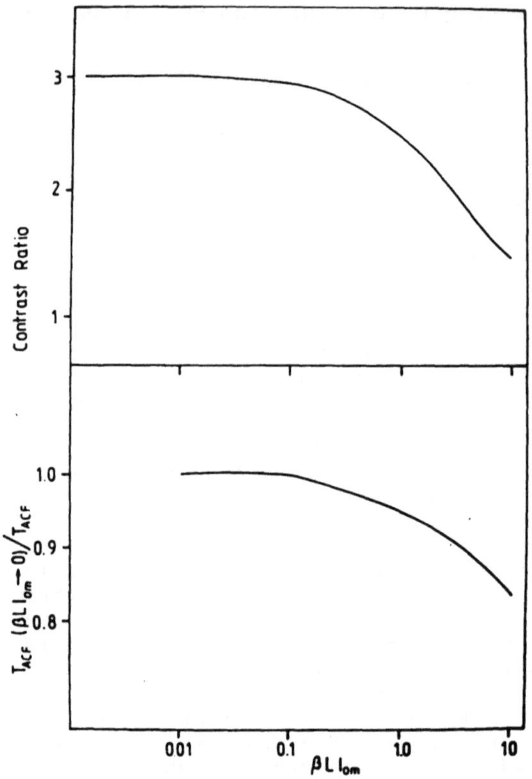

Fig.4: Contrast ratio and factor by which the measured autocorrelation width T_{ACF} has to be multiplied to get the true value T_{ACF} $(\beta L I_o \to 0)$ as a function of the pulse intensity

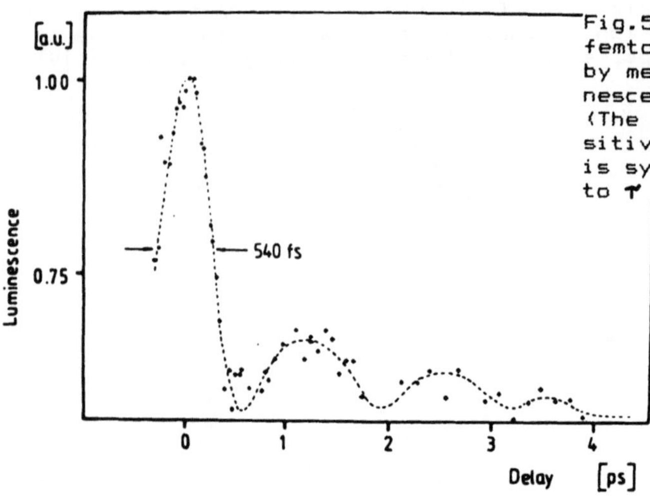

Fig.5: Autocorrelation of UV femtosecond pulses recorded by means of two-photon luminescence in CsI:Na
(The curve is shown for positive time delay τ only, but is symmetrical with respect to $\tau = 0$)

References

[1] J.H.Glowina, G.Arjavalingam, P.P.Sorokin J.E.Rothenberg: Opt.Lett. 11, 79(1986)

[2] S.Szatmari, B.Racz, F.P.Schäfer: OptCommun. 62, 271(1987)

[3] E.S.Kintzel, C.Rempel: Appl.Phys. B **42**, 91(1987)
[4] M.Morita, T.Yajima: Appl.Phys. B **28**, 25(1982)
[5] S.L.Shapiro, (Ed.): Ultrashort Light Pulses, Springer New York (1977)
[6] N.Bloembergen: IEEE J.Quant.Electr. **QE-10**, 375(1974)
[7] E.D.Aluker, W.W.Gavrilov, R.G.Deich, S.A.Tschernov: Ultrafast Processes stimulated by Radiation in Alkali-Halides (in Russian), Riga (1987)
[8] W.Dietel, E.Döpel, V.Petrov, C.Rempel, W.Rudolph, B.Wilhelmi, G.Marowsky, F.P.Schäfer: Appl.Phys B **46**, 183 (1988)
[9] B.Dick, S.Szatmari, B.Racz, F.P.Schäfer: Opt.Commun. 62, 277 (1987)

Picosecond Resolution in Optical Coherent Transient Experiments with Nanosecond Sources

A. Débarre, J.-C. Keller, J.-L. Le Gouët, and P. Tchénio

Laboratoire Aimé Cotton, CNRS II, Bât. 505, Université Paris-Sud, F-91405 Orsay Cedex, France

Abstract. First we recall how, using broadband correlated laser pulses, one can monitor atomic dipole evolution on a time scale much shorter than the pulse duration. We analyse the influence of the field strength in this technique. We show that in the strong field regime, the destruction of atomic memory is surprisingly fast.

Second, we propose a new technique to read structures engraved in an inhomogeneous line. This technique also takes avantage of the high time resolution which can be achieved with broadband correlated pulses.

1. Introduction

In the time domain, the usual way to investigate ultrafast optical phenomena (<1ps) like atomic dipole relaxations is to perform experiments with femtosecond laser sources. However, an alternative approach has been proposed recently. It makes use of broad bandwidth laser pulses, with a correlation time τ_c which is much smaller than the pulse duration τ_l. Then, it is shown that the dipole evolution can be monitored with a time resolution τ_c in time-delayed four wave mixing experiments [1]. In these experiments, the atomic dipole evolution is monitored as a function of the time delay between two broadband correlated pulses which are obtained by splitting a single initial pulse. Thus, subpicosecond T_2-type decay rates have been measured with nanosecond lasers which are much less expensive and easier to handle than femtosecond sources.

The key to the temporal resolution of these experiments is the combination of two factors. First, the delay between two correlated broadband pulses is well defined (with a resolution τ_c) . Second, in weak field regime, atomic dipoles freely evolve during this interval. But when the field becomes strong, it perturbs dipole memory. We show that, when the field strength increases , the memory time of the dipoles decreases surprisingly fast.

The high temporal resolution of the experiments performed with broadband correlated pulses can be applied in domains other than relaxation time measurements. We propose a new technique which takes avantage of this property to read a structure engraved in a broad inhomogeneous line.

Springer Proceedings in Physics, Vol. 49 **Ultrafast Phenomena in Spectroscopy**
Editors: E. Klose · B. Wilhelmi

2. Monitoring a dipole evolution with broadband correlated fields.

The electric field of a broad band pulse fluctuates with a characteristic time τ_c, which identifies with the inverse of the spectral width of the light. Such a pulse can be regarded as a train of elementary components of duration τ_c without phase relations. The second pulse is a time delayed replica of the first one. So, to each elementary component of the first pulse corresponds a twin elementary component in the second pulse. The time interval between a pair of twin components is the same for all the pairs. This defines the delay t_{12} between the two pulses with a resolution τ_c . Simple evidence of this resolution is that the contrast of an interference pattern produced by the two pulses vanishes when their time interval exceeds τ_c (Figure 1a).

Atomic memory allows one to take avantage of this well-defined delay. Consider the simple experiment where two correlated angled beams resonantly interact with an atomic sample. The first pulse creates atomic dipoles. The second pulse converts them into a spatial population grating. As the dipoles created by a particular elementary component retain the memory of its phase, a population grating can be carved even when the time delay between pulses exceeds τ_c. This grating reflects the quasi-free evolution of the dipoles, specifically their relaxation during the time interval t_{12} between a pair of twin components. This population grating can be probed by a delayed third pulse. Recording the diffracted energy as a function of the time delay between the first two correlated beams gives direct access to dipole relaxation.

The developed picture is valid in the weak field regime. But to increase signal intensity, one can perform the experiment in strong field conditions. Then the experiment raises a question: how long can an atomic system irradiated by a strong stochastic field retain the memory of its interaction with a given elementary component, and recognize its time delayed replica to ensure the time resolution of the experiment ?

We have investigated both theoretically and experimentally the diffraction of a weak broadband pulse by the transient grating carved by two broadband correlated pulses, when one of the correlated pulses is allowed to be strong [2,3]. To focus on the field effect, we have chosen a "model" atom which doesn't relax during the pulse sequence. Now, recall the relevant time parameters of the problem. The first one is the characteristic evolution time T of the atomic system in the presence of the radiation field. The strong field regime is obtained for $T \ll \tau_l$, where τ_l is the pulse duration. The last parameter is the correlation time τ_c which is assumed to be shorter than any other time parameter in the problem. The diffracted signal intensity is recorded as a function of the time delay t_{12} between correlated pulses. The delay t_{12} is varied around 0 in a domain which is much narrower than τ_l and than T_2 which is the dipole relaxation time. In the weak field regime, atomic dipoles can keep memory of a given interaction with the field during their lifetime T_2. So, the recorded signal depends on the correlation between the first two pulses all over the explored domain of t_{12} [3,4]. When one of the correlated pulses is strong, the atomic memory time is dramatically reduced. Then, we have shown that except for a narrow domain of width τ_c located at

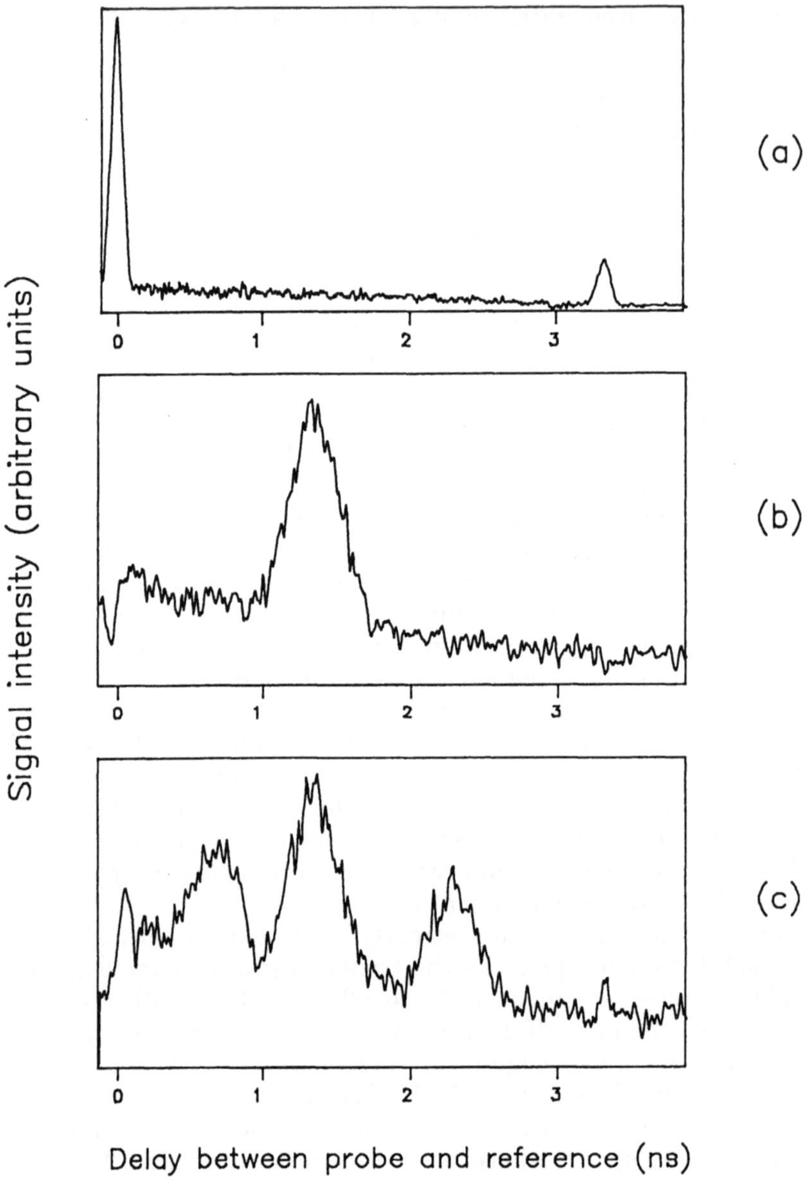

Figure 1. Experiment in calcium vapor. The contrast of a spatial interference pattern is recorded as a function of the delay between probe and reference pulses (correlated pulses).(a) interference between the probe and the reference beams: autocorrelation profile of the probe pulse. The time interval between the two peaks corresponds to the cavity round trip. (b)&(c) interference between the signal and the reference: amplitude cross-correlation profiles between the diffracted signal and the reference pulse. The engraved structure in (b) is sinusoidal. In (c), it is the superposition of three sinusoidal structures.

t_{12}=0, the signal identifies with the signal which would be produced by uncorrelated pulses. This result is somewhat surprising. Indeed, when $\tau_c \ll t_{12} \ll T$, it means that the atomic quantities involved in the signal lose memory of the correlation between excitation pulses while the atomic system scarcely evolves during the time interval between two twin components. This very fast destruction of atomic memory was not evidenced in the previous reports on strong field effects [5]. Recent theoretical results predict that, when the signal is produced by *two* strong correlated pulses, the memory obliteration of the correlation between excitation pulses occurs on the same time scale τ_c [6].

3. Reading a structure engraved in an inhomogeneous absorption line.

So far only the energy of the diffracted signal was considered. But in an inhomogeneous broadened material, when the probe pulse is short, the time behavior of the signal gives a direct access to new information, namely the shape of the spectral domain excited by the first two pulses. More precisely, if a short pulse of area θ probes the grating at time t_0, the radiated field $E_s(t,\mathbf{r})$ reads

$$E_s(t,\mathbf{r}) = \theta\ \xi_s(t-t_0)\ \exp(i\Omega t-\mathbf{k_s}.\mathbf{r}) + \text{c. c.} \qquad (1)$$

where Ω is the central frequency of the absorption line. $\xi_s(t)$ reads

$$\xi_s(t) = \begin{cases} \displaystyle\int_{-\infty}^{+\infty} d\Delta\ n(\Delta)\ \exp((i\Delta-\gamma)t) & \text{if } t\geq 0 \\ 0 & \text{if } t\leq 0 \end{cases} \qquad (2)$$

where γ is the dipole decay rate and $n(\Delta)$ is the amplitude of the grating made of excited atoms with a given resonance frequency $\Omega+\Delta$. As soon as the relaxation can be neglected during the emission time interval, the field shape $\xi_s(t)$ reproduces that of the Fourier transform of the structure $n(\Delta)$ which is engraved in the inhomogeneous line.

The characteristic variation time τ_s of the signal $\xi_s(t)$ is determined by the inverse of the spectral width of the structure. When the structure is carved all over the inhomogeneous width, τ_s ranges from less than 100fs to 1ns. Specifically, τ_s is short in photochemical materials of interest for optical data storage which exhibit very broad inhomogeneous widths as large as 100cm^{-1}. As no dectector is fast enough, it has been proposed to determine the signal intensity $|\xi_s(t)|^2$ by a sampling method. The intensity cross-correlation between the signal and a short reference pulse has been measured as a function of the time delay between the reference pulse and the probe pulse [7]. This method requires a reference pulse and a probe pulse shorter than τ_s, the inverse of the spectral width of the structure.

Now, we address the problem of reading the engraved structure with pulses much longer than τ_s. Then, the temporal behavior of the signal intensity reflects the temporal shape of the light pulse and not that of the Fourier transform $|\xi_s(t)|^2$ of the spectral structure. In order to determine the response $\xi_s(t)$ we propose a new sampling method which takes avantage of the well defined delay between two correlated broadband pulses. In this method a broadband pulse probes the grating. A fraction of this probe pulse is sent in a variable delay line. It constitutes the reference . The *amplitude* cross-correlation function between the diffracted signal and the reference pulse is obtained through the contrast of a spatial interference pattern which is detected on a linear CCD. This contrast is measured as a function of the delay between the probe and the reference pulses. We show that the resulting profile reproduces $|\xi_s(t)|$.

The electric field of the probe pulse can be written as

$$E(t,\mathbf{r}) = \varepsilon(t)\ \exp\left[\,i\,(\Omega t-\mathbf{k}\mathbf{r})\,\right] + c.c. \tag{3}$$

where the envelope $\varepsilon(t)$ has a width τ_l and fluctuates with a characteristic time τ_c. The pulse is assumed to be long and broadband. Thus the following inequalities hold: $\tau_c << \tau_s << \tau_l$. An elementary slice of the field in the interval $[t_0,t_0+dt]$,with $dt< \tau_c$, can be regarded as a short probe pulse with an area $\varepsilon(t_0)dt$. It stimulates an emission given by the equations (1,2). The total radiated field $R(t,\mathbf{r})$ is the superposition of all elementary responses in weak field regime.$R(t,\mathbf{r})$ reads

$$R(t,\mathbf{r}) = \int_{-\infty}^{+\infty} dt'\ \varepsilon(t')\ \xi_s(t-t')\ \exp\left[i\left(\Omega t-\mathbf{k}_r\mathbf{r}\right)\right] + c.c. \tag{4}$$

To measure the amplitude cross-correlation between the signal $R(t,r)$ and the reference pulse $E(t-T,r)$ we make them interfere. As they have different wave vectors, respectively $\mathbf{k}$ and $\mathbf{k}_r$, a fringe pattern is created. It exhibits a contrast $g(T)$ which, according to equations (3,4), reads

$$g(T) \propto \left| \int_{-\infty}^{+\infty} dt'\ \xi_s(t')\ A(t'-T) \right| \quad \text{with} \quad A(t)=\int_{-\infty}^{+\infty} dt'\ \varepsilon^*(t')\ \varepsilon(t'-t) \quad . \tag{5}$$

$A(t)$ is the autocorrelation function of the probe field and it has a width τ_c. As the scale of variation of $\xi_s(t)$ is $\tau_s >> \tau_c$, $A(t)$ can be considered as a δ-function with regard to $\xi_s(t)$. So, $g(T)$ reads

$$g(T) \propto |\xi_s(T)| \quad . \tag{6}$$

Measuring the fringe contrast as a function of the reference pulse delay T gives direct access to the Fourier transform of the engraved structure.

Experimentally we have demonstrated this technique. The experiment was performed in Ca vapor. The inverse of the inhomogeneous (Doppler) width of the excited transition is τ_s=0.66ns. The transient engraved structures are probed by a broadband pulse of duration τ_l=7ns, with a correlation time τ_c=50ps. To determine the correlation time τ_c the autocorrelation function of the probe pulse is measured (Figure 1a) [8] . Figures 1b&1c represent the variation of the fringe contrast as a function of the time delay of the reference pulse for two different engraved structures.

In Figure 1b, the structure results from the excitation of the sample by two angled correlated beams with a time delay t_{12}. The carved structure can be written as

$$n(\Delta)= n_0(\Delta) \exp(-i\Delta t_{12}) \qquad (7)$$

where $n_0(\Delta)$ is the equilibrium shape of the line. As expected, Figure 1b reproduces the Fourier transform of this structure which is simply the Fourier transform of the equilibrum line shape delayed by t_{12} with regard to the probe pulse.

In Figure 1c a more complex stucture has been engraved. The excitation is provided by four correlated pulses P1, P2,P3,P4 with a time delay t_{1i} between P1 and Pi. The pulse P1 (write pulse) builds a spatial grating in the sample with each one of the three pulses P2,P3,P4 (data pulses), which propagate along the same direction. The carved structure is the superposition of three sinusoidal modulations and it can be written as

$$n(\Delta)= n_0(\Delta) [a_2 \exp(-i\Delta t_{12}) + a_3 \exp(-i\Delta t_{13}) + a_4 \exp(-i\Delta t_{14})] \qquad (8)$$

where the a_i set contains the relative amplitudes and phases of the pulses. Figure 1c exhibits the three peaks expected in the Fourier transform profile of the engraved structure.

References.

1. S. Asaka, H. Nakatsuka, M. Fujiwara and M. Matsuoka: Phys. Rev. A **29** 2286 (1984). N. Morita and T. Yajima: Phys.Rev.A 30 2525 (1984).
2. P. Tchénio, A. Débarre, J.-C. Keller, J.-L. Le Gouët: Phys.Rev. A **39** 1970(1989).
3. P. Tchénio, A. Débarre, J.-C. Keller, J.-L. Le Gouët: Phys.Rev.Lett. **62** 415(1989).
4. M. Defour, J.-C. Keller, J.-L. Le Gouët: J.Opt.Soc.Am.B 3 544 (1986).
5. R. Beach, D. De Beer, S. R. Hartmann: Phys.Rev. A **32** 3467 (1985) . M. Defour, J.-C. Keller, J.-L. Le Gouët: Phys.Rev. A **36** 5226 (1987) .
6. V. Finkelstein and P. R. Berman: to be published.
7. A. Rebane, J. Aaviksoo and J. Kuhl: Appl.Phys.Lett. **54** 93 (1989)
8. A. Débarre, J.-C. Keller, J.-L. Le Gouët, A. Richard, P. Tchénio: Opt.Comm. to be published

Enhancement of Second Harmonic Generation from a Silver Island Film by Self-Polarization

F.R. Aussenegg, A. Leitner, and J.D. Pedarnig

Institut für Experimentalphysik der Karl-Franzens-Universität, Universitätsplatz 5, A-8010 Graz, Austria

"Island films" are metal layers consisting of droplet-like submicroscopic particles on a solid substrate. In such structures localized electron plasma oscillations can be driven resonantly by incident light, causing an increased field inside and at the surface of the particle and thus an enhancement of second harmonic generation (SHG) created in the surface of the particles. A further enhancement of SHG by a factor 10 is observed when the island film is positioned parallel to a smooth mirror layer. This can be attributed to a coupling of the particle's dipole with its image dipole ("self-polarization"). As the described system is thermally damaged at an energy density exceeding 2.5 mJ/cm^2 picosecond pulses have to be used for irradiation.

When an oscillating dipole is positioned close to a conducting metal surface the image dipole induced in the metal surface reacts on the dipole and modifies the optical properties, a phenomenon called "self-polarization" /1/. Recently a system was investigated consisting of a layer of submicroscopic silver particles (a so-called "island film") as the dipoles, separated by a dielectric quartz layer from a flat silver surface /2/. The enhanced absorption of this system compared with the absorption of the free island film was explained by self-polarization.

In this work we focus on the question whether second harmonic generation (SHG) at silver island films can be enhanced by the self-polarization effect too.Silver island films show SHG due to symmetry breaking of their surface which is in principle a property of every surface /3/. Even particles of highly symmetric shape possess a non-symmetric polarization situation at the surface due to the different dielectric function of the substrate they are affixed to and the air at the other side, respectively /4,5/.

Fig. 1 shows the investigated sample geometry. A 200 nm smooth silver layer was deposited on a flat glass substrate and acts as a bulk metal with a flat surface. Atop this a quartz layer with stepwise changing thickness was placed. Finally the silver islands were deposited by slow thermal evaporation technique (for details see /4/).

The incident light drives electron-plasma oscillations in the spheriodlike particles. Every particle's dipole induces an image dipole in the 200 nm "bulk" metal layer. At a certain distance (quartz layer thickness) the phase shift between the particle's dipole and its image

Springer Proceedings in Physics, Vol. 49 **Ultrafast Phenomena in Spectroscopy**
Editors: E. Klose · B. Wilhelmi

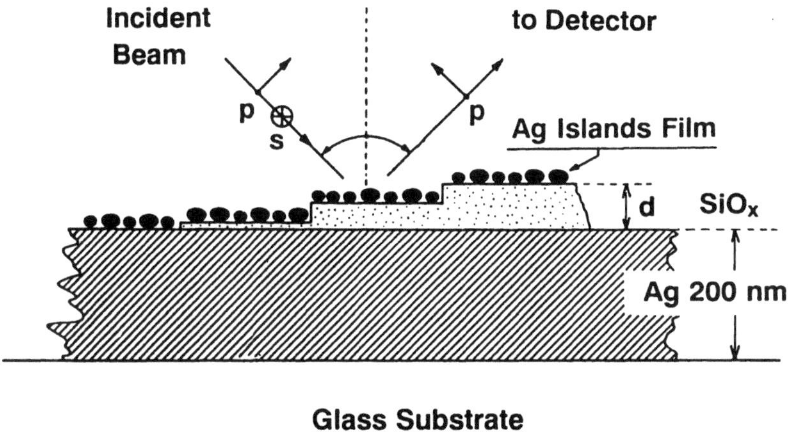

Fig. 1: Sample geometry; s,p: polarization directions

is optimum for supporting the dipole oscillation. Following /1/, when approximating a particle as a point-dipole in its center the induced dipole moment μ_{ind} in an island particle is given by the (free) particle's polarizability α_0 and the sum of two fields: the primary field E_p, given by the incoming light in the absence of the particle, but in the presence of the solid surface, and the image field E_{image} of the dipole:

$$\mu_{ind} = \alpha_0(E_p + E_{image}) \ . \qquad (1)$$

The image field is given by the dipole moment μ_{ind}, the smooth mirror's field reflection coefficient γ (in our case assumed to be unity) and a distance dependent phase function f(n2d), where n is the refraction index of the interlayer and d is the distance to the metal surface:

$$E_{image} = \mu_{ind}\, \gamma\, f(n2d) \ . \qquad (2)$$

Combining eq. (1) and eq. (2) and with $\alpha_{eff} = \mu_{ind}/E_p$ for the effective polarizability of a particle we get

$$\alpha_{eff} = \alpha_0 / [1 - \gamma\, \alpha_0\, f(n2d)] \ . \qquad (3)$$

In the same way the nonlinear term of the polarizability $\beta_{eff}(2\omega;\omega,\omega) = \mu_{ind}^{(NL)} / E_p^2$ can be deduced:

$$\beta_{eff}(2\omega;\omega,\omega) = \frac{\beta_0(2\omega;\omega,\omega)}{[1 - \gamma\, \alpha_0(2\omega)\, f(2nd)]\,[1 - \gamma\, \alpha_0(\omega)\, f(2nd)]^2} \qquad (4)$$

where $\mu_{ind}^{(NL)}$ is the induced nonlinear dipole, $\beta_0(2\omega;\omega,\omega)$ is the second-order particle

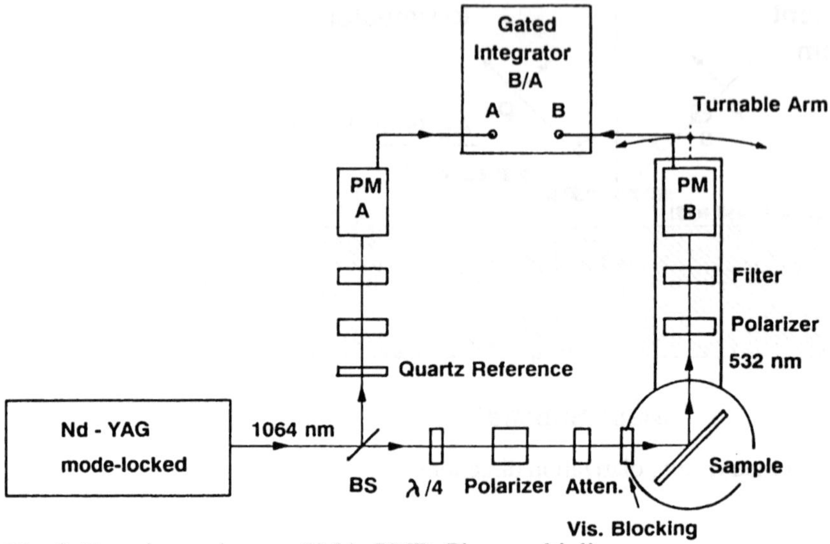

Fig. 2: Experimental setup; PMA, PMB: Photomultipliers

polarizability for the SHG process and $\alpha_0(\omega)$ and $\alpha_0(2\omega)$ are the linear polarizability of the particle at the fundamental and SH frequency, respectively. From this an enhancement of SHG by self-polarization is expected when $\alpha_0(\omega)$ or $\alpha_0(2\omega)$ gets comparatively large.

Fig. 2 shows the experimental setup. The fundamental was produced by a passively mode-locked Nd-YAG laser. The fundamental beam was incident at 45^0 to the surface normal. For both polarization directions of the fundamental parallel (p) and vertical (s) with respect to the plane of incidence the SH was reradiated in the reflection direction and was p-polarized.

As there is a non-negligible absorption of the irradiated fundamental light (1064 nm) in silver particles (the oscillating electron plasma represents an electric current and causes ohmic losses) the metal is heated and therefore the energy density applicable to the sample is limited. Assuming that the absorbed energy cannot flow from the metal to the substrate during the laser pulse but that this energy is dissipated completely before the next pulse is absorbed the temperature increase Δt is given by $\Delta t = Q\,A\,/\,(\,C\,\rho\,a\,d_m\,)$ where Q is the pulse energy, A the absorption of the film, C the specific heat of the metal, ρ its density, a the area of the laser spot at the sample and d_m the mass thickness of the island film. As the temperature of the metal must not exceed 130^0C in order to prevent changes in particle morphology /4/ for our films (4.0 nm mass thickness, A = 0.04 at 1064 nm) a maximum pulse energy density of about 2.5 mJ /cm^2 results.

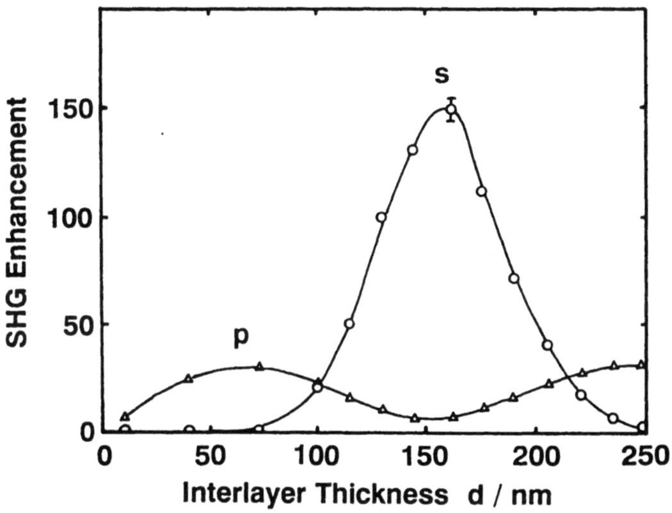

Fig. 3: SH intensity of 4 nm mass thickness silver island film vs interlayer thickness (normalized to SH intensity without smooth metal surface).Polarization direction of the incident fundamental parallel (p) and perpendicular (s) to the plane of incidence, respectively

In order to get a measurable SHG signal we need 25×10^6 W/cm^2 fundamental intensity as a minimum. Therefore laser pulses shorter than 100 ps are absolutely required to investigate the enhancement of optical nonlinearity enhanced by self-polarization.

Fig. 3 shows the measured SH intensity as a function of the distance between an island film (4 nm mass thickness) and a smooth metal surface normalized to the SH intensity of the same island film without mirror. Discussing the results we have to take into account the increase in the fundamental's field strength by reflection at the flat metal surface. The maximum increase in SH intensity caused by this effect is a factor 16. So we have to attribute approximately a factor 10 of SHG enhancement for s-polarized fundamental to the self-polarization effect.

References:

1. S. Efrima and H. Metiu, J. Chem. Phys., 70, 1602 (1979)
2. R.W. Holland and D.G. Hall, Phys. Rev. Lett., 52, 1041 (1984)
3. J.Rudnick and E.A.Stern, Phys. Rev. B 4, 4274 (1971)
4. F.R. Aussenegg, A. Leitner and J.D. Pedarnig, Appl. Phys. B49, 279 (1989)
5. A. Wokaun, J.G. Bergmann, J.P. Heritage, A.M. Glass, P.F. Liao and D.H. Ohlson, Phys. Rev. B 24, 849 (1981)

Spectral Holography of Ultrashort Pulses and Photochronography of Ultrafast Processes with Pico- to Femtosecond Time Resolution

Yu. T. Mazurenko

State Optical Institute, 199034 Leningrad, USSR

Abstract. We consider the methods of registration, reconstruction and transformation of ultrashort light pulses based on interference, diffraction, holography and nonlinear optics of spectral decomposition waves.

1. Spectral Holography

The longitudinal dimensions of the wave packets formed from ultrashort light pulses are substantially less than the typical dimensions of the elements of optical systems. Because of this, diffraction of ultrashort pulses may result in the change of not only the spatial but also the temporal structure of radiation. These features underlie the principles of holography of ultrashort light pulses in the spectrally nonselective media [1,2,3,4,5].

From the methods considered the most promising one is the holography of time-varying signals, which includes the spatial spectral decomposition of light [2,3,4,5]. The wave formed during the spatial spectral decomposition of a short light pulse may be called a spectral decomposition wave. The formation of such a wave is possible if the pulse length t is much shorter than the reciprocal T of the frequency resolution of the spectral instrument.

The spectral decomposition wave exists during the time T and is restricted in space by the near diffraction zone of the plane of spectrum. It contains the complete set of the spaced Fourier-transform components of the light pulse, these components having the appropriate amplitude, phase and frequency. So the spectral decomposition wave is actually a physical realization of the Fourier transform of a temporal signal in optics.

If the time distance between two pulses is less than T, then the interference of their spectral decomposition waves can exist. The result of registration of the interference pattern can be called a spectral hologram. In general, the spectral hologram contains the registration of a signal pulse relative to a reference pulse.

The signal pulse can be reconstructed from the spectral hologram in the form of a real temporal signal. The reconstruction occurs by the filtration of the spectral decomposition wave of the reference pulse with the spectral hologram and the subsequent composition of the newly formed spectral components by the use of the spectral instrument operating on the backward waves. Besides the reconstructed signal pulse the time-reversed copy of it is generated.

An ultrashort wave packet can contain the spatial modulation in the direction perpendicular to the spectral decomposition direction. In the most general case it corresponds to the 1-D motion picture. Such a wave packet forms the 2-D spectrum of temporal and spatial frequencies on the plane of spectrum of the instrument. The 2-D spectra of two pulses can interfere on the plane of spectrum. The result of registration of the interference pattern is the 2-D spectral hologram.

Springer Proceedings in Physics, Vol. 49 Ultrafast Phenomena in Spectroscopy
Editors: E. Klose · B. Wilhelmi

By the use of a 2-D spectral hologram we can register and reconstruct the 1-D motion picture. Besides, such a hologram allows us to separate spatially the direct and time-reversed copy of the signal pulse on reconstruction [4].

The volume or 3-D spectral hologram can be obtained by the registration of the 3-D interference pattern of oppositely travelling spectral decomposition waves of the signal and the reference wave packets. A 3-D spectral hologram in comparison with a 2-D one has the same type of advantages as those of an ordinary 3-D hologram in comparison with a 2-D one, namely, high efficiency and possibility of discrimination of undesirable signals (in our case these signals are the time ones) during the reconstruction. We may obtain from a 3-D spectral hologram only direct or only time-reversed copies of a signal wave packet at will [3].

2. Holographic Spectral Filtration

If we interpret the reconstruction of a spectral hologram as the complex spectral filtration of a temporal optical signal, we easily see that this filtration makes it possible to obtain light pulses of any amplitude and phase structure whose length does not exceed T.

Note that the holographic spectral filtration allows us to obtain phase filtration of the spectrum of a temporal signal by the use of an amplitude transparency (a 2-D spectral hologram).

By using the spectral hologram as a temporal analog of a Vander-Lugt filter, we can realize recognition and optimal filtering of temporal optical signals [6].

If the spectral hologram is reconstructed as an ordinary Fourier-transform hologram, a coherent image is formed which is an exact analog of the temporal signal written on the hologram. The inverse process is also possible: a holographic synthesis of a spectral hologram and thus of a temporal signal from a given image. Besides, the spectral hologram of a desired signal can be produced artifically on the basis of calculation.

Note the possibility of compensation of the arbitrary phase modulation (the arbitrary chirp) of ultrashort pulses by the use of holographic spectral filtration.

Overall, spectral holography can be the universal means of linear transformations of ultrashort light pulses [6].

3. Spectral Nonlinear Optics

We will consider the nonlinear-optical phenomena in the spectral decomposition waves. These waves exist for a time T which is much more than the length of dispersed pulses t, and hence the interaction time between the radiation and the nonlinear medium can be increased by several orders of magnitude. The more important feature of these interactions is the possibility for basically new transformations of optical pulses [7].

In addition to spectral decomposition waves we also consider monochromatic waves incident normally on the spectrum plane and which produce in this plane certain amplitude images which can be assumed to correspond the spectra of spatial frequencies of certain initial 1-D images.

We fill a region of diffraction depth near the spectrum plane with a nonlinear medium and we obtain a wave of nonlinear polarisation generated by interaction of some number of spectral decomposition waves and monochromatic waves.

The wave of nonlinear polarisation produces the light wave which can be a new spectral decomposition wave or a new monochromatic wave.

The newly generated spectral decomposition wave can be converted into a new wave packet by a spectral instrument operating on the backward waves.

Hence, the nonlinear interaction of spectrally dispersed wave packets and coherent images expanded in a spatial-frequency spectrum leads to generation of a new wave packet or a new 1-D image. In consequence, mutual conversions of the time signals, conversions of spatial signals into time signals and of time signals into spatial signals are possible. The generated spectrum of spatial or temporal frequencies is the product of the initial spectra. Therefore the generated time signals or spatial signals are connected with the initial time and spatial signals by convolution-type conversion.

Apparently, four-wave nonlinear interactions of the type widely used for spatial transformation of monochromatic wave fields provide the greatest possibilities for realizing the principles considered. In this case we are concerned essentially with extending the principles of self-diffraction and dynamic holography to spectral decomposition waves. By the four-wave interaction we can obtain under dynamic conditions compression, stretching and time-inversion of optical pulses. One can also perform the dynamic space-time and time-space transformation of optical signals.

We note the possibility of using these space-time conversions for extremely broad-band modulation and subsequent demodulation of light beams. Conversion of the image into the time signal can be regarded as modulation, and the inverse transformation can be regarded as demodulation. If the signal which is being modulated is a sequence of ultrashort pulses, then when it is modulated a frequency band with a wide order of inverse pulse lengths may be covered. The speed of response of the modulating device is determined by the requirement to change the image at the pulse repetition frequency, and in the optimum case the pulse repetition frequency is equal to the frequency resolution of the spectral dispersion. Hence, multiplication of the modulation frequency band by a factor of T/t occurs. The reverse situation occurs on demodulation. Of course, the advantage obtained must be ensured by the number of independently modulated image elements equal to T/t [7].

4. Superfast Scanning of Light Beams and Possibilities for Pico-Femtosecond Chronophotography

Principles for spatial scanning of optical radiation based on the use of four-wave nonlinear interaction with the participation of spectral decomposition waves of ultrashort pulses were put forward [8]. The interaction between the spectral decomposition wave and the counterpropagating plane monochromatic wave in dynamic light-sensitive medium leads to the formation of the rotating holographic grating. The longitudinal structure of this grating is typical for the reflective holograms, and, therefore, this grating can be used as a holographic analog of a rotating mirror. This mirror is formed using neither direct nor indirect macroscopic shifts of optical hardware. Because of this, there are no mechanical restrictions on the time resolution when scanning.

As in the case of a usual rotating mirror the time resolution is limited by the scanning time of the diffraction angle which corresponds to the mirror aperture. It can be shown [8] that the time resolution equals the length of a light pulse used to form the hologram. Because of this, the time resolution of a single diffraction element of an image when scanning the optical radiation turns out to be principally limited only by a value lasting several light oscillation periods [8].

At present, light pulses are obtained which contain less than ten oscillations. For this reason, the practical limits of resolution are consistent with the fundamental ones, and it is possible now to achieve the pico-femtosecond time resolution of scanning. This leads to the possibility of chronophotography of ultrafast processes with pico-femtosecond time resolution [8].

References

1. Yu.T.Mazurenko: Opt. Spektrosk. **56**, 583-584 (1984)
2. Yu.T.Mazurenko: Opt. Spektrosk. **57**, 569-571 (1984)
3. Yu.T.Mazurenko: Opt. Spektrosk. **59**, 608-614 (1985)
4. Yu.T.Mazurenko: Kvant. Elekronika 12, 1235-1241 (1985)
5. Yu.T.Mazurenko: Usp. Fiz. Nauk **148**, 537-539 (1986)
6. Yu.T.Mazurenko: Opt. Spektrosk. **57**, 8-11 (1984)
7. Yu.T.Mazurenko: Opt. Spektrosk. **59**, 57-61 (1985)
8. Yu.T.Mazurenko: Opt. Spektrosk. **60**, 551-558 (1986)

Subfemtosecond Studies of Four-Photon Scattering in Dye Solutions

V.L. Bogdanov, A.B. Evdokimov, G.V. Lukomskij, and B.D. Fainberg

State Optical Institute, 199034 Leningrad, USSR

Abstract. This paper contains some results of experimental studies of resonance four-photon scattering of ultrashort laser pulses in dye solutions using precision time-delay lines. A short delay (≈50fs) between scattering signals in different directions has been observed, which was explained by the peculiarities of nonstationary self-diffraction. Modulation of scattering signals with a period equal to the period of the optical oscillations in the laser light (1.95 fs) and equal to half the period of the optical oscillations (<1fs) has been found. These results were explained by interference of two light fields - four-photon and Rayleigh scatterings. The possible application of such interference to the analysis of ultrashort pulses is discussed.

1. Experimental

Investigation of resonance four-photon scattering is a powerful method for the determination of ultrafast relaxation times for electronic transitions in different media. This paper is devoted to the study of such scattering for ultrashort pulses in a dye solution, i.e. DODCI in ethylene glycol. The experimental scheme is shown in Fig.1. A jet dye laser was synchronously pumped by ML argon laser pulses. We usually obtained spectrally limited pulses

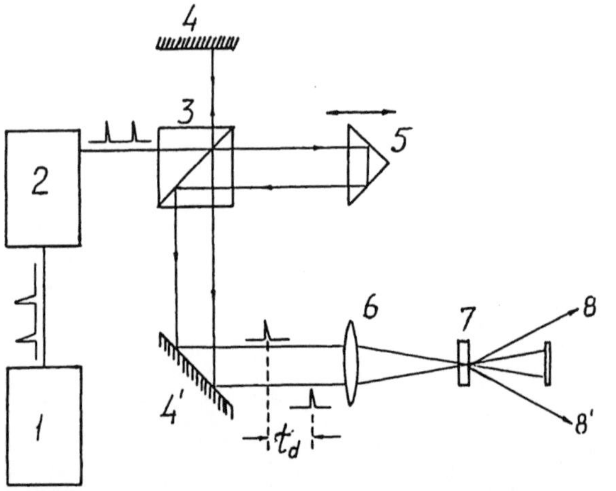

Fig.1. Experimental set-up: 1 - ML argon laser, 2 - jet dye laser, 3 - glass cube, 4, 4' - mirrors, 5 - delay line, 6 - lens, 7 - dye solution, 8, 8' - photodetectors, t_d-delay time.

Springer Proceedings in Physics, Vol. 49 Ultrafast Phenomena in Spectroscopy
Editors: E. Klose · B. Wilhelmi

with durations (t_p)≈2ps only with rhodamine 6G active media in ethylene glycol. However, the addition of the DODCI solution to the active media for passive synchronization of the modes made it possible to generate pulses with t_p≈300 fs. Such pulses, as well as more prolonged and non-spectrally-limited ones (coherence time≈100 fs), have been used in our experiments.

Pulses were introduced in the Michelson interferometer with a precision delay line. The rate of delay line CW scanning may be as slow as 1 fs/s. A pair of space-separated and time-scanned pulses was focused in a cell with the DODCI solution. The angle between the directions of the laser pulses (k_1 and k_2) was 3°, the thickness of the cell being ≈ 100 µm. The scattered signals were recorded in directions $k_3=2k_1-k_2$ and $k_4=2k_2-k_1$ after space selection by means of two photomultipliers and self-recorders.

2. The delay between signals

It was found that the maxima of the signals in directions k_3 and k_4 were available non-simultaneously. Usually we observed that the delay between maxima was 50-100 fs. Similar results were obtained earlier by Yajima et al. /1/. On observing the delay between the maxima of four-photon scattering in different directions, the authors of Ref. /1/ concluded that such a delay was initiated by the dephasing process with a duration of about 50 fs. We suppose that this delay is a result of the pulse self-diffraction by a grating formed from the excited molecules with a long (ns) lifetime. Such a conclusion is made on the basis of analysis of transient self-diffraction for complex molecules with strong broadened vibronic spectra and non-Markovian relaxation /2/. As was shown in /2/, the apparent influence of the dephasing process on the delay of a scattered pulse may be observed for pulses of $t_p<T'$, where the irreversible dephasing time $T'=(t_s/\sigma_{2s})^{1/3}$; t_s is the decay time of the correlation function of the low frequency molecular vibration system, σ_{2s} is the contribution of such a system to the second central moment of the absorption spectrum. Typical values of T' for complex molecules in solution are 20 fs /2/, i.e. $T'\ll t_p$ for our experiments. In this case the delay of pulses can only be connected with the dynamics of the formation of these pulses by a transient population grating /2/.

3. Modulation of scattering signals

Using the slow rate of delay line scanning, we observed the modulation of intensity for scattered signals (Fig.2). It appeared that the envelope duration of the signals was due to coherence time rather than exciting pulse duration. The forms of the upper and lower envelopes were different. In particular, two minima were observed for the lower envelope.

On the envelope sides the modulation period was equal to the laser exciting light period (Fig. 2b), but close to the signal maxima the modulation period was equal to the half-period of the exciting light (Fig. 2c). It was found that the modulation periods are linear functions of the exciting light wavelength.

Such data can be interpreted as a result of the interference between Rayleigh and four-photon scattering fields. As shown in /3/ the resulting signals consist of 4 contributions with different dependences on time delay between k_1 and k_2 pulses:

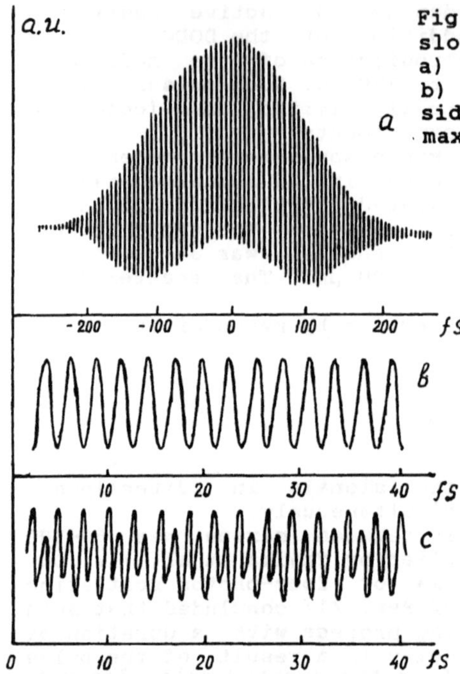

Fig.2. Scattering signals registered at slow velocity of delay-line scanning: a) - signal envelopes, b) - modulation on the envelopes' sides, c) - modulation close to the maxima of the envelopes.

1 - not depending on t_d, induced by Rayleigh scattering components solely, 2 - slowly depending on t_d, induced by four-photon scattering alone, 3 - oscillating with laser light frequency, induced by interference between Rayleigh components of k_3 and k_4 directions and between Rayleigh and four-photon scattering, 4 - oscillating with double laser frequency, induced by interference between Rayleigh and four-photon scatterings only. The above picture of signal formation allows us to interpret all of the experimental data.

The observed beats, as well as beats in the interferometric method based on second-harmonic generation /4/, may be used for the analysis of the ultrashort laser pulses (form and phase modulation of pulses). The interference of different fields of scattering may supply additional information compared to the method of /4/. This results from the use of higher-order interaction of light with matter in our case (the sixth order of light field amplitude) in comparison with the generation of the second harmonic (the fourth order of light field amplitude).

References

1. T.Yajima, Y.Ishida, Y.Taira: In Picosecond Phenomena II, ed.by R. Hochstrasser, W.Kaiser, C.V.Shank (Springer,Berlin, Heidelberg, 1980),p. 190-194.
2. B.D.Fainberg; In Abstracts of the XX All-Union Congress on Spectroscopy, pt. I (Kiev, 1988) p. 288(in Russian).
3. V.L.Bogdanov, A.B.Evdokimov, G.V.Lukomskij, B.D.Fainberg: Pisma v GETF, **49**, 135 (1989)(in Russian).
4. J.C.Diels, J.J.Fontaine, I.C.McMichael, F.Simoni: Appl.Opt.,**24**, 1270 (1985).

Theoretical Study of Coherent Artifacts in Pump-and-Probe Experiments

E. Döpel, W. Rudolph, G. Werner, and B. Wilhelmi

[1]Sektion Physik, Friedrich-Schiller-Universität Jena, Max-Wien-Platz 1, DDR-6900 Jena, German Dem. Rep.

Abstract. We present results of a general numerical treatment of coherent pulse interaction in pump-probe experiments for arbitrary pulse shape, energy and duration in an arbitrary two-level system which is assumed to be homogeneously broadened. In particular we investigate the influence of phase relaxation time, detuning, pulse shape and chirp on the expected signals in the probe direction (coherent artifact) and background-free direction.

1. Introduction

Several methods of pump-probe techniques have become a standard tool for measuring ultrafast relaxation times of matter. The two (or more) necessary laser pulses are often produced by beam splitting of the output of a single source and therefore they are coherent. The interference of the two incident beams (which usually include a small angle) in the probe induces a grating which causes diffraction. This gives rise to a coherent artifact in the measured signal and, moreover, light is diffracted into new (the background-free) directions.

On one hand such coherent effects are used to measure phase relaxation times /1,2,3/. On the other hand the coherent coupling may lead to difficulties in the interpretation of pump-probe experiments near zero delay /4,5,6,7/. The exact calculation of signals formed by coherent interaction is therefore of great importance.

Rate equations were applied to deal with this problem for optically thin media /8,9/. The influence of parallel or orthogonal polarisation directions of the different beams was discussed in /5/ and /9/. In /6/ the different interactions of self-induced amplitude and phase gratings was investigated. Recent experiments made it necessary to take into account the phase relaxation time of the sample and its influence on the coherent coupling in the probe direction /10/ and in the background-free direction /11/. The solution found in /11/ allows us to consider arbitrary pulse shapes, but the influence of a detuning of the laser from the transition center frequency is not discussed. The analytical solution derived in /10/ is valid for some special cases only, which are not very significant for the experiment. The influence of pulse chirp has not been considered yet.

The present paper presents results of a numerical treatment of the system of density matrix and wave equations. The influence of the phase relaxation time, of pulse chirp and detuning of laser and transition frequency is discussed for an absorbing atomic two-level system.

2. Theoretical description

We assume two plane waves E_L and E_T (parallel polarisation direction) intersecting with an angle θ at an absorbing medium described by a homogeneously broadened two-level system (see fig.1). The direction of the angle's bisection is z. As a result of the beam interference a concentration grating arises (i.e. regions with a different number of excited states /4/) which we assume to be thin. The direction of the grating vector is parallel to the x-direction and its absolute value is $k_g = 2 * c * \theta / \lambda$. λ is the wavelength of the incident light.

The problem is governed by the density matrix equations for the atomic system and the wave equation. The incident field E is the superposition of the two waves E_L and E_T and for small θ it can be written as:

$$E = \{ [A_T * \exp(1/2\, i\, k\, \theta\, x) + A_L * \exp(-1/2\, i\, k\, \theta\, x)] /2 \} * \exp[i(\omega t - k z)] + c.c. \qquad (1)$$

(k : wave number)

The diagonal elements ρ_{ii} of the density matrix representing the occupation density of the two levels can be written as a sequence of Fourier elements in x-direction:

$$\rho_{ii} = \rho^{0}_{ii} + \sum_{m=1}^{\infty} \rho^{m}_{ii} * \exp(i\, k\, m\, \theta\, x) + c.c. \qquad (2)$$

With the ansatz (1),(2) one can derive equations for the transmitted as well as for the diffracted components of the electrical field which can be solved numerically.

In the present paper we show calculations for the following conditions:

- the energy relaxation time T_1 is much larger than the pulse duration T_L
- weak saturation S, with $S = \sigma \int I\, dt = 0.08$
  (σ : cross section, I: photon flux density)
- small absorption (small signal transmission 97%)
- energy of probe pulses (with amplitude A_T) 50 times less than pump pulses (with amplitude A_L)
- cos - shaped grating (i.e. $m \leq 1$)

From the last point it follows that the maximum error of all results for S=0.08 is less than 2 % .

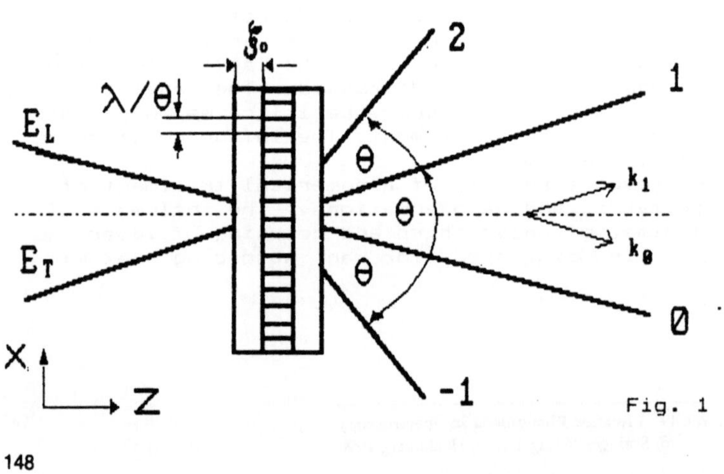

Fig. 1 : Beam geometry

3. Results

The influence of phase relaxation time T_2 on the coherent artifact in the probe direction (direction 1 in fig.1) is depicted in fig.2. The artifact can clearly be seen only for a small ratio $t_2=T_2/T_L$. Further calculations show a strong dependence of the artifact on pulse shape. A comparison with the analytical results of /10/ obtained for double sided exponential pulses exhibits a weaker decrease of the peak with increasing phase relaxation time than for sech-pulses used in our calculations. This is of importance because the sech-pulse shape is often assumed in experiments.

Next we want to discuss the dependence of the signal's amplitude in the background-free direction on T_2, which is depicted in fig.3. The parameter is the detuning Δ of the laser frequency ω from the transition frequency ω_{21} of the two-level system: $\Delta=(\omega-\omega_{21})\ /\ (\omega_{hom}/2)$ (ω_{hom}:homogeneous line width), where $\mu_{21}^2 * (\omega_{hom}/2)\ /\ [(\omega-\omega_{21})^2+\omega_{hom}^2/4\ ] = \text{const.}$ (μ_{21}:atomic dipole moment)

If there is no detuning the behaviour can be fitted by a single exponential function. The decrease is caused by a slow build-up of

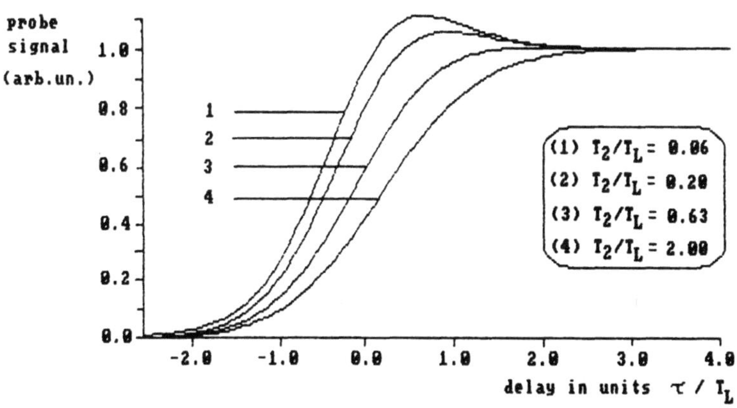

Fig.2 : Probe signal of a sech-pulse for different phase relaxation times (no chirp and no detuning)

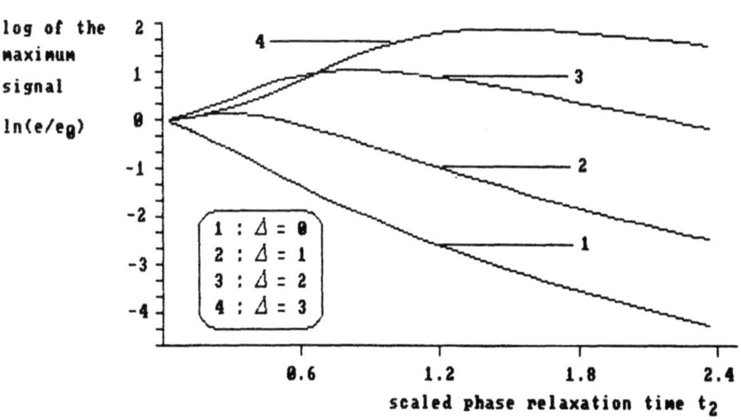

Fig.3 : Amplitude of the background-free signal as a function of the scaled phase relaxation time t_2 for various detunings (Gaussian pulses without chirp, normalized to equal values at $t_2=0$)

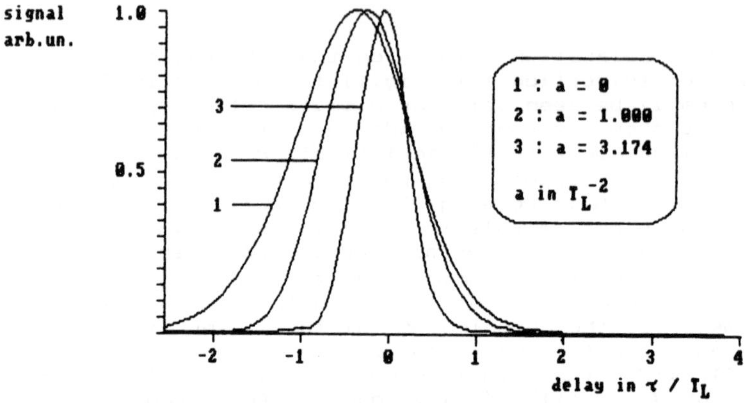

Fig.4 : Background-free signal as a function of linear chirp (normalized to equal values for the maxima), T_2/T_L=0.06, no detuning

the matter's polarisation and therefore of the self-induced grating with time constant t_2. The reason for the normalized signal being larger for $\Delta\neq0$ as compared with $\Delta=0$ is that the pulse absorption increases, though the small signal absorption coefficent for cw light at the pulse's mid frequency was kept constant.

Now we want to consider the signals in the background-free directions -1 and 2 (see fig.1) from another point of view. In general the position of the maximum of the signal as a function of delay τ is not zero (fig.4). In the two background-free directions the difference to zero delay has the opposite sign but the same absolute value, which is equal to half of the so-called peak shift. In our case the peak shift depends on the coherence length of the incident light pulses and their pulse duration T_L as well as on T_2 (the coherence length is determined by a phase term exp(i*a*t²) in the expression for the complex pulse amplitude that causes linear chirp, a:chirp parameter). This is opposed to /12/ where a rate equation analysis was chosen to determine the homogeneous line width of an inhomogeneously broadened sample when using long pulses with stochastic phase distribution and where the peak shift vanishes for pure homogeneous broadening. Accordingly, our method can be used to measure T_2 in the case of homogeneous broadening if the pulse duration is not too long.

References

1. T.Hattori, T.Kobayashi, Chem. Phys. Lett. **133**, 230 (1987)
2. K.Kurokawa, T.Hattori, T.Kobayashi, Phys. Rev. **A36**, 1298 (1987)
3. A.M.Weiner, S.De Silvestri, E.P.Ippen, Ultrafast Phenomena IV, ed. by D.H.Auston and K.B.Eisenthal, Springer Verlag (1984), 230
4. C.V.Shank, D.H.Auston, Phys. Rev. Lett. **34**, 479 (1975)
5. Z.Vardeny, J.Tauc, Phys. Rev. Lett. **46**, 1223 (1981)
6. S.L.Palfrey, T.F.Heinz, JOSA **B2**, 674 (1985)
7. M.J.Rosker, F.W.Wise, C.L.Tang, Appl. Phys. Lett. **49**, 1726 (1986)
8. B.Wilhelmi, J.Hermann, Sov.Journ.Quantum Electron.**10**, 1082 (1980)
9. B.S.Wherrett, A.L.Smirl, T.F.Boggess, IEEE Journ. Quant. Electr. **QE19**, 680 (1983)
10. M.W.Balk, G.R.Fleming, Journ. Chem. Phys. **83**, 4300 (1985)
11. J.G.Fujimoto, T.K.Yee, Appl. Phys. **B34**, 55 (1984)
12. S.Asaka, H.Nakatsuka, M.Fujiwara, M.Matsuoka, Phys. Rev. **A29**, 2286 (1984)

Part III

Light Pulse Shaping and Nonlinear Optical Phenomena

Soliton Generation from c.w. Diode Laser Signals

E.J. Greer, D.M. Patrick, P.G.J. Wigley, and J.R. Taylor

Femtosecond Optics Group, Physics Department, Imperial College,
Prince Consort Road, London SW7 2BZ, UK

Abstract. Using an input c.w. diode laser signal, solitons as short as 200fsec have been generated through stimulated Raman amplification, at the diode laser wavelength, provided by a mode locked Nd:YAG laser at 1.32μm. In addition soliton-like structures have been generated through cross phase modulation on a c.w. diode laser signal from an intense pump signal at 1.06μm from a c.w. pumped mode locked Nd:YAG laser.

1.1 Soliton generation via Raman amplification of diode laser signals

One of the simplest schemes to generate solitons in single mode fibres involves the launch of relatively long pulses of around 100ps into relatively short fibre lengths (~100's of meters) at modest average powers (50mW-1W) [1]. Through stimulated Raman scattering a Raman continuum containing several single solitons is generated, and simply by spectrally selecting in the continuum, pulses of around 200fsec and peak powers in the kilowatt regime can be observed. When the pump pulse is in the normal dispersion regime and the Raman signal in the anomalously dispersive region the threshold for soliton generation is substantially greater than when the pump pulse experiences anomalous dispersion [2]. In this latter case modulational instability [3] acts as the precursor of the soliton generation scheme. In the centre of the pump pulse, where the intensity is greatest, modulational instability breaks up the central region of the pump into soliton-like structures. Through modulational instability the structures grow in intensity and also experience stimulated Raman amplification on transmission. Due to the amplification process the solitons compress and experience intra pulse Raman scattering, or the soliton self frequency shift, as it is most commonly referred to [4]. As a consequence, the solitons experience a frequency down shift, the shift being strongly dependent on the pulse duration $\Delta\gamma \propto 1/\tau^4$ [5]. Since the solitons at the peak of the pump pulse are formed first, as they experience the highest gain, they will then move towards the rear of the pump pulse due to the combined effect of soliton self frequency shifting and group velocity dispersion. This results in inelastic soliton collisions, giving rise to further pulse shortening and spectral shifting resulting in the spectrally broad ensemble of solitons.

The formation mechanism gives rise to some problems in the application of the generated solitons since there is no real control over their spectral extent, apart from a coarse manipulation through the pump power. Similarly, the number of solitons under the pump envelope is also not precisely controllable. One means to overcome this is to seed the process using either precise wavelengths from a c.w. diode laser or pulses from a mode-locked semi-conductor laser or even noise bursts, from the same source. We have demonstrated single soliton evolution from all these sources through seeded Raman amplification processes and will report on the former process here.

Springer Proceedings in Physics, Vol. 49 Ultrafast Phenomena in Spectroscopy
Editors: E. Klose · B. Wilhelmi

1.2 Experimental

The experimental scheme used to generate solitons from a c.w. diode laser signal via stimulated Raman amplification is shown in Figure 1. A c.w. pumped mode locked Nd:YAG laser operating at 1.32μm was used to provide the Raman gain signal. This laser generated pulses of 100psec with average powers of up to 2W. However, the required experimental power levels were significantly lower.

The continuous seed signal was obtained from a 500μm long InGaAsP laser diode in an external cavity arrangement, with one of the plane facets anti-reflection coated which permitted wavelength tuning, using a diffraction grating in the region 1.33μm to 1.38μm. An average output of 1mW was obtained from the laser diode system. Using a dichroic beam splitter the pump and signal beams were launched into a 1.98km length of single mode fibre with a dispersion minimum at 1.34μm and a loss of 0.5dB/km at 1.3μm. From the known fibre parameters, the soliton power for a 250fs pulse at 1.36μm was calculated to be 62.6W, which corresponded to an average power of 1.6mW at 100MHz repetition rate.

Figure 2(a) shows the spectrum obtained at the fibre output for an average pump power of 160mW at 1.32μm in the fibre. It can be seen that a low level Raman band centred around 1.38μm was present, but for the fibre length used, the pump power level was insufficient to permit soliton generation and no short pulse evolution was detected within this band. Initially, the diode laser was tuned to 1.36μm, where it can be seen from Figure 2(a) no spectral component was present at the 160mW pump power. On introduction of 60μW average power of diode signal, rapid spectral enhancement via Raman gain was observed at this wavelength, see Figure 2(b).

When the pump power was increased to 180mW, in the absence of a seed signal, the pump power level was sufficient for soliton generation within a distinct band centred around 1.38μm . However, at 1.36μm no solitons were present and seeding at this wavelength still gave rise to rapid soliton generation via the modulated Raman gain on the c.w. diode laser signal. Soliton collisions and soliton self frequency shifting also extended the Raman band to 1.42μm.

In the time domain, background free autocorrelations revealed 250fs solitons when a 1.36μm bandpass filter was used to examine the Raman enhanced band under experimental conditions similar to those of Figure 2(b). A typical autocorrelation trace is shown in Figure 3.

From spectral filtering of the feature at 1.36μm we inferred an average power of 6.8mW. This corresponded to 1.8mW being in the soliton component, which is equivalent to 72.7W peak power and would imply that, to within the experimental error, operation at these pump and signal power levels gave rise to single fundamental soliton evolution with a defined spectral band.

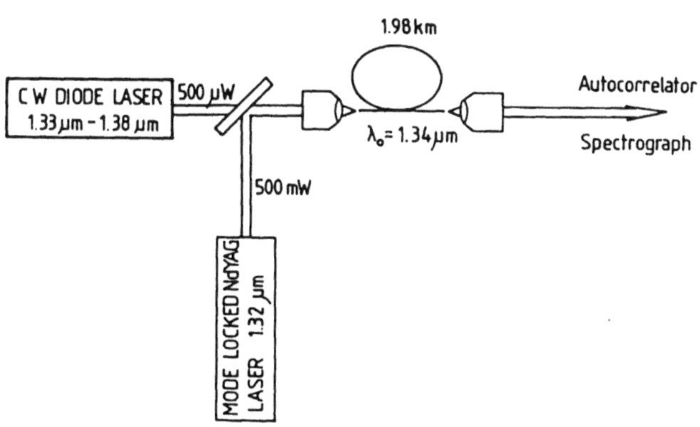

Figure 1. Schematic of experimental arrangement used in soliton generation from a c.w. diode laser signal using preferential Raman amplification.

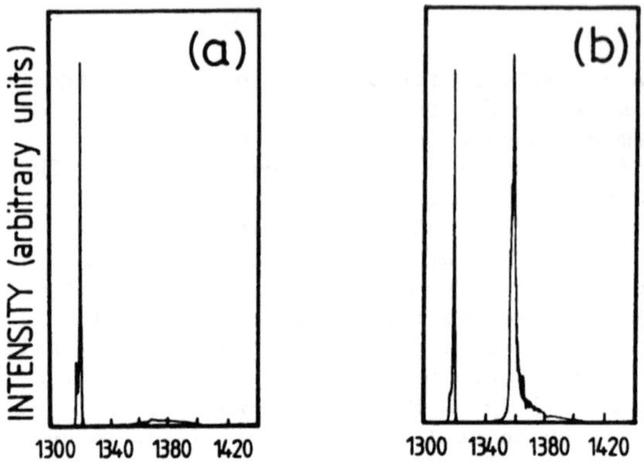

Figure 2. Output spectra at an average pump power of 160mW at 1.32μm plus (a) no signal input (b) 60μW average signal power at 1.36μm.

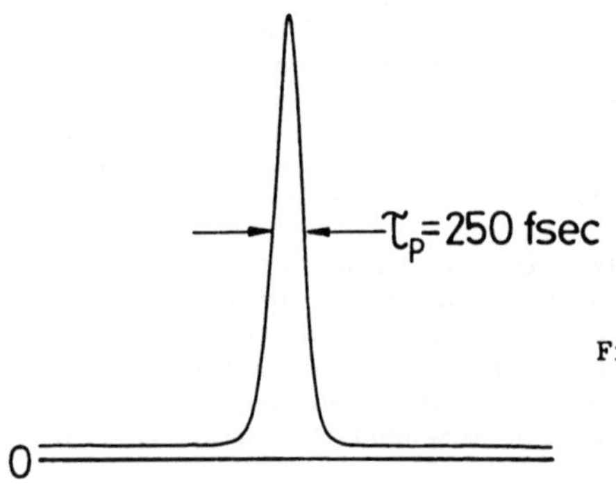

Figure 3. Background free autocorrelation of soliton generated at 1.36μm from c.w. diode laser signal via seeded Raman amplification.

However, as the pump power was increased, the pedestal component of the autocorrelation trace increased in intensity and the measured pulsewidth increased. This behaviour is similar to that observed in the soliton evolution from noise, caused by soliton collisions towards the front end of the fibre and is associated with the formation of the soliton continuum consisting of solitons with uncontrollable wavelength and exhibiting considerable temporal jitter.

Through operating on threshold pump powers for Raman generation it was possible to generate single fundamental solitons throughout the complete tuning band of the c.w. diode laser. Typically these solitons were relatively independent of the signal powers and were in the range of 250fsec. Varying the pump power had a much greater effect on the generated soliton pulsewidths.

2. Generation of pulses from a c.w. diode laser in the anomalous dispersion regime through cross phase modulation

When a continuous signal is launched into a fibre together with an intense pulsed pump, the intensity dependent refactive index change due to the pulsed beam gives rise to a frequency chirp on the c.w. signal. If the signal is in the anomalous dispersion regime, then the effect of dispersion on the cross phase modulated signal experiencing a frequency down shift to the front of the disturbance and an up shift to the rear will tend to give rise to soliton-like shaping. Where the pump and signal beams can be propagated with similar group velocities this shaping mechanism should give rise to a pulsed output on the input c.w. signal.

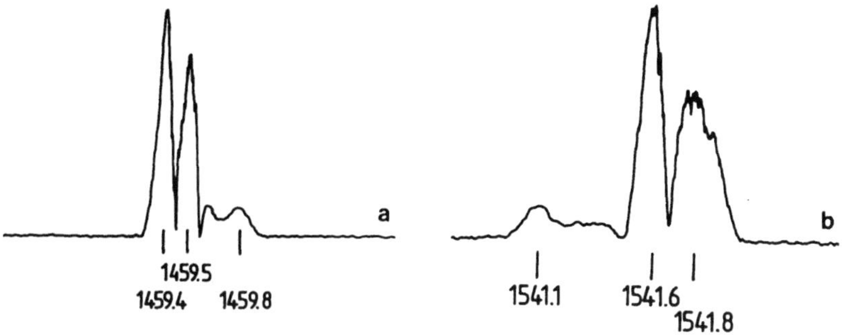

Figure 4. Output spectra generated through cross phase modulation by a 1.06μm mode locked laser signal for input wavelengths of (a) 1459nm and (b) 1541nm from a c.w. diode laser.

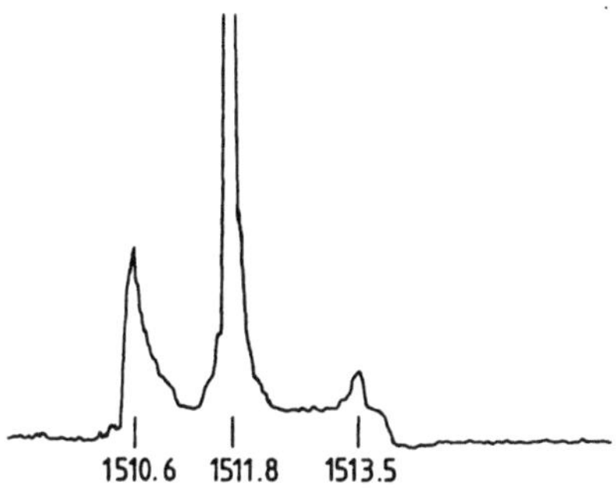

Figure 5. As in Figure 4, but for an input signal wavelength of 1511nm.

The mathematical modelling of this process was first carried out by Schadt & Jaskorzynska [6]. Using a system of two coupled equations describing the pump and signal amplitudes, they also showed that for readily achievable experimental parameters, pulse structures on the c.w. signal could be achieved with durations up to forty times shorter than the pump signal at input.

We have investigated this pulse shaping mechanism using an experimental arrangement similar to that in Figure 2. The pump signal was provided by a c.w. pumped mode locked Nd:YAG laser, operating at 1.06μm while the c.w. signal was derived from a grating tuned InGaAsP diode laser tunable from 1.43μm to 1.56μm, with average powers of up to 500μW. Signal and pump beams were coupled into a 1.2km length of single mode fibre, which had a dispersion minimum at 1.27μm.

Initially, the spectral broadening due to the wavelength dependence of the cross phase modulation was investigated. Figures 4(a) and (b) show the spectra obtained for the launched diode wavelengths of 1459.5nm and 1541.6nm respectively. The asymmetry observed in the spectra was a consequence of the group velocity dispersion mismatch between the pump (1.06μm) and signal. Both spectra shown were recorded for average pump power levels of 480mW and a signal level of 60μW.

When tuned to 1511.8nm a much more symmetrical cross phase modulated signal spectrum was obtained, as shown in Figure 5, which corresponded to an optimum matching of the group velocities of the pump and signal beams. The spectral enhancement through operating at the matching point is clearly seen from Figure 6, which shows the variation of the measured signal bandwidth with wavelength for a constant average pump power of 250mW. At average pump power levels around 480mW further spectral enhancements could also be observed due to cross phase modulation from pump signals which were chirp distorted due to the enhancement caused by stimulated Raman scattering.

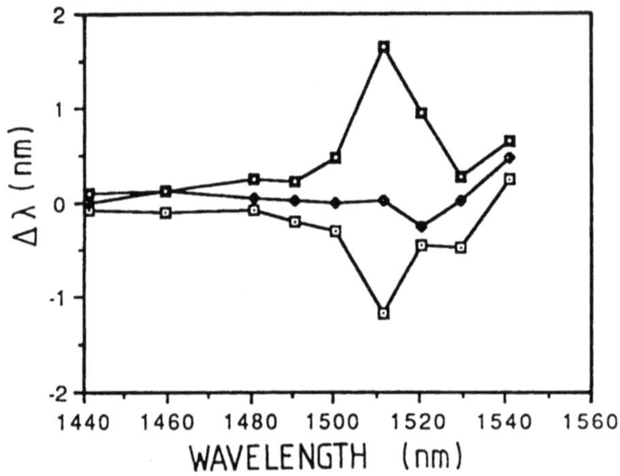

Figure 6. Variation of the spectral broadening through cross phase modulation with input wavelength.

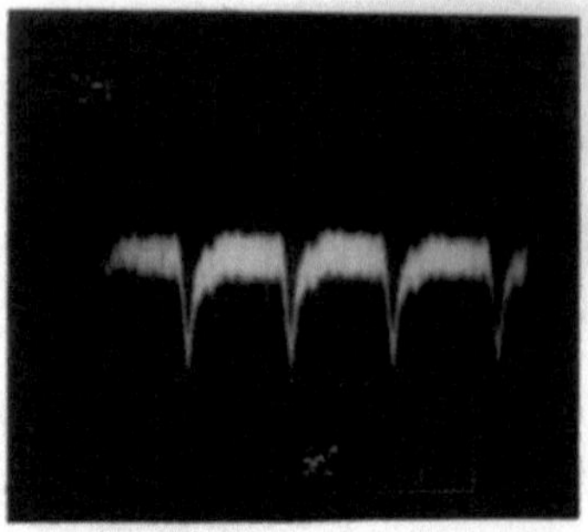

Figure 7. **Temporally modulated** signal from a c.w. diode laser through cross phase modulation.

In the time domain, a photodiode-oscilloscope combination revealed a pulsed output from the c.w. laser diode which was enhanced when operating around 1512nm. At present, autocorrelation traces have not been obtained due to the low level nature of the signals and temporal resolution of the generated pulses has been limited to 50ps, the resolution limit of the diode-oscilloscope combination. A portion of a pulse train generated using this cross phase modulation technique is shown in Figure 7.

It has also been indicated by Jaskorzynska and Schadt [7] that this technique can be applied to compress weak pulses in the regime of negative group velocity dispersion.

3. Conclusion

We have shown that from c.w. diode lasers signals in the anomalous dispersion regime it is possible to generate femtosecond solitons through pulsed Raman gain and soliton-like pulses via cross phase modulation from pump signals in the normal dispersion regime.

Acknowledgement

The major financial support for this work from British Telecom Research Laboratories is gratefully acknowledged, as is the contribution of SERC.

References

[1] A.S. Gouveia-Neto, A.S.L. Gomes & J.R. Taylor, Optics Lett. 12, 1035 (1987).
[2] A.S. Gouveia-Neto & J.R. Taylor, Electronics Lett. 24, 1544 (1988).
[3] A. Hasegawa & W.F. Brinkman, IEEE J. Quantum Elect. QE-16, 694 (1980).
[4] F.M. Mitschke & L.F. Mollenauer, Optics Lett. 11, 659 (1986).
[5] J.P. Gordon, Optics Lett. 11, 662 (1986).
[6] D. Schadt & B. Jaskorzynska, Electronics Lett. 23, 1090 (1987).
[7] B. Jaskorzynska & D. Schadt, IEEE J. Quantum Elect. QE-24, 2117 (1988).

Nonlinear Propagation of Ultra-short Optical Pulses in Activated Fibers

E.A. Vanagas [1] *and A.I. Maimistov*

[1]Semiconductor Physics Institute of Lithuanian Academy of Sciences, Pozelos 52, 232600 Vilnius, Lithuania
[2]Moscow Engineering-Physics Institute, Kashirskoje shosse 31, 115409 Moscow, USSR

Abstract. Both the coherent and quasi-coherent amplification of ultra-short pulses in single-mode optical fibers activated by resonant impurities have been studied via computer simulations. We found that the autosoliton can be created in an optical fiber with both inverted resonant impurities and linear losses. It is shown that the main parameters of this single wave are independent of the initial pulse. In addition, the single wave spectrum shift in the quasi-coherent case is found to depend on the sign of the group-velocity dispersion.

1. Introduction

In the last ten years intense research into the formation and propagation processes of nonlinear single waves in optical fibers has been done. Here under certain conditions the appearance of optical solitons is possible [1], as was experimentally demonstrated in the investigation [2]. Using solitons for the transmission of information through optical fiber communication lines presupposes their periodical amplification, as the fiber generally possesses losses. Except for amplification due to stimulated Raman scattering, for these purposes it is possible to use a fiber containing resonant impurities with inverted population, which is analogous to the traditional laser amplifier. In addition, the presence of resonant impurities increases the nonlinear dielectric permeability of the medium where the optical pulse propagates [3]. As has become obvious, the fibers containing erbium Er^{+3} as impurities [4-6] are suitable for amplification purposes.

The theoretical analysis of optical pulse evolution in such activated fibers is being impeded from the point of view of analytical calculations. In general, the nonlinear equations describing pulse amplification appear to be non-integrable by the inverse spectral transform method - a fairly efficient means of investigating this kind of problem.

In this report the results of the numerical simulation of coherent and quasi-coherent amplification of short optical pulses in single-mode weakly nonlinear fibers with both inverted resonant impurities and linear losses are presented.

2. Model equations

The evolution of the envelope of a short optical pulse electric field and the state of resonant impurities under definite conditions can be described by both the nonlinear Schrödinger equation and the Bloch system [7-9]:

$$i\partial_{,z}q + \varepsilon\partial^2_{,t}q + \mu|q|^2q + i\Gamma q + a\langle\sigma\rangle=0,$$

$$\partial_{,t}\sigma = i\delta\sigma + ifqn - \gamma\sigma, \qquad (1)$$

$$\partial_{,t}n = 2if(q^*\sigma - q\sigma^*),$$

where $z = Z/L_g$, $t = (T-Z/v_g)t_p^{-1}$, $L_g = 2\beta t_p^2/|\alpha_2|$ is the dispersion length. β is the light-guide propagation constant, $v_g = [d\beta/d\omega]^{-1}|_{\omega=\omega_0}$ is the group velocity, $\alpha_2 = v_g^{-2}[1- \beta dv_g/d\omega]|_{\omega=\omega_0}$, $q = A/A_0$ is the normalized envelope of the electric field, A_0 is the peak electric-field intensity, σ and n describe the polarization and the population inversion of resonant levels, respectively. The interaction of the radiation with the resonant impurities is characterized by the dimensionless constant $f = (\bar{d}A_0/\hbar)t_p$. Here $\bar{d}$ is the element of the dipole transition between the resonant states. The coefficient μ is equal to the ratio of the dispersion length to the Kerr length and is given by

$$\mu = \frac{4\pi\omega_0^2 \int d\vec{\rho}\ \vec{U}_0^4(\vec{\rho})\ |\chi_3(\vec{\rho})|}{c^2|\alpha_2| \int d\vec{\rho}\ \vec{U}_0^2(\vec{\rho})},$$

where $\chi_3(\vec{\rho})$ is the Kerr susceptibility, $\vec{U}_0(\vec{\rho})$ is the light-guide mode function, and $\vec{\rho}=(x,y)$. The coefficient a is equal to the ratio of the dispersion length to the length of the resonant gain and is given by

$$a = \frac{4\pi(\omega_0 t_p)^2 \int d\vec{\rho}\ N_A(\vec{\rho})\ \bar{d}\ \vec{U}_0^2(\vec{\rho})}{A_0 c^2|\alpha_2| \quad \int d\vec{\rho}\ \vec{U}_0^2(\vec{\rho})},$$

where $N_A(\vec{\rho})$ is the concentration of the resonant impurities in a nonlinear fiber. The coefficients Γ and γ describe the linear losses and the gain saturation, respectively. The angle brackets in (1) denote summation over all the normalized frequency detunings δ from the center of the inhomogeneously broadened gain line of resonant impurities, and t_p is the optical pulse duration.

3. Numerical results

Let us discuss the case of coherent amplification of optical pulses in an activated fiber (γ=0). It has been found by numerical investigations that for a definite concentration of resonant impurities and linear losses, the autosoliton appears and its shape does not depend on the initial pulse. It is noteworthy that the existence of autosolitons in nonlinear optics was first predicted by Grigorian [10] and was numerically investigated in [11]. All the parameters of this single wave are determined by the medium characteristics: the gain coefficient a and the loss coefficient Γ. The total power of this single wave is determined by the empirical formula

$$U_* = \int_{-\infty}^{+\infty} |q|^2 dt = a / 2\Gamma .$$

It is worth mentioning that the shape of this single wave considerably depends on the value of U_*. Fig. 1 shows the evolution of an ultra-short pulse in an activated fiber when a=4, Γ=4.

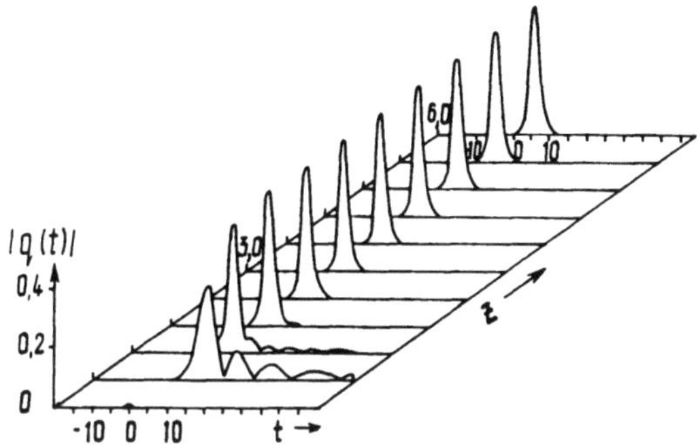

Fig. 1 Spatial evolution of initial pulse $q(z=0,t)=10^{-3}\text{sech}(t)$ for $a=4$, $\Gamma=4$, $\mu=2$, $\delta=0$, $f=1$, and $\varepsilon=1$.

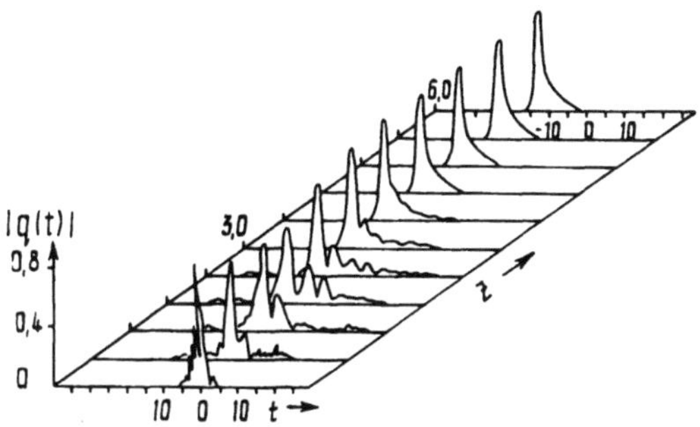

Fig. 2 Spatial evolution of initial pulse $q(z=0,t)=\text{sech}(t)\exp(i\varphi)$ for $a=2$, $\Gamma=1$, $\mu=2$, $\delta=0$, $f=1$, $\varepsilon=1$, and $D(\varphi)=1$. Here $D(...)$ is the dispersion of the random function φ.

Evidently, in the distance $Z \geq 1.8L_g$ a symmetrical single wave is formed. The total power of this single wave equals $U_* = 0.5$. A change in concentration of resonant impurities or linear losses in the fiber, which causes an increase in U_*, leads to the deformation of the single wave shape. Fig. 2 shows the evolution of a stochastically modulated initial pulse and the formation of an asymmetrical single wave when $a=2$, $\Gamma=1$, and $U_*=1$. By means of numerical experiments, it has been determined that the stationary single wave only forms when $a \leq 2\Gamma$. The non-stationarity of a single wave when $a>2\Gamma$ is connected with the self-phase modulation of the pulse in a nonlinear fiber. The energy of this pulse is not equal to U_*.

The scenario of nonlinear propagation of optical pulses in an activated fiber in the case of quasi-coherent amplification ($\gamma \simeq 1$) differs considerably from the previous case. We have found that, depending on the sign of group-velocity dispersion, the spectrum of the amplified pulse shifts either to the red side of the spectrum, for $\text{sgn}(\varepsilon) = 1$, or to the blue side, for $\text{sgn}(\varepsilon) = -1$. Fig. 3 presents both the evolution and the spectrum of the pulse in the anomalous dispersion region of the fiber for $\text{sgn}(\varepsilon)=1$ and when $a=2$, $\Gamma=0.25$, and $\gamma=1$. The red shift of the pulse spectrum arises simultaneously with pulse amplification. Let us point out that if $a \gg 1$ the compression of the pulse takes place concurrently with its amplification (Fig.4). However, in the distance $Z \geq 2.4L_g$ the pulse amplitude decreases and it decays into two separate single waves. Similar pulse amplification took place in the normal dispersion region of the fiber. Fig. 5 describes the pulse evolution and its spectrum when $\text{sgn}(\varepsilon)=-1$, $a=2$,

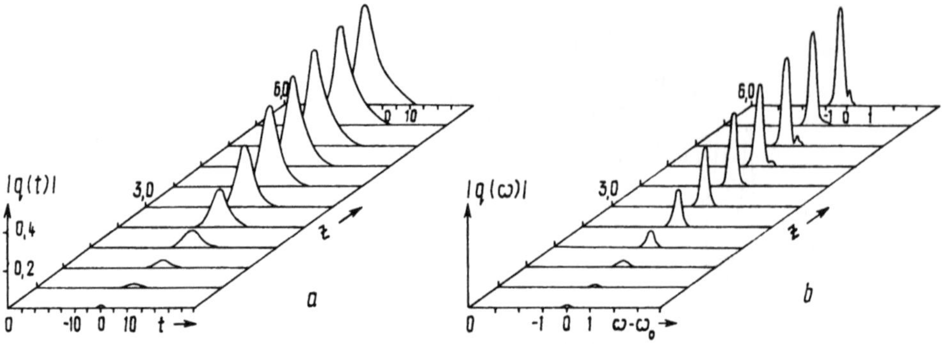

Fig.3 Spatial evolution (a) and frequency spectrum evolution (b) of initial pulse $q(z=0,t)=10^{-2}\,\mathrm{sech}(t)$ for $a=2$, $\Gamma=0.25$, $\mu=2$, $\delta=0$, $f=1$, $\gamma=1$, and $\varepsilon=1$.

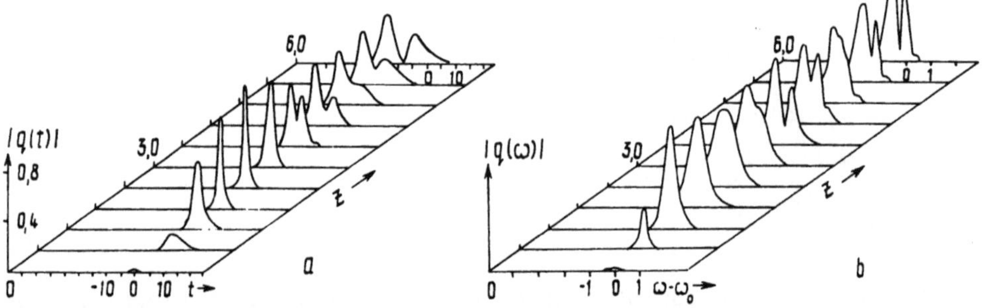

Fig.4 Spatial evolution (a) and frequency spectrum evolution (b) of initial pulse $q(z=0,t)=10^{-2}\,\mathrm{sech}(t)$ for $a=8$, $\Gamma=2$, $\mu=2$, $\delta=0$, $f=1$, $\gamma=1$, and $\varepsilon=1$.

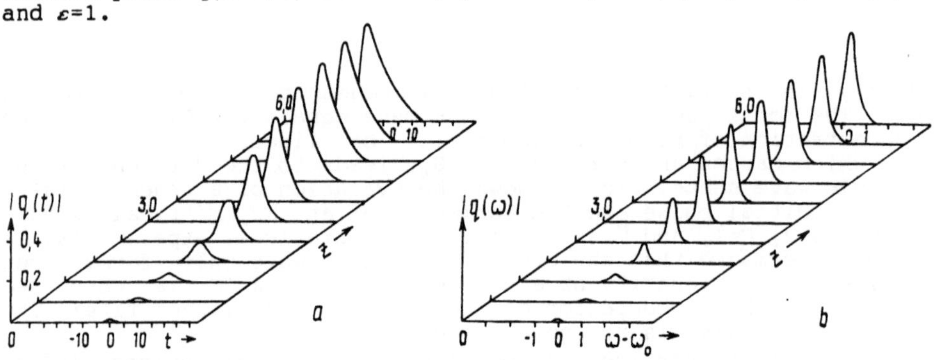

Fig.5 Spatial evolution (a) and frequency spectrum evolution (b) of initial pulse $q(z=0,t)=10^{-2}\,\mathrm{sech}(t)$ for $a=2$, $\Gamma=0.25$, $\mu=2$, $\delta=0$, $f=1$, $\gamma=1$, and $\varepsilon=1$.

$\Gamma=0.25$, and $\gamma=1$. However, in this case a blue shift of the single wave spectrum occurs.

Let us note that in the case of quasi-coherent amplification we observed only non-stationary single waves. The non-stationarity of these waves is connected with the dynamical shift of their spectrum in relation to the amplification line center of resonant impurities.

4. Conclusions

The results of numerical investigations allow us to draw the following conclusions:
(1) In the case of coherent amplification of ultra-short optical pulses in a single-mode fiber with both inverted resonant impurities and linear losses, the formation of stationary single waves - autosolitons - is possible.
(2) In the quasi-coherent case of amplification, the single waves are non-stationary, and the shift in their spectrum depends on the group-velocity sign.

Acknowledgments

The authors wish to thank Dr. V.S. Grigorian, Dr. A.B. Grudinin and Dr. D.V. Korobkin,Jr., for helpful discussions.

References

1. A. Hasegawa and F. Tapert, Appl. Phys. Lett., 23, 142 (1973).
2. L.F. Mollenauer, R.H. Stolen and J.P. Gordon, Phys. Rev. Lett., 45, 1095 (1980).
3. E.A. Manykin, I.I. Surina and A.I. Maimistov, Propagation of optical pulses in fibers: a review, CNIIatominform, Moscow (1987).
4. R.J. Mears, L. Reekie, I.M. Jauncey and D.N. Payne, Electron. Lett., 23, 1029 (1987).
5. M. Nakazawa, Y. Kimura and K. Suzuki, Electron. Lett., 25, 199, (1989).
6. E.M. Dianov, A.B. Grudinin, I.Y. Khrushchev and D.V. Korobkin,Jr., Abstracts of 4th International Symposium :" Ultrafast Phenomena in Spectroscopy ", Neubrandenburg, 1989, p. PTh 36.
7. A.I. Maimistov and E.A. Manykin, Sov. Phys. JETP, 58, 685 (1983).
8. E.A. Vanagas, E.A. Manykin and I.I. Surina, preprint of I.V. Kurchatov Institute of Atomic Energy N-4627/1, Moscow (1988).
9. E.A. Vanagas, SPIE Proceedings, 1085 (1989) (in press).
10. V.S. Grigorian, Pis'ma v Zh. Eksp. Teor. Fiz., 44, 447 (1986).
11. V.S. Grigorian, A.I. Maimistov and Yu.M. Skliarov, Zh. Eksp. Teor. Fiz., 94, 174 (1988).

Influence of Various Nonlinear Optical Processes on Femtosecond Light Pulse Propagation in Fibers

J. Herrmann and B. Wilhelmi

Zentralinstitut für Optik und Spektroskopie, Akademie der Wissenschaften,
Rudower Chaussee 6, DDR-1199 Berlin, German Dem. Rep.

This talk presents a study of chirp generation for ultrashort light pulses in fibers with saturable nonlinearity and positive group velocity dispersion (GVD). Using a variational approach the minimum output pulse duration has been calculated for fiber-grating compressors. Stimulated Raman scattering (SRS) of ultrashort pulses in optical fibers and the stabilization of pump pulses by SRS are considered.

1. Introduction

Recently the propagation of ultrashort light pulses through single-mode fibers became the subject of many theoretical and experimental investigations. This interest originates from the application of fibers in generating, shaping and compressing ultrashort pulses.

In particular we mention the compression of pulses in fiber-grating compressors, and the generation and propagation of solitons which might play a role in optical communication technology.

In our talk we study the influence of various nonlinear optical processes on the pulse propagation in fibers, such as the chirp generation in fibers with saturable nonlinearity and the effect of stimulated Raman scattering on self-phase modulation and pulse shortening in fiber-grating compressors.

2. Chirp generation in fibers with saturable nonlinearity

In fiber-grating compressors a bandwidth-limited pulse is sent through a single-mode fiber with positive GVD where self-phase modulation is produced by light induced changes of the refractive index. After the pulse has passed sufficiently long distances in fibers with Kerr-type nonlinearity, it obtains a nearly rectangular shape which is independent of its initial shape, and it gains a positive chirp that is constant over nearly the entire pulse, which means that the frequency increases linearly with time (/1/,/2/,/3/). This broadened and chirped pulse can be compressed very favourably by passing it through a linear medium with negative group velocity dispersion (e.g. a grating compressor). The resulting minimum pulse duration at the output of the compressor is given by

$$\tau_L \approx \sqrt{k_L''/\varkappa}\ \frac{1}{|A_0|} \qquad (1)$$

Springer Proceedings in Physics, Vol. 49 **Ultrafast Phenomena in Spectroscopy**
Editors: E. Klose · B. Wilhelmi

where K_L'', Ao and $\varkappa$ are the group velocity dispersion, the input peak field strength and the nonlinear coefficient, respectively.

Equ.(1), derived from the nonlinear Schrödinger equation (NSE) with a Kerr-type nonlinearity, yields unlimited pulse shortening with increasing nonlinearity and input field strength. From this equation one would expect that materials with high nonlinear coefficients, e.g.semiconductors, semiconductor-doped glasses, or organic polymers, are very favorable and could give shorter pulses than achieved up to now. The field-induced change of the refractive index, however, is influenced by higher order nonlinearities and, as a consequence, this change becomes saturated at higher field strengths in most cases. The field strengths where saturation occurs depend on the particular physical processes that cause the nonlinear behaviour.

Therefore it is interesting to study chirp production in fibers with saturable nonlinearity, described by the basic equation

$$i \frac{\partial A}{\partial z} + \frac{k_L''}{2} \frac{\partial^2 A}{\partial \eta^2} + \frac{\varkappa |A|^2}{1 + \gamma |A|^2} = 0 \qquad (2)$$

for the slowly varying amplitude $A(z,\eta)$ of the field strength where $\eta = t - z/v$ Equation (2) differs from the NSE by the saturation of nonlinearity, which is taken into account in analogy to systems with resonant interaction and is characterized by the parameter γ. Such a model can be used e.g.for the description of semiconductor-doped glasses with highly nonlinear optical properties (/4/). Moreover, many other nonlinear materials show a similar saturation behaviour, and for these equation (2) can be used at least in the sense of a model equation.

An approximate solution of equ.(2) can be found using a variational approach with the trial function

$$A(z,\eta) = B(z) \exp(i\varphi(z)) \exp(i\beta(z)\eta^2) f(\eta/\tau_L(z)) \qquad (3)$$

where the amplitude B(z), the phase $\varphi(z)$, the chirp parameter ß(z) and the pulse width $\tau_L(z)$ are free functions which depend on the propagation distance z and will be determinded by variational procedures. For the shape function $f(y)$, $y = \eta/\tau_L(z)$ we can use a rectangular pulse profile. We note that the results do not depend sensitively on the chosen shape, e.g. we got almost the same behaviour for a hyperbolic secant function.

After the pulse has passed a critical distance in the fiber, the pulse duration is determined by the following relation (the details of the calculations are presented in /5/):

$$\tau_L^F(z) = z \left(\frac{24\varkappa k_L''}{\gamma} \left[1 - \frac{1}{\gamma\, |B_o|^2} \ln(1 + \gamma |B_o|^2) \right] \right)^{1/2} . \qquad (4)$$

Thus the duration increases linearly with the propagation distance z.

For the chirp parameter we get

$$\beta(z) = (2k_L'' z)^{-1} . \qquad (5)$$

After chirp compensation in a linear optical device with negative group velocity dispersion the output pulse duration is given

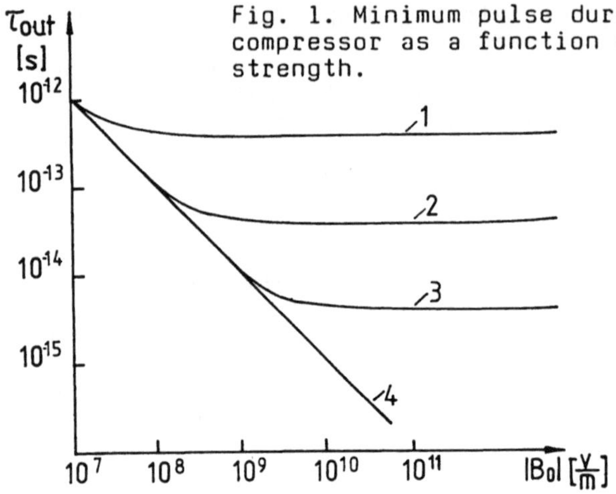

Fig. 1. Minimum pulse duration for a fiber-grating compressor as a function of the input field strength.

by

$$\tau_{out} = \frac{2.78}{\beta \tau_L^F} = 0.8 \left(\frac{k_L'' \gamma}{\varkappa} \right)^{1/2} \left[1 - \frac{1}{\gamma \; |B_o|^2} \ln(1+\gamma |B_o|^2) \right]^{-1/2} , \quad (6)$$

which does not depend on the fiber length z. For small field strength Bo of the pulse maximum at the fiber entrance relation (6) is approached by equ. (1). With $\gamma |B_o|^2 > 1$, however, we get a sharp limit for the output pulse duration. In Fig.1 the output pulse duration is depicted as a function of the field strength for different saturation intensities; the saturation behaviour of the output pulse duration is obvious.

From (6) and with the maximum value of light induced refraction index change $\Delta n_{sat} = \varkappa n_o / \gamma k_L$ there results

$$\tau_{out}^{min} = 0.8 \; (k_L'' c / \; \omega_L \Delta n_{sat})^{1/2} \quad (7)$$

for the minimum possible pulse duration. With typical values for n_{sat} in glass fibers of about 10^{-4} we obtain values for the minimum pulse duration of about 5 fs.

We mention that in some semiconductors Δn_{sat} is as high as about 0.1. However, whether shorter light pulses can be produced with such materials has to be carefully studied taking account of the real linear and nonlinear optical frequency response and the relaxational processes.

3. Stimulated Raman scattering and self-phase modulation in fibers

Now we consider the influence of stimulated Raman scattering (SRS) on the chirp production in fibers. Extensive experimental investigations of SRS in fibers with a length of a few walk-off lengths were made, e.g. by Johnson et al. (/6/) and Gomes et al. (/7/). In recent experiments Heritage et al. (/8/) and Kuckartz et al. (/9/) investigated SRS in long fibers (some 100 m) and obtained the result that SRS is not always unfavourable in pulse shortening experiments as assumed before. On the contrary, SRS-influenced pulse propagation highly stabilizes the pump pulses and provides high compression factors for the pump pulses.

These observations are connected with rather complex processes, which require some new theoretical investigations. Numerical studies of the basic equations for SRS, taking into account self-phase modulation, pump depletion and pulse walk-off, were done in /10,11,12/. Neglecting GVD however, an analytical solution of the basic equation can also be found (/13/):

$$I_S(z,\eta) = \frac{I_{SO}\,\exp[\,2Q\int_0^z I_{LO}(\eta - Dt)dt]}{1 + 2QI_{SO}\left[\int_0^z dt\ \exp[\,2Q\int_0^t I_{LO}(\eta - Dt')dt']\right]} \qquad (8)$$

$$I_L = \frac{1}{2Q}\frac{d}{dz}\ln I_s$$

$$\varphi_{S,L} = -\varkappa\int_0^z [\,2I_{L,S}(\eta,t) + I_{S,L}(\eta,t)]\,dt$$

where $A_{S,L} = (I_{S,L})^{1/2}\exp(i\varphi_{S,L})$, $\eta = t - z/v_S$, $D = 1/v_S - 1/v_L$.

$v_{S,L}$ are the group velocities of pump and Stokes pulse, respectively, and Q is the Raman gain coefficient.

On the basis of the solution (8) one can describe the main characteristic parameters of the pump pulse and the Raman pulse for fiber length of a few walk-off lengths $L_v = \tau_{LO}/|D|$, e.g. pulse duration, peak intensity and chirp. Fig.2 exhibits the evolution of the pump pulse and the Stokes pulse, where the fragmentation of the pump pulse is obvious.

In Fig. 3 the chirp parameter of the pump pulse and the Stokes pulse is represented as a function of fiber length and gain. After two walk-off lengths the Stokes chirp increases faster than that of the laser pulse and surpasses it up to three times. This fact is in good agreement with the experimental findings in /6/ and /7/.

Now we consider the case of large fiber lengths, where GVD cannot be neglected. For an approximate description we divide the fiber into two pieces. The first piece ends at twice the walk-off length ($z=2L_W$), where GVD can be neglected and the solution (8) can be used. For a hyperbolic secant input pulse shape we obtain for the output pump pulse

$$I_L(\eta,z) = \frac{I_{LO}\,ch^{-2}(\eta/\tau)}{1 + g\,\frac{I_{SO}}{2I_{LO}}\,\exp(\,g(1-th(\eta/\tau)]\ E_1[\,g(1-th(\frac{\eta+Dz}{\tau}))]} \qquad (9)$$

where

$$E_1(x) = \int_x^\infty e^{-t}\,t^{-1}\,dt\ ,\quad g = 2Q\tau I_{LO}|D|^{-1}\ ,\quad \tau_L = 1.76\,\tau\ .$$

At $z=2L_W$ pump and Raman pulses are separated. The further evolution of both pulses can be described independently by the respective NSE with account of GVD and with the asymmetric pulse intensity (9) as input and the corresponding input chirp. Under the influence of GVD the chirp of the pump pulses is linearised at

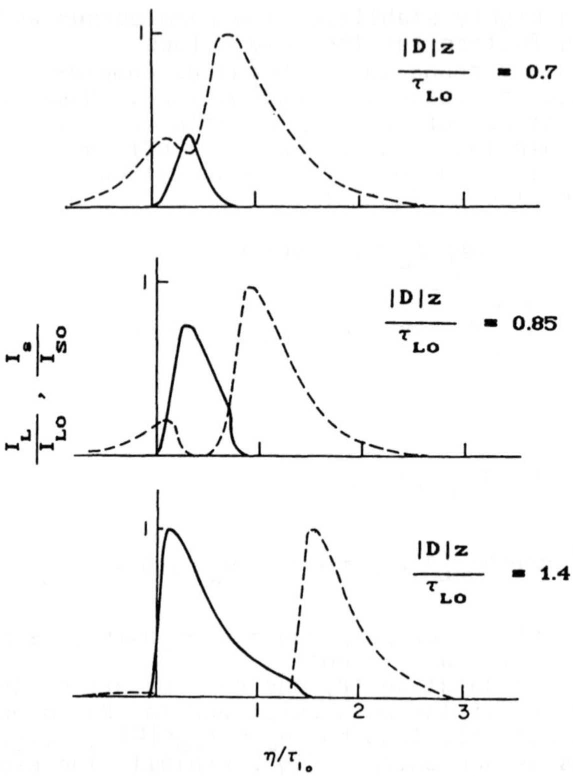

Fig. 2. Temporal and spatial evolution of the pump (---) and the Stokes (——) pulse intensities for a gain coefficient g = 25.

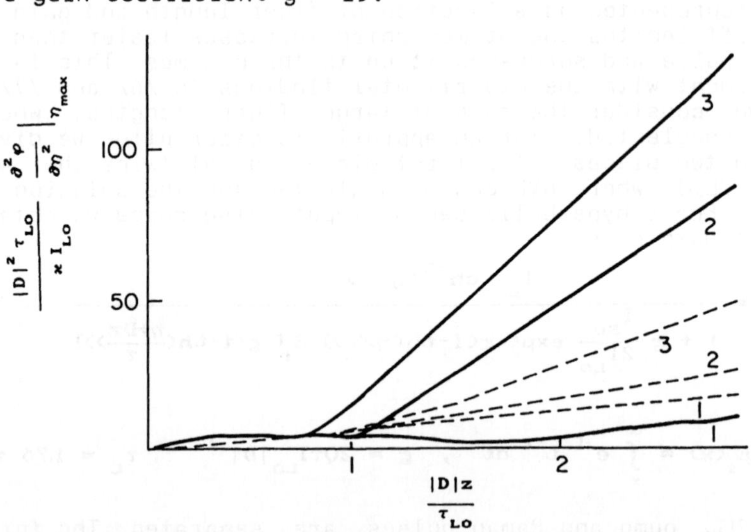

Fig. 3. Chirp at the pump and Stokes pulse maxima as a function of the fiber length for different gain coefficients g (indexes 1,2 and 3 refer to g = 15, 20 and 25, respectively).

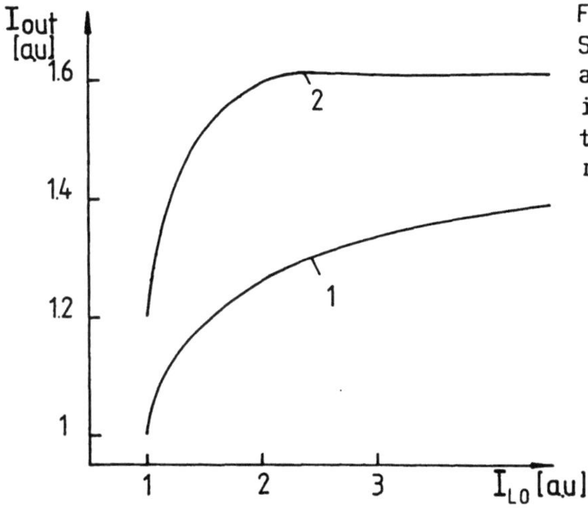

Fig. 4. Output intensity of SRS-influenced pump pulses as a function of the input intensity. (indexes 1,2 refer to $z=4L_W$ and $z=8L_W$, respectively).

the end of the fiber and the shape becomes symmetric. The self-stabilization of fluctuating pump pulse intensities results mainly from the action of SRS in the first piece of the fiber.

In Fig. 4 the output intensity of the pump pulse is depicted as a function of the input intensity. Obviously, the fluctuations of the input intensity are highly suppressed at the end of the fiber.

The pulse duration is stabilized in similar manner. As already mentioned such stabilization of fluctuating pulse parameters was observed by Heritage et al. (/8/) and Kuckartz et al. (/9/), where a pulse shortening from 70 ps pulses to 550 fs pulses and a stabilization of peak power and pulse width below 1 % were achieved.

References.

/ 1/ D. Grischkowski, A.C. Balant. Appl.Phys.Lett. 41, 1(1982)

/ 2/ R. Meinel, Opt.Comm. 47, 343 (1983)

/ 3/ W.J. Tomlinson, R.H. Stolen, C.V. Shank J.Opt.Soc.Am.B, 1, 139 (1984)

/ 4/ Th. Peschel, Phys. stat. solid. (b) 150, 657 (1988)

/ 5/ J. Herrmann to be published

/ 6/ A.M. Johnson, R.H. Stolen, W.M. Simpson, Ultrafast Phenomena V, edited by G.R. Fleming and A.E. Siegman, Berlin, Springer-Verlag (1986)

/ 7/ A.S. Gomes, V.L. da Silva, J.R. Taylor, J.Opt.Soc.Am. B, 5, 373 (1988)

/ 8/ J.P. Heritage, A.M. Weiner, R.J. Hawkins, Opt.Comm. 67, 367 (1988)

/ 9/ M. Kuckartz, R. Schulz, H. Harde, J.Opt.Soc.Am. B, 5, 1353 (1988)

/10/ D. Schadt, B. Jaskorzynska, U. Osterberg, J.Opt.Soc.Am. B 3, 1257 (1986)

/11/ M. Kuckartz, R. Schulz, H. Harde, Opt.Quant.Electron. 19, 237 (1987)

/12/ D. Schadt, B. Jaskorzynska, J.Opt.Soc.Am. B, 4, 856 (1987)

/13/ J. Herrmann, J. Mondry, J.Mod.Optics 35, 1919 (1988)

Distortion of Femtosecond Laser Pulses in Optical Systems and Their Applications

Z. Bor and G. Szabó

JATE University, Department of Optics and Quantum Electronics,
H-6720 Szeged, Dóm tér 9, Hungary

The pulse front of femtosecond pulses can be distorted by several picoseconds upon traversing a lens. The effect is due to the fact that in optical materials the group velocity of light is typically 2-20% smaller than the phase velocity. The paper presents equations describing the pulse front delay in singlet lenses, achromats and compound lenses. The pulse front delay is several orders of magnitude larger than the broadening caused by group velocity dispersion in the lens material. Delays occurring in Fresnel-type zone plates is also described.

1. Introduction

In dispersive media the speed of propagation of a light pulse is the group velocity. The difference between the phase and group velocity is determined by the first derivative of the refractive index with respect to the wavelength [1, 2]. The second derivative is responsible for pulse broadening due to group velocity dispersion (GVD) [3].

The propagation effects occurring upon refraction on plane surfaces (e.g. in prism systems) have been studied in detail and are used successfully for intra and extracavity pulse compression [2, 4-7]. However, attention to the propagation effects occurring in lenses and lens systems has been paid only very recently [8-13]. In this paper the temporal and spatial distortion of a pulse front is described upon traversing a lens or lens system. The equations are valid for thin lenses in the paraxial approximation.

2. Propagation time of a pulse front through a lens

Due to the difference between the phase and group velocity, the pulse front - which is defined as the surface coinciding with the peak of a pulse - is delayed with respect to the phase front. This delay can be calculated as [1-3, 9, 10]

$$\Delta T(r) = \frac{l}{c}\left(-\lambda\frac{dn}{d\lambda}\right), \tag{1}$$

where l is the path length of light in the medium having a refractive index $n(\lambda)$.

Springer Proceedings in Physics, Vol. 49 Ultrafast Phenomena in Spectroscopy
Editors: E. Klose · B. Wilhelmi

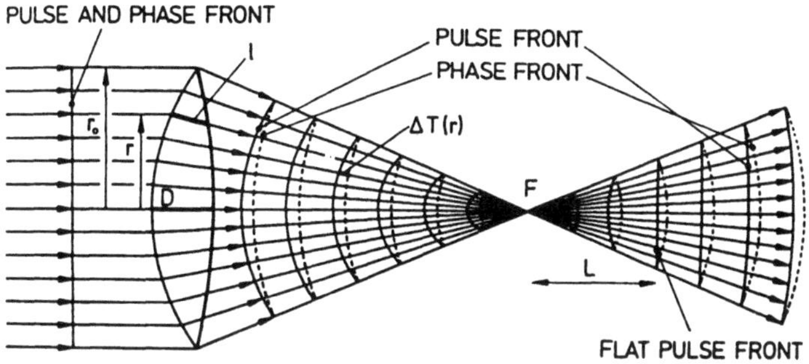

Fig. 1 Due to the difference between the group and phase velocity, behind a lens the pulse front is delayed with respect to the phase front

Using Fermat's principle and applying analytical ray tracing in the paraxial approximation it can be shown that the delay between the phase and pulse front can be calculated as

$$\Delta T(r) = \frac{r_0^2 - r^2}{2cf(n-1)} \left(-\lambda \frac{dn}{d\lambda}\right), \tag{2}$$

where r and r_0 are the input paraxial radii of an arbitrary and the marginal ray, respectively (Fig. 1). The above effect is proportional to $\lambda(dn/d\lambda)$ and will be referred to as propagation time difference (PTD). In equation (2) the delay for the marginal ray (i.e. $r = r_0$) is assumed to be 0.

For optical materials of practical interest, $dn/d\lambda$ has a negative value in the spectral range of transparency. The axial part of the beam is, therefore, delayed with respect to the marginal part.

The effect of PTD is illustrated in Fig. 2, which shows the locations of the pulse fronts at the near vicinity of the focus at different moments. The calculations have been carried out for a fused silica lens having f = 150 mm, r_0 = 40 mm, n = 1.50799 [16], $\lambda(dn/d\lambda)$ = -0.1375 at the wavelength of the KrF excimer laser (249 nm). The equations describing the pulse fronts are

$$x = c\cdot\left(t - \Delta T(r)\right)\cos\alpha, \tag{3}$$

$$y = c\left(t - \Delta T(r)\right)\sin\alpha, \tag{4}$$

where x and y are the spatial coordinates measured from the focal point, $\alpha = \mathrm{arctg}(r/f)$ and the origin of time t is chosen so that the marginal ray propagating with phase velocity arrives at the focus at t = 0.

A peculiar feature of the pulse front is the formation of a loop for times $0 < t < T(r_0)$. The reason for it is that for $0 < t < T(r_0)$ the part of the pulse propagating marginally has already passed through the focal point F, while the axially propagating part did not reach

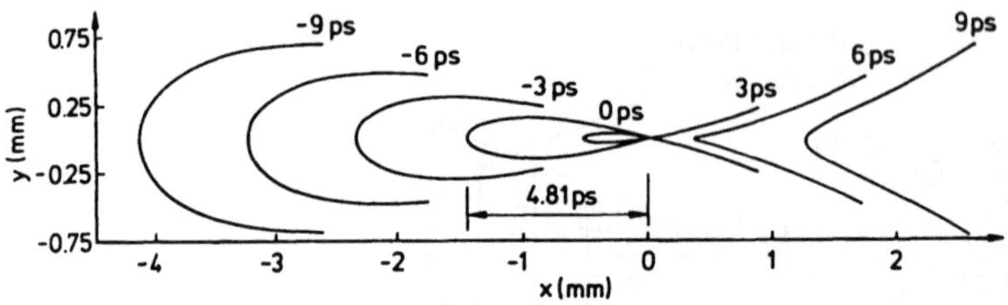

Fig. 2 Behaviour of the pulse and phase front in the vicinity of the focal plane (geometrical optical approximation). The temporal delay between the marginal and axial ray is 4.81 ps. The formation of the loop means that for $0 < t < 4.81$ ps the pulse propagating along the marginal ray has passed the focus, while the axial part has not reached it

it yet. The present analysis makes use of the geometrical optical approximation. It is, therefore, not valid in the proximity (about a few hundred μm) of the focus. In this range the shape of the pulse could have been calculated by solving the Kirchhoff-Fresnel integrals. Such calculations have been carried out previously by Martinez [14] for a much simpler case (i.e. for plane waves tilted by gratings or prisms).

Fig. 2 demonstrates clearly the importance of PTD for ultrashort pulses. A photodetector located at the focal plane of the lens would detect a signal stretched by $T(r = 0) - T(r = r_0)$ which amounts to 4.81 ps for the lens specified above.

It is interesting to note that the curvature of the pulse front changes from convex to concave behind the focus. From simple geometrical consideration it can be shown that at

$$L = \frac{f}{n-1}\left(-\lambda\frac{dn}{d\lambda}\right) \qquad (5)$$

the pulse front is flat, while the phase front is a sphere having a radius of L. This property of the lens can be used to avoid the unwanted effect of the radius dependent propagation time T(r). For a silica lens at 249 nm this distance is at $L = 0.27 \cdot f$ behind the focus F.

The pulse front may be also affected by the spherical aberration of the lens. The study of this effect is beyond the scope of this paper.

Calculations show [9, 10] that for an achromatic doublet the propagation time from plane A to the focal point F is

$$T(r) = \frac{d_1}{c}\left(n_1 - \lambda\frac{dn_1}{d\lambda}\right) + \frac{d_2}{c}\left(n_2 - \lambda\frac{dn_2}{d\lambda}\right) + \frac{f}{c}, \qquad (6)$$

where the meaning of the symbols is shown in Fig. 3. In the derivation of (6), Fermat's principle and the condition of achromatism obtained from $df/d\lambda = 0$ was used [10]. Thus for an achromat, in contrast to singlet lenses, the propagation time is the

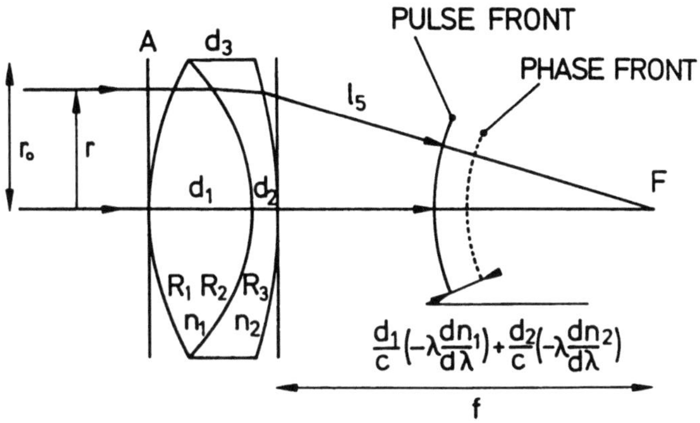

Fig. 3 For an achromat the delay between the pulse and phase fronts is constant over the cross-section of the lens

same for any input radius r (6). In most experiments, this is equivalent to no delay between the two fronts.

3. Pulse broadening due to group velocity dispersion

For a singlet lens the broadening of the pulse due to GVD can be calculated as [10]

$$\Delta\tau(r) = \frac{\lambda}{c}\frac{d^2n}{d\lambda^2}\,\Delta\lambda\; l(r)\;, \tag{7}$$

where l(r) is the path length of a ray in the lens material (Fig. 1).

It is convenient to express the bandwidth of the pulse as $\Delta\lambda = \lambda/(2\cdot N)$, where N is the number of coherent optical cycles in the pulse. (For a transform-limited pulse, where the form factor does not differ far from unity, $N = c\tau/\lambda$.)
A detector or a target located at the focal plane perceives both PTD (2) and pulse broadening due to GVD (7). Therefore it is instructive to express their ratio

$$\frac{\Delta T(r)}{\Delta\tau(r)} = \frac{2}{A}\,N\;, \tag{8}$$

where A is a quantity expressing the proportionality between the second and first derivative of the refractive index [15]. A is slightly wavelength dependent and has a typical value between 2.6 and 4.0. (Higher values of A correspond to shorter wavelength towards spectral cutoff [15]). Using the polynomial given for the refractive index of fused silica, we obtain that at 249 nm, A = 3.775 [15, 16]). From (8) we obtain that for a 100 fs long pulse for fused silica lens at 249 nm, the exact value of the PTD is 64 times larger than broadening due to GVD.

For a singlet lens one can, in general assume that for any input radius r, the effect of delay between pulse and phase fronts is approximately N times larger than the broadening of the pulse due to GVD of the lens material. The above approximation is correct with the precision of a factor of 2/A.

For an achromat, the pulse broadening due to GVD observed in the focal plane can be calculated as

$$\Delta\tau = \frac{dT(r)}{d\lambda}\,\Delta\lambda\ , \tag{9}$$

where T(r) is given by (6). Taking the derivative of (6) with respect to λ we obtain

$$\Delta\tau(r) = \frac{\lambda^2}{2cN}\left(d_1\,\frac{d^2n_1}{d\lambda^2} + d_2\,\frac{d^2n_2}{d\lambda^2}\right) . \tag{10}$$

Thus for an achromat the broadening of the pulse due to GVD is independent of the input radius r. Therefore it can be compensated by grating [17] or prism compressors [2, 4-6].

4. Pulse front delay in a telescope

Pulse front delay occurs in lens systems as well. For example in telescopes the propagation time also depends on the input radius r. Using (2) for both lenses of the telescope the delay between the phase and pulse front $T(r_2)$ is obtained as

$$\Delta T(r_2) = \frac{r_{02}^2 - r_2^2}{2cf_2(n-1)}\left(-\lambda\frac{dn}{d\lambda}\right)\left(1 + \frac{1}{M}\right) . \tag{11}$$

Here $M = f_2/f_1$ is the magnification factor of the telescope and r_2 is the output radius (i.e. the distance of an arbitrary ray at the output side of the lens from the optical axis). For the derivation of (11) it was assumed that the lenses both were made from the same material.

The temporal front delay in a telescope was measured by time-of-flight interferometry with an accuracy of 20 fs [12, 13]. These measurements proved the validity of (2) and (11).

5. General expression for the temporal front delay

In [10] a more general expression for the delay between the phase and pulse fronts has been derived,

$$\Delta T(r) = \frac{r_0^2 - r^2}{2cf^2}\,\lambda\frac{df}{d\lambda}\ , \tag{12}$$

where $df/d\lambda$ is the longitudinal chromatic aberration of the lens. (12) may be very useful for the calculation of PTD because the longitudinal chromatic aberration is usually specified by the suppliers of lenses. In the paraxial approximation, (12) is valid for singlet lenses, achromatic doublets, not fully compensated achromats and, as it was shown in [9], also for a Fresnel-type focussing zone plate.

It is somewhat surprising that (12) is valid also for a Fresnel-type focussing zone plate in which the delay between the pulse and phase fronts has quite a different origin compared to a usual refractive lens. This is a hint that (12) has a more fundamental physical meaning. It can be shown [15] that the general expression (12) is the manifestation of the uncertainty principle for a spectrograph consisting of a lens with chromatic aberration and a pinhole in the focus of the lens.

6. Conclusions

Due to the difference between the phase and group velocity the pulse front is delayed with respect to the phase front (Fig. 1). The delay is a parabolic function of the input beam radius (see (2)). The delay was also experimentally demonstrated in [11-13].

The delay may amount to several picoseconds in large aperture UV optics. The delay may considerably affect the power deposition into a target located at the focus.

For positive lenses, the pulse front is flat at a distance of L behind the focal plane given by (5). For negative lenses this plane is before the focal plane.

For singlet lenses the pulse broadening caused by GVD is in practice several orders of magnitude smaller than the delay due to the difference between the phase and group velocity (see (8)).

The general equation expressing the delay through the longitudinal chromatic aberration (12) turned out to be valid for a Fresnel-type focussing zone plate as well.

Achromats have the valuable property that the delay between the pulse and phase fronts is constant over the lens cross section (see(6) and Fig.3). In most experiments, this is equivalent to no delay between the two fronts. Thus, the application of achromats for most experiments can eliminate the unwanted effects of pulse front distortion.

For achromats, the pulse broadening caused by GVD is larger than for singlet lenses. However, it is constant over the cross section of the lens (see (10)). Such spatially homogeneous pulse broadening can be compensated by pulse compressors.

The consideration of effects described above is particularly important in designing large aperture, high F number UV optics for inertial confinement fusion experiments and x-ray laser pumping.

Acknowledgements

This work has been supported by OTKA Foundation of the Hungarian Academy of Sciences and the German-Hungarian Exchange Program.

References

1. M.R. Topp, G.C. Orner: Opt. Commun. **13**, 276 (1975)
2. Z. Bor, B. Rácz: Opt. Commun. **54**, 165 (1985)
3. L.G. Cohen, C. Lin: Appl. Opt. **16**, 3136 (1977)
4. O.E. Martinez, J.P. Gordon, R.L. Fork: J. Opt. Soc. Am. **A1**, 1003, (1984)
5. R.L. Fork, O.E. Martinez, J.P. Gordon: Opt. Lett. **9**, 159 (1984)
6. J.D. Kafka, T. Baer: Opt. Lett. **6**, 401 (1987)
7. Z. Benkő, Z. Gogolák, Z. Bor, G. Szabó: Conference on Ultrafast Phenomena in Spectroscopy, Neubrandenburg, 1989, ibid
8. H. Staerk, J. Ihlemann, A. Helmbold: Laser und Optoelektronik **20**, 6 (1988)
9. Z. Bor: Opt. Lett. **14**, 119 (1989)
10. Z. Bor: J. Modern Optics **35**, 1907 (1988)
11. S. Szatmári, G. Kühnle: Opt. Commun. **69**, 60 (1988)
12. Z. Bor, Z. Gogolák, G. Szabó: Opt.Lett. **14**, 862 (1989)
13. Z. Bor, Z. Gogolák, G. Szabó: Conference on Lasers and Electro-Optics, Baltimore, 1989, Paper PD 32
14. O.E. Martinez: Opt. Commun. **59**, 229 (1986)
15. Z. Bor, B. Rácz: Appl. Opt. **24**, 3440 (1985)
16. I.H. Malitson: J. Opt. Soc. Am. **55**, 1205 (1965)
17. E.B. Treacy: IEEE. J. Quantum Electron. **QE-5**, 454, (1969)
18. Z. Bor: (to be published)

Optical Pulse Shaping by Spectral Hole Burning Filters

R. Kaarli, P. Saari, and H. Sõnajalg

Institute of Physics, Estonian Academy of Sciences, Riia 142,
202400 Tartu, Estonia, USSR

Abstract.The shaping of ultrashort light pulses with arbitrary amplitude and frequency modulations by the filters synthesized via spectral hole burning, is investigated both theoretically and experimentally.

1.Introduction

The pulse shaping device is a crucial component of any optical communication or optical processing system as well as of a time-resolved spectroscopy setup. In addition to various devices based on fibers, gratings and interference filters, the application of the filters synthesized via spectral hole burning (SHB) [1-3] seems to be perspective [4-8]. It is because of the following reasons:

1.SHB is a unique method for spectral and spatial 'engineering' of the dielectric permittivity of an optical element.

2.SHB filters can be prepared not only by exposing the plate to proper light pulses but also via synthesizing by preprogrammed scanning of monochromatic irradiation over the given spectral range.

3.The limits of the temporal range of SHB filters (10^{-7}-10^{-13}s) are determined by the homogeneous width of the impurity absorption line and by the width of the inhomogeneously-broadened absorption band, respectively, and these do not depend upon the size of the filter.

4.The fact that the photoactive impurity molecules have different orientations makes it possible to control also the polarization variations of output pulses.

In this work, we have investigated the shaping of linearly polarized light pulses by SHB filters.

2. Theoretical considerations

Let us ask which is the complex response $r(\nu)$ of an SHB filter with the transparency spectrum $T(\nu)$. In general, the response of a linear optical system is not determined by the transparency spectrum together with the causality as an *a priori* knowledge [9]. But in the case when the propagation of light through the system may be described by a single first-order wave equation (i.e. the optical system is isotropic and the dielectric permittivity has no sharp spatial modulations), the complex response is given as [8,9]

$$r(\nu)=\exp\left[0.5*(1+i\mathcal{H})[\ln T(\nu)]\right], \qquad (1)$$

where the symbol $\mathcal{H}$ means the Hilbert transformation. A relevant feature of the responses $r(\nu)$, given by eq.(1), is the absence of zeros on the complex half-plane where $r(\mathrm{Re}\nu+i*\mathrm{Im}\nu)$ is analytic. In our particular case, eq.(1) is valid if the SHB filter is synthesized by making use of the circularly (or linearly) polarized irradiation. Therefore, the class of responses available with the isotropic SHB filters is restricted. However, in order to find out the suitable response to be realized according to eq.(1), the investigation of $r(\mathrm{Re}\nu+i*\mathrm{Im}\nu)$ on a complex half-plane becomes indispensable.

The considerations are illustrated by computer simulations (fig.1). First of all, imagine two optical systems, whose temporal responses consist of two (the amplitudes are denoted with a_1 and a_2) pulses attenuated and retarded by the same time interval. The responses differ from each other by the temporal order of their components: the pulse with the amplitude a_1 is followed by the one with the amplitude a_2 or vice versa. These optical systems have different phase characteristics but identical transparency spectra. If a SHB filter with this transparency spectrum is made, then the front pulse of the response is more intense compared with its second pulse (fig.1a). However, a holographic approach enables the play-back or shaping, in principle, of any kind of signals but only after an intense nondiffracted read-out pulse (fig.1b). As is

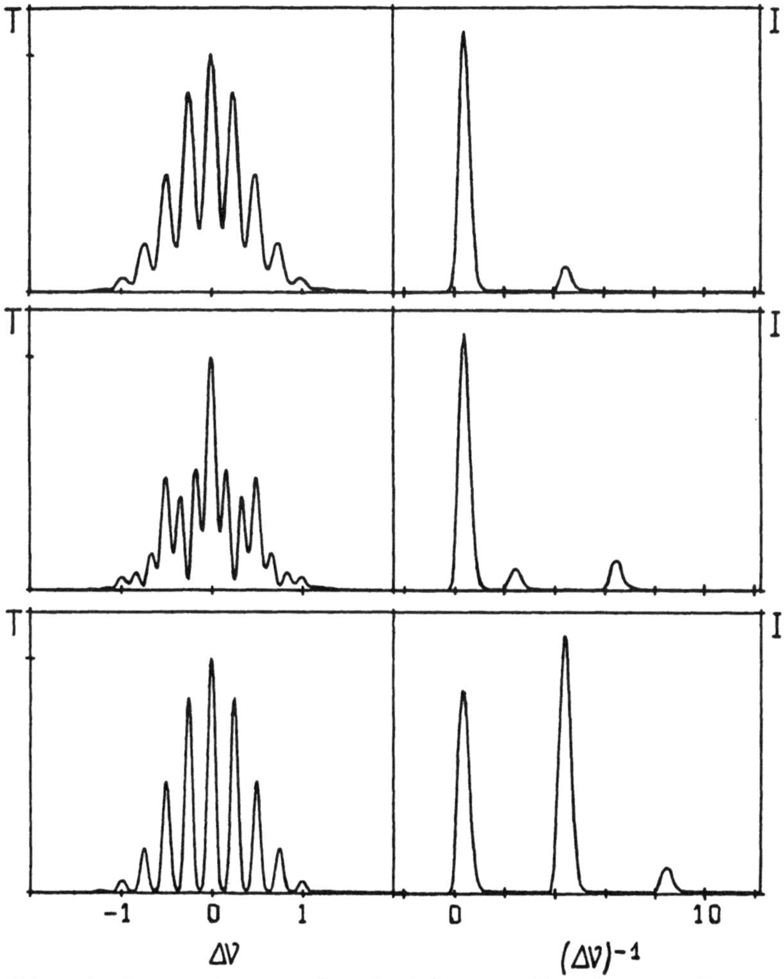

Fig.1.Computer simulation of the pulse shaping by SHB filters: the transparency spectra (on the left) and the intensity profiles of the output pulses.

shown in fig.1c, responses whose intensity profiles do not coincide precisely but are quite similar to the ones not allowed by eq.(1), may also be found.

In what follows, we will discuss the properties of anisotropic SHB filters burnt in with elliptically polarized irradiation. The response of the filter is given by a symmetrical (2×2) matrix [10]. Our theoretical analysis shows that the nondiagonal element of this matrix, which describes the rotation of the plane of light polarization by $\pi/2$ is not governed by the limitation connected with the absence of complex zeros. Consequently, one may get a pulse-shaping device by putting a proper anisotropic SHB filter between two perpendicularly orientated polarizers. The amplitude and the phase characteristics of this device are independently controllable in the limits set by strict causality.

3. Experiment

To prepare the SHB filters [8] an octaethylporphin-doped polystyrene (OEP-PS) [11] plate at 1.8K was irradiated with a monochromatic light from a single-frequency Rhodamine 6G ring laser. Temporal responses were measured by passing picosecond pulses from a synchronously pumped CW laser through the SHB filters. In fig.2, the transparency spectra T(ν) of the SHB filters (upper insert), the corresponding

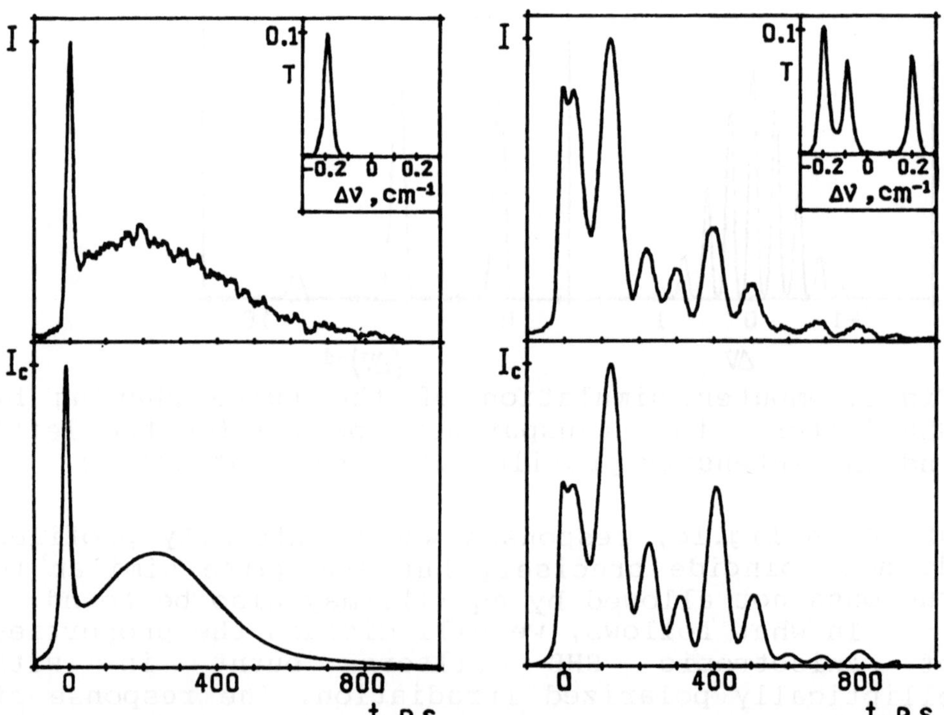

Fig.2. Transparency spectra of SHB filters T(ν) (ν_0=16140cm^{-1}, upper insert), the corresponding streak-camera images I(t) (upper main frame) and the calculated responses I_c(t) (bottom).

streak-camera images (upper main frame) and the responses calculated from $T(\nu)$ according to eq.(1) (bottom) are shown. As one can see, the calculated responses coincide well with the experimental ones.

The experiments with anisotropic SHB filters are in progress.

4. Conclusions

In this paper, the results of the investigation of the shaping of linearly polarized ultrashort light signals with arbitrary amplitude and frequency modulations by using photochemically burnt-in spectral filters, are presented. The pulse shaping device consisting of an anisotropic SHB filter and two polarizers, is offered to extend the class of available output pulses.

The authors are indepted to A.Gorokhovskii, V.Palm and M.Rätsep for their participation in the experiments and to I.Renge for chemical preparation of the sample.

References

1. A.A.Gorokhovskii, R.K.Kaarli, and L.A.Rebane, JETP Lett.20, 216 (1974).
2. B.M.Kharlamov, R.I.Personov, and L.A.Bykovskaya, Opt. Commun.12, 191 (1974).
3. W.E.Moerner, Ed., *Persistent spectral hole-burning: science and applications* (Springer, Berlin, 1988).
4. G.C.Bjorklund, U.S.Patent 4 306 771 (1981).
5. J.H.Eberly, S.R.Hartmann, and A.Szabo, Phys.Rev.A 23,2502, (1981).
6. A.Rebane and R.Kaarli, Chem.Phys.Lett.110, 317, (1983).
7. P.Saari, R.Kaarli, and A.Rebane, J.Opt.Soc.Am.B3, 527, (1986).
8. H.Sõnajalg, A.Gorokhovskii, R.Kaarli, V.Palm, M.Rätsep, and P.Saari, Opt.Commun.71, 377, (1989).
9. J.S.Toll, Phys.Rev. 104, 1760, (1956).
10. R.M.A.Azzam and N.M.Bashara, *Ellipsometry and polarized light* (North-Holland Publishing Company, Amsterdam, 1977).
11. A.A.Gorokhovskii, in *Zero-phonon lines and spectral hole burning in spectroscopy and photochemistry*, edited by O.Sild and K.Haller (Springer, Berlin, 1988).

High Time Resolution with Incoherent Light in the Raman Fringe Decay

K. Wynne, M. Müller, and J.D.W. van Voorst

Laboratory for Physical Chemistry, University of Amsterdam, Nieuwe Achtergracht 127, 1018 WS Amsterdam, The Netherlands

Abstract. The use of a new method to measure vibrational dephasing, the Raman Fringe Decay (RFD), is proposed that can be used with unamplified non-transform-limited dye-laser pulses. The advantages of the method are a higher signal-to-noise ratio compared to low repetition-rate experiments and a signal that decays with the vibrational dephasing rate $\gamma = (T_2)^{-1}$ without any significant "coherent artifacts". The results of the experiments on the 524 cm^{-1} mode of liquid methyl-iodide and the 656 cm^{-1} mode of liquid carbon-disulfide are presented, which demonstrate the power of this technique.

1. Introduction

A number of recent studies by various groups has shown that the time resolution in time-resolved degenerate four wave mixing [1,2,3] and coherent Stokes Raman scattering [4,5,6] (CSRS) is determined by the coherence-time of the pulses involved rather than their widths. All these studies [7] have been performed with low repetition-rate nanosecond pulsed lasers and require careful comparison with theory to determine the relevant physical parameters.

In contrast to this, the Raman Fringe Decay (RFD) [8] technique uses unamplified non-transform-limited picosecond dye laser pulses. Due to the different experimental set-up of the RFD, as compared to incoherent CSRS, the former technique has a number of advantages. First, the higher repetition rate provides a better S/N ratio. Second, due to the nature of the technique the signal decays with the vibrational dephasing rate, $\gamma = (T_2)^{-1}$ - rather than $\gamma = (T_2/2)^{-1}$, without a 'spike' on top of the signal, which does appear in incoherent CSRS. This enables unambigious interpretation of the results. Finally, for a polarization condition where the excitation pulse pair is polarized perpendicular to the probe pulse pair, the experiments are not limited to the measurement of dephasing times shorter than the pulse width, but can be extended smoothly to the measurement of longer dephasing times, without any effect due to the limited pulse width.

2. Theory

The difference between CSRS and the method discussed in this paper is that here *both* the excitation and the probe pulse pairs consist of a Stokes and a laser pulse. After laser and Stokes have been made (nearly) resonant with the Raman active vibrational mode of interest and the probe pulse pair has been passed through a variable delay, the beams are made collinear and are focused into the sample. The signal is then detected at the anti-Stokes frequency. A theoretical description of this situation using perturbation theory and assuming

Springer Proceedings in Physics, Vol. 49 **Ultrafast Phenomena in Spectroscopy**
Editors: E. Klose · B. Wilhelmi

(Gaussian-Markovian) stochastic laser pulses, leads to an expression with fringe terms oscillating at the laser and Stokes difference frequency as a function of the delay τ. Formally writing the average RFD intensity as

$$I \propto 4\alpha \cos\{(\omega_L-\omega_S)\tau\} + 4\beta \sin\{(\omega_L-\omega_S)\tau\} \ , \qquad (1)$$

it is found that

$$\alpha = e^{-\gamma|\tau|} \{ X \cos \Delta\tau + Y \sin \Delta|\tau| \} + Z e^{-\Gamma|\tau|} \ , \qquad (2a)$$

$$\beta = e^{-\gamma|\tau|} \{ Y \frac{\tau}{\tau} \cos \Delta\tau + X \sin \Delta|\tau| \} - Y \frac{\tau}{\tau} e^{-\Gamma|\tau|} \ , \qquad (2b)$$

where

$$X = \frac{\Gamma \{(\Gamma-\gamma)(\Gamma+3\gamma) + \Delta^2\}}{\gamma|\Gamma-\phi|^2|\Gamma+\phi|^2} \ , \qquad Y = \frac{4\Gamma\Delta}{|\Gamma-\phi|^2|\Gamma+\phi|^2} \ ,$$

$$Z = \frac{(\Gamma-\gamma)(\Gamma^2-\gamma\Gamma-4\gamma^2) + \Delta^2(\Gamma+4\gamma)}{\gamma|\Gamma-\phi|^2|\Gamma+\phi|^2} \ , \qquad (2c)$$

$\phi = \gamma+i\Delta$, Δ is the detuning from Raman resonance, ω_L and ω_S are the laser and Stokes frequencies, respectively, and $\Gamma = (\tau_c)^{-1}$ the inverse coherence time of the laser field. The amplitude of the fringes decays with the vibrational dephasing rate γ, has no prominent coherence spike term and no background. Note that it follows from eqs. (1) and (2) that in general the RFD signal oscillates at both the Raman transition frequency and the difference frequency of laser and Stokes. This can alternatively be expressed by stating that beats appear for nonzero detuning from resonance.

3. Experimental

The above-described experiment was performed on the 524 cm^{-1} mode in liquid methyl-iodide and the 656 cm^{-1} mode in liquid carbon-disulfide at room temperature.

The inset of figure 1 shows part of the raw data, oscillating at the Raman transition frequency, obtained by scanning the probe pulse pair relative to the excitation pulse pair. It has been checked that the oscillating part of the signal depends on all four beams. The raw data is digitally filtered and the fringe amplitude, averaged over 10μm ($\cong$ 66.6 fs), as a function of delay is determined. Figure 1 shows the RFD measured in methyl-iodide; although the coherence time of the laser is only just shorter than the vibrational dephasing time of methyl-iodide, it is unambiguously found that the signal decays with the expected [9,10] T_2= 2.4 ps. The solid line in the figure is a least squares fit of our result, eq. (1), to the data that yields the expected dephasing and coherence time. It should be noted that the interpretation is not hampered by a prodigious coherence spike and that no baseline had to subtracted as in incoherent CSRS [4].

If laser and Stoke are detuned from 656 cm^{-1} to 645 cm^{-1} in carbon-disulfide, the result of figure 2 is obtained, which shows a beat with a period of 3 ps. However, since the dephasing time in carbon-disulfide is longer (T_2= 20 ps [11]) than the pulse widths, the interpretation is not unambiguous.

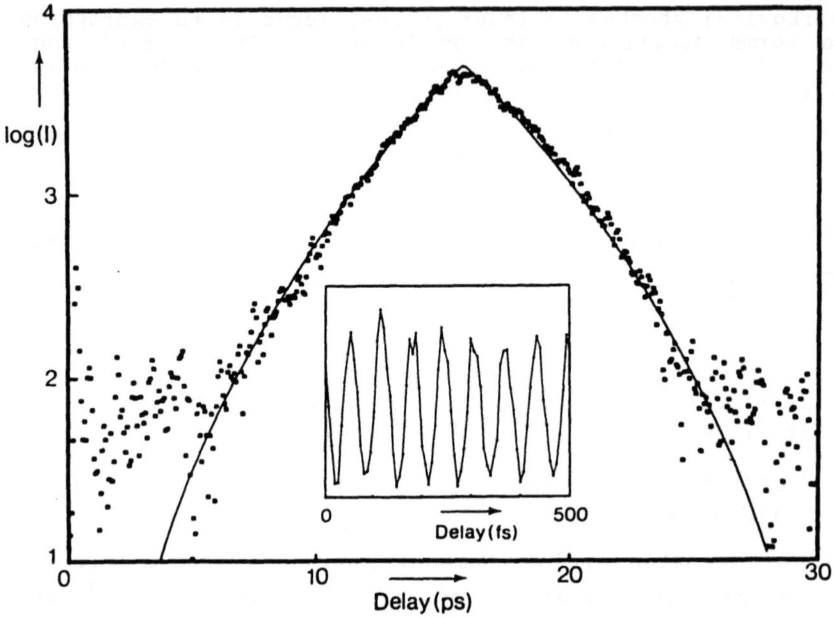

Figure 1. Raman fringe decay in room-temperature methyl-iodide at two-photon resonance at 524 cm^{-1}. The smooth curve is a least-squares fit to the theory with T_2= 2.4 ps. The inset shows part of the raw data.

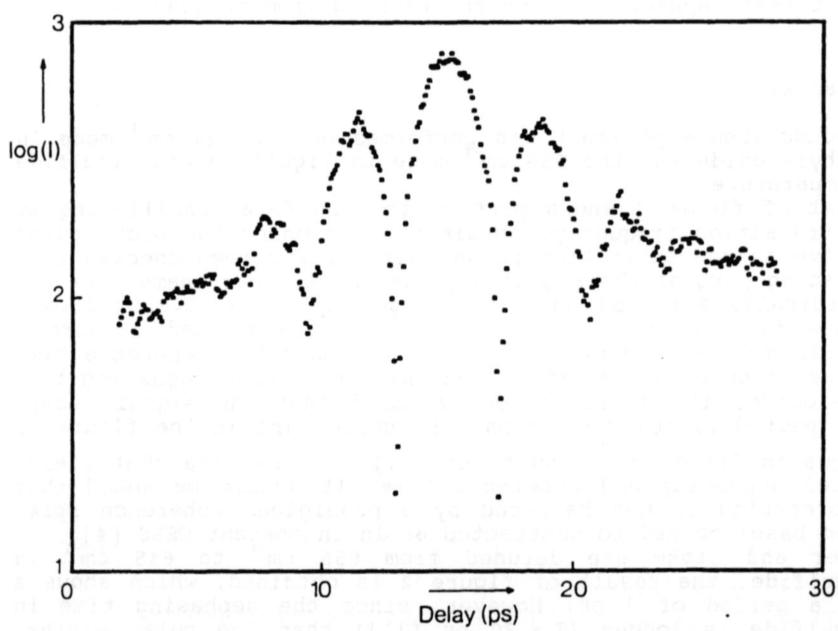

Figure 2. Raman fringe decay in room-temperature carbon-disulfide with two-photon detuning from 656 to 645 cm^{-1}.

4. Conclusion

In conclusion, it has been shown that it is possible to perform experiments using incoherent pulses with CW mode-locked lasers in electronically nonresonant Raman scatterers. The described method pairs a high signal-to-noise ratio to high time resolution. The results of this new approach to the use of incoherent light compares favourably with other incoherent techniques, e.g. incoherent CSRS.

Acknowledgement. The research reported here was partially supported by SON (Stichting Scheikundig Onderzoek Nederland) and NWO (Stichting Nederlands Wetenschappelijk Onderzoek).

References

1. N. Morita, T. Yajima, Phys. Rev. **A30**, 2525 (1984)
2. R. Beach, D. DeBeer, S.R. Hartmann, Phys. Rev. **A32**, 3467 (1985)
3. N. Morita, T. Tokizaki, T. Yajima, J. Opt. Soc. Am. **B4**, 1269 (1987)
4. T. Hattori, A. Terasaki, T. Kobayashi, Phys. Rev. **A35**, 715 (1987)
5. T. Kobayashi, T. Hattori, A. Terasaki, K. Kurokawa, Rev. Phys. Appl. **22**, 1773 (1987)
6. T. Kobayashi, A. Terasaki, T. Hattori, K. Kurokawa, Appl. Phys. **B47**, 107 (1988)
7. The single exception is: S. Asaka, H. Nakatsuka, M. Fujiwara, M. Matsuoka, Phys. Rev. **A29**, 2286 (1984)
8. K. Wynne, M. Müller, J.D.W. van Voorst, Phys. Rev. Lett. **62**, 3031 (1989)
9. N. Kohles, A. Laubereau, Chem. Phys. Lett. **138**, 365 (1987)
10. R.B. Wright, M. Schwartz, C.H. Wang, J. Chem. Phys. **58**, 5125 (1973)
11. M. van Exter, A. Lagendijk, Opt. Comm. **56**, 191 (1985)

Temporal and Spatial Evolution of Picosecond Phonon-Polariton Pulses in Crystals

F. Vallée, G.M. Gale, and C. Flytzanis

Laboratoire d'Optique Quantique du CNRS, Ecole Polytechnique, F-91128 Palaiseau Cedex, France

Abstract. Propagation and relaxation of picosecond phonon-polariton pulses in cubic and non-cubic crystals are investigated by a time and space resolved CARS technique. Temperature and frequency dependent measurements of the polariton dephasing time allow a detailed discussion of the polariton relaxation mechanisms.

1. Introduction

Ultrashort pulses of collective propagating excitations in crystals are very sensitive probes of dynamical and structural crystalline properties, yielding new information on crystal behavior via their temporal and spatial evolution. The creation and measurement of such pulses constitutes an important new problem in solid state physics which has stimulated the development of new experimental techniques for both bulk and surface excitations [1-5]. In particular, in the far infrared domain, the recent demonstration of a nonlocal nonlinear technique, namely time and space resolved CARS, has allowed the probing of new propagation and relaxation features of Raman active bulk phonon-polariton pulses in crystals, without any restriction on polariton frequency range [2]. Although a femtosecond technique has also been used to perform coherent time-domain spectroscopy of infrared and Raman-active modes [4], this method is restricted to low-frequency polaritons and has only a limited selectivity.

One of the simplest ways to study phonon-polaritons in noncentrosymmetric crystals is near-forward spontaneous Raman scattering, which is particularly well adapted to the investigation of the polariton dispersion curve and scattering intensity [6]. Polariton relaxation is much more difficult to address with this or similar local techniques, because of the large variation of polariton frequency with scattering angle and related propagation effects. The picosecond nonlocal CARS technique avoids this problem and allows the proper separation of temporal and spatial polariton features.

2. Experimental technique

The time and space resolved CARS method reposes on a "time of flight" type technique; its principle is schematically shown in Fig.1. Two space and time coincident optical picosecond pulses of frequencies ω_L and ω_S, with $\omega_L - \omega_S = \omega_\pi(k_\pi)$, the frequency of the polariton of wavevector k_π, create by coherent scattering at time $t = 0$ and at $X = 0$ a polariton wave-packet. As the polariton is an admixture of mechanical (phonon) and electromagnetic (photon) excitation, its coupling with the excitation fields occurs through both of these components and can be seen as a mixture of coherent Raman scattering and parametric amplification. When the excitation process has terminated, this polariton wave-packet propagates freely inside the crystal with its group velocity $V_g(\omega_\pi)$, along a direction

Springer Proceedings in Physics, Vol. 49 **Ultrafast Phenomena in Spectroscopy**
Editors: E. Klose · B. Wilhelmi

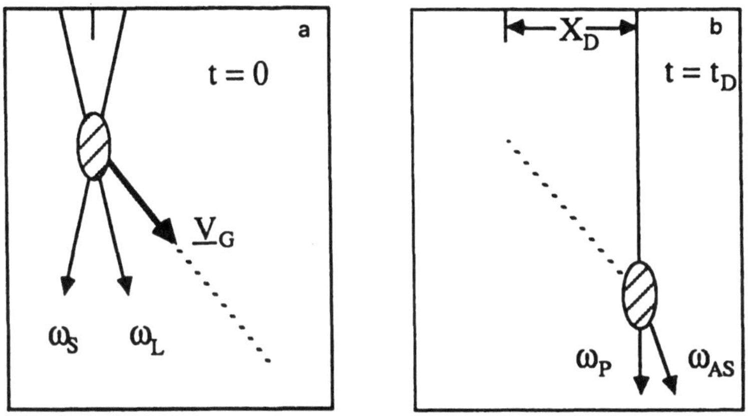

Fig.1: Principle of the time and space resolved CARS technique: (a) coherent polariton excitation, (b) coherent polariton detection

fixed by overall wave-vector conservation. The temporal and spatial evolution of the propagating excitation is followed by phase-matched coherent anti-Stokes scattering of a third picosecond probe pulse of frequency ω_P, displaced with respect to the excitation in space (by X_D) and in time by t_D (see Fig.1).

After an initial regime (typically for $t_D \leq t_P$ the duration of the pulses) due to the finite temporal width of the excitation process and to the finite size of the excitation zone, a linear dependence of the spatial displacement X_D where the signal is maximum, on time delay t_D is predicted. The slope of this "long" time linear dependence measures the polariton group velocity V_g. For the parameters of our experimental system and for the investigated samples, spatial and temporal spreading of the polariton wave packet is small, and the intensity decrease of the coherent signal with probe time delay when the polariton wave-packet is followed inside the crystal (i.e. X_D and t_D are varied simultaneously) gives for the first time a direct access to the polariton dephasing time as a function of polariton frequency.

The three independent picosecond pulses necessary for the experiment, ω_P, ω_L, and ω_S, which is continuously tunable, are produced by frequency conversion of a single 5 ps infrared pulse provided by a passively mode-locked Nd^{3+}:glass laser. After polarization adjustment, the three beams are focused into the cooled sample using a noncollinear geometry and the generated anti-Stokes signal is detected after spatial, spectral and polarization selection. The temporal resolution of our system is routinely of 0.3ps and its high sensitivity allows the observation of the coherent signal over several orders of magnitude, which strongly improves the spatial resolution (~ 20µm) compared to the focal diameter (~100µm). Samples long (~1cm)compared to the length of the excited zone were used, which practically obviates any influence of polariton escape from the crystal on the measured damping.

3. Polariton pulses in a cubic crystal: ammonium chloride

3.1 Group velocity measurement: ν_4 polariton

The first demonstration of the time and space resolved CARS technique was performed on the polariton associated with the ν_4 polar mode in the cubic crystal of ammonium chloride

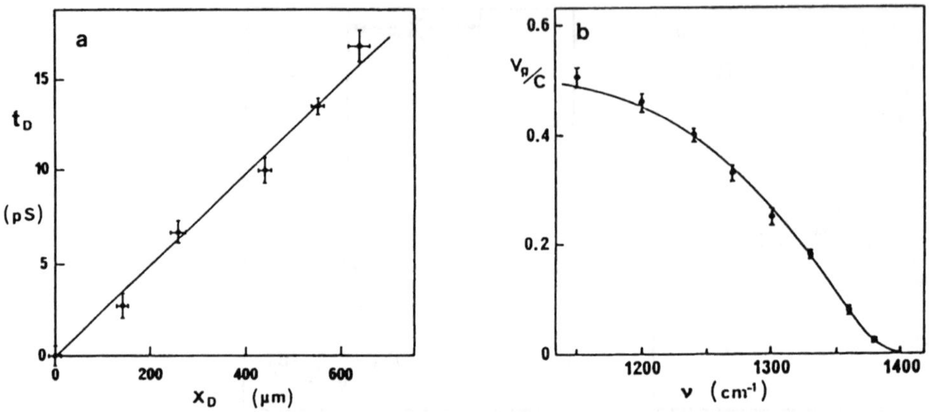

Fig.2(a): Group velocity measurement for the 1300 cm^{-1} polariton in NH_4Cl. (b)Frequency dependence of the ν_4 polariton group velocity in ammonium chloride.

(NH_4Cl) [2]. An example of the group velocity measurements performed on this polariton is shown in Fig.2a, where we have plotted the measured position X_D of the peak coherent signal as a function of probe delay t_D, for a polariton frequency of 1300 cm^{-1}. Note that all the points correspond to the resonant polariton response, except the first one, which is due to the non-resonant third order nonlinear response of the crystal. This response is dominant whenever the three beams spatially and temporally overlap, precisely fixing the temporal and spatial "origin " in the experiment, but decreasing very rapidly with time and space delay, following the system response function. The experimental points are well described by a straight line of slope p, passing through the origin since the resonant small delay regime is completly masked by the non-resonant response in this experiment. The group velocity may thus be calculated from

$$V_g = p / (\sin\theta + pn\cos\theta/c)$$

where n is the crystal refractive index at the probe wavelength, θ defines the polariton propagation direction (see Fig.1), and c is the speed of light in vacuum.

The group velocity measured for various polariton frequencies is shown in Fig.2b. As the polariton dispersion curve has been extensively studied in ammonium chloride by near forward spontaneous Raman scattering [7], we can independently estimate the polariton group velocity by differentiating the measured dispersion curve, which gives the full line in Fig.2b. Agreement with the experimental points is excellent (there is no fitting parameter) which gives an indirect confirmation of our results. The access to polariton dephasing time using this technique was also demonstrated in this experiment. In particular, the polariton dephasing was investigated in the vicinity of the ammonium chloride order/disorder phase transition, pointing up the strong influence of crystal disorder on polariton dynamics.

3.2 Polariton Fermi resonance

It is particularly interesting to use this new technique to investigate polariton interaction with other collective excitations of the crystal, of which a typical example is polariton Fermi resonance. Polariton Fermi resonance is produced whenever a polariton branch crosses a many-phonon band, most often two-phonon, of same symmetry as the polariton. This

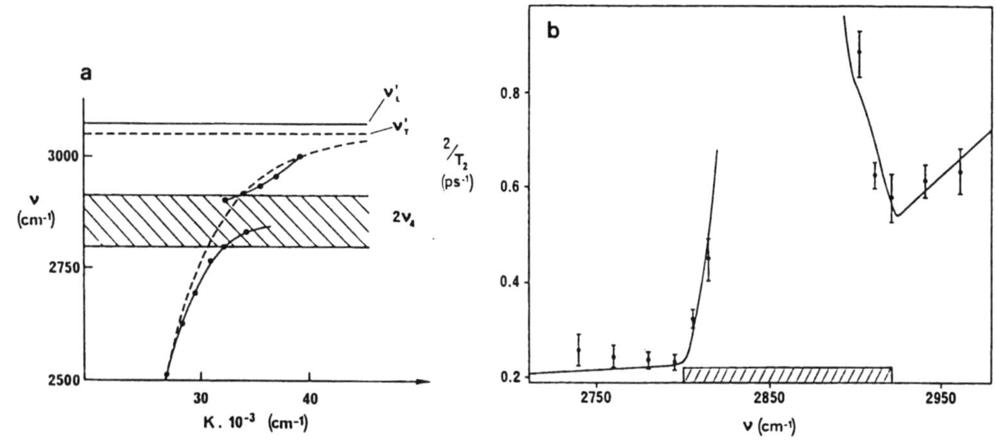

Fig.3: The polariton dispersion curve (a) and the frequency dependence of the polariton dephasing rate (b) in the vicinity of the $2\nu_4$ two phonon polar band (shaded region). The full line of Fig.3(b) is calculated (see text).

situation is frequently encountered because both polariton and two-phonon states span large regions in frequency and wave-vector space. For sufficiently strong anharmonic coupling, this crossing leads to a drastic alteration of polariton properties, formation of an additional gap in the polariton dispersion curve, and strong, frequency dependent damping of the polariton line. Due to the intrinsic difficulties associated with polariton linewidth measurements by spontaneous Raman techniques, this last problem has not yet been addressed in detail.

We have performed the first time and space resolved study of polariton Fermi resonance in ammonium chloride [3], in a frequency region where the polariton is in strong interaction with the polar $2\nu_4$ two-phonon band (see Fig.3a) [7]. From a theoretical analysis, it can be shown that under conditions of polariton Fermi resonance the polariton line shape is still Lorentzian on the edge of the two-phonon band (i.e. for relatively weak two-phonon density of states), but is strongly broadened by the opening of a new direct relaxation channel into the two-phonon band [3]. This mechanism leads to a contribution directly proportional to the two-phonon density of states, which is clearly evidenced in our experiments.

The dramatic frequency dependence of the polariton dephasing rate for polariton frequencies close to the edge of the $2\nu_4$ band is depicted in Fig.3b (crystal temperature is 78K). The rapid increase of the damping rate inside the $2\nu_4$ band is a clear indication that polariton disintegration into two isoenergetic free phonons constitutes a very efficient polariton relaxation channel, whose contribution essentially reflects the two-phonon density of states which, by definition, strongly increases at the edges of the two-phonon band. Outside the two-phonon continuum, this direct relaxation channel is closed, but polaritons can still preferentially relax into this band with the assistance of a low energy lattice phonon, in an up- or down-conversion process for polaritons respectively below and above the continuum. Taking into account these two main mechanisms and a smaller contribution due to higher order processes, we obtain the full line in Fig.3b, which correctly describes the frequency dependence of the measured damping rate. This interpretation of polariton relaxation in the $2\nu_4$ frequency region receives additional support from our temperature dependent measurements of the polariton dephasing rate.

4. Polariton in an uniaxial crystal: Lithium Iodate

We have generalized this technique to non-cubic crystals where the birefringence allows to circumvent the wave-vector restrictions associated with Raman techniques in cubic crystals [6]. In uniaxial crystals, using suitable crystal orientation and polarization of the beams, the entire polariton dispersion curve can be observed [8]. We have performed time and space resolved CARS experiments in the uniaxial $LiIO_3$ crystal, where we have investigated the upper and lower ordinary E_1 polariton, on both sides of the highest frequency Reststrahlen band (ω_{TO}=768cm^{-1}, ω_{LO}=843cm^{-1}).

Our group velocity measurements for the ordinary polariton in $LiIO_3$ are in very good agreement with the indirect value obtained from the derivative of the polariton dispersion curve [8]. The frequency dependence of the measured polariton dephasing rate is depicted in Fig.4a, for a fixed crystal temperature of 80K, showing a minimum of the dephasing rate in the vicinity of the Reststrahlen band.

In ordered crystals, polariton dephasing is mainly due to anharmonic interactions with many phonons states, and hence the frequency dependence of the polariton dephasing rate mainly reflects the density of states of the accessible many phonons bands. In lowest order, two processes intervene: down-conversion ($\omega_\pi \rightarrow \omega + \omega'$) into two lower energy phonons and up-conversion ($\omega_\pi + \omega \rightarrow \omega'$) due to scattering by thermal phonons. We have estimated the number of channels for these two mechanisms in $LiIO_3$ using the frequency of the zero wave-vector phonons [9], energy conservation and symmetry requirements. The results are plotted at the bottom of Fig.4a, and show a good correlation between the number of accessible states and the measured polariton damping rate. Below the Reststrahlen band both down and up conversion processes are possible. The former correspond to polariton splitting into two phonons of comparable energy, and the latter to polariton conversion into the ω_{TO}-ω_{LO} frequency region with absorption of a thermal lattice phonon (~100cm^{-1}). As the

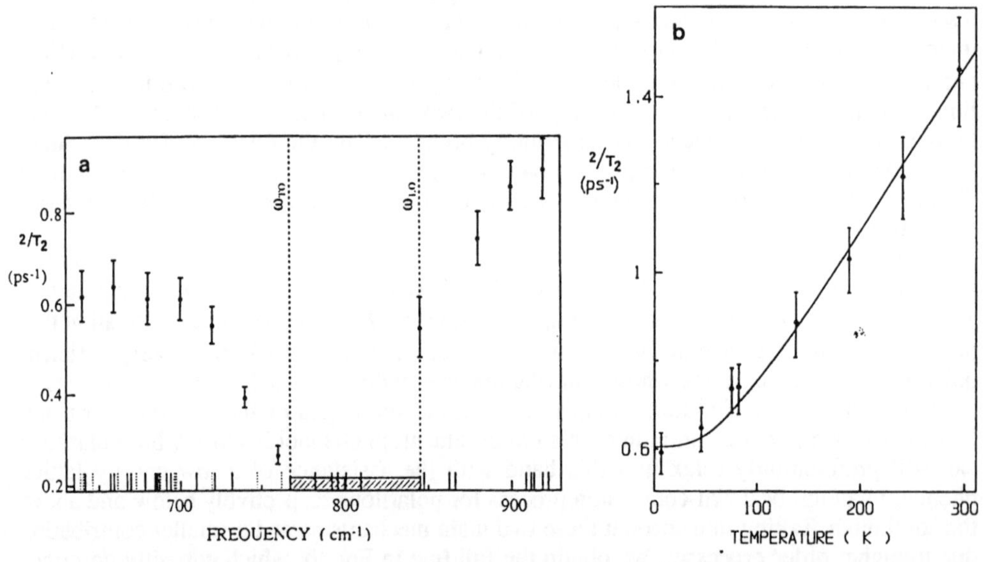

Fig.4(a): Polariton dephasing rate as a function of frequency in $LiIO_3$ at T=80K. The full and dashed lines indicate respectively the allowed down and up conversion processes. (b) Temperature dependence of the 880 cm^{-1} polariton dephasing rate (full line is calculated).

Reststrahlen band under investigation corresponds to the highest one phonon frequency region in $LiIO_3$, up-conversion is impossible for polaritons above the longitudinal mode, so that their dominant relaxation channel is down conversion into the ω_{TO}-ω_{LO} region with emission of a lattice phonon. Although the physical processes are different, it is interesting to note that the ω_{TO}-ω_{LO} frequency region plays a role similar to the two-phonon band in the case of polariton Fermi resonance (see above), acting as a reservoir of accessible states for phonon assisted up- or down-conversion of the polariton.

As for the polariton Fermi resonance in NH_4Cl, the temperature dependence of the damping rate is imposed by the previous interpretation of polariton relaxation, which then can be tested by temperature dependent measurements. This is exemplified in Fig.4b for the 880cm^{-1} polariton. For this particular polariton frequency the dephasing rate can be written

$$\Gamma(880) = \gamma(880)[1+n(100)+n(780)]$$

where γ is a frequency dependent average anharmonic coupling factor for the down-conversion process, and $n(\omega)$ the Bose factor for the phonon of frequency ω. As γ is fixed by low temperature measurements there is no real fitting parameter in this expression. Nevertheless, very good agreement between predicted (full line of Fig.4b) and measured temperature dependence of the dephasing rate is obtained. Similar agreement is obtained over the whole frequency range studied, on both sides of the Reststrahlen band comfirming our interpretation of polariton relaxation.

5. Conclusion

We have demonstrated that polariton dynamics can be directly investigated in cubic and noncubic noncentrosymmetric crystals using a time and space resolved CARS technique, which gives access to polariton relaxation mechanisms in a wide variety of systems. Our results clearly show that this technique constitutes a powerful tool for the continuous probing of crystal anharmonicity as a function of frequency, which, together with temperature dependent measurements, opens up new possibilities for the investigation of anharmonic processes and phonon relaxation in solids.

Furthermore, this method allows some fundamental problems concerning polaritons to be addressed, such as polaritons scattering by crystal disorder or defects, or polariton reflection or transmission at an interface. Similar methods could be applied to the investigation of other types of propagating collective excitation of solids, and in particular to surface phonon-polaritons.

References

[1] Y.Masumoto,S.Shionoya and T.Takagahara, Phys.Rev.Lett. **51**,923 (1983).
[2] G.M.Gale,F.Vallée and C.Flytzanis, Phys.Rev.Lett. **57**,1867 (1986).
[3] F.Vallée,G.M.Gale and C.Flytzanis, Phys.Rev.Lett. **61**,2102 (1988).
[4] D.M.Auston and M.C.Nuss, IEEE J.Quantum Electron. **24**,184 (1988).
[5] M.van Exter and A.Lagendijk, Phys.Rev.Lett. **60**,49 (1988).
[6] Y.N.Polivanov, Sov.Phys.Usp. **21**,805 (1978).
[7] V.S.Gorelik,O.P.Maximov,G.G.Mitin,M.M.Sushchinskii, Sol.St.Com. **21**,615 (1977).
[8] L.A.Kulevsky,Y.N.Polivanov and S.N.Poluektov, J.Ram.Spectr. 3,239 (1975).
[9] F.E.A.Melo and F.Cerdeira, Phys.Rev. **B26**,720 (1982) and references therein.

Part IV

Applications to Solid-State Physics

Femtosecond Studies of Nonlinear Optical Processes in Wide-Gap II-VI Semiconductors

J. Puls[1], *F. Henneberger*[1], *W. Rudolph*[2], *and D. Lab*[2]

[1]Humboldt-Universität zu Berlin, Sektion Physik, Bereich Halbleiteroptik, Invalidenstr. 110, DDR-1040 Berlin, German Dem. Rep.

[2]Sektion Physik, Friedrich-Schiller-Universität Jena, Max-Wien-Platz 1, DDR-6900 Jena, German Dem. Rep.

Abstract : We report on a study of various nonlinear optical processes in wide-gap II-VI semiconductors on the fs time scale. For above-gap excitation of CdSe we have studied the state- and bandfilling kinetics in the high-density carrier system. For below-gap excitation of CdS_xSe_{1-x} we have directly observed and resolved in time the gap-shrinkage . By light-induced grating measurements we were able to deduce the diffusion constant for a highly degenerate electron-hole plasma.

1. Introduction

The linear optical properties of wide-gap semiconductors are dominated by the electron-hole Coulomb interaction. It leads to discrete exciton resonances below the gap and enhanced continuum transitions above. When the optical power is increased, the density of electron-hole pairs eventually becomes so large that many-body effects are of importance [1], which cause a shifting of the various energy levels. This effect is dramatic for the continuum states and results in a strong shrinkage of the energy gap. In contrast, the neutral exciton is scarcely affected by the surrounding carriers, so that the exciton levels stay practically constant with rising excitation. As a consequence, the band-edge crosses the lowest exciton level at a certain density and, above that, excitons do not exist (Mott transition). At the same time, the exciton oscillator strength is decreased by screening, exchange, and phase space occupation. In addition to these nonlinear optical changes due to the vanishing of the exciton resonances, there appears nonlinearity related to bandfilling in the free carrier plasma. The relative contribution of these two processes to the total nonlinear response depends on whether the onset of bandfilling is below or above the Mott transition. For the wide-gap II-VI-compounds one has specifically $n_{BF} \approx 10\ n_{Mott}$. This order-of-magnitude difference leads to a clear separation of the mechanisms: At low intensities the nonlinearity is exciton-related and at higher intensities, above the Mott transition, the nonlinear response is dominated by state- and bandfilling.

Previous studies have revealed very large nonlinear absorption [2] and refraction [3] of the visible band-gap II-VI compounds at room-temperature based on the above-discussed exciton process (for a recent summary see [4]). They have enabled demonstrations of optical bistability and self-diffraction using the mW-output of the Ar-ion laser [5]. Excite-and-probe measurements on the ps time scale [6] have shown band-to-band recombination controlled recovery of the nonlinearity within ≈200 ps. However, switch-on occurred instantaneously with the pump pulse of 6 ps. Therefore, a study of the switch-on transients and the physical processes behind them requires fs pulses.

2. Experimental

For generation of the fs pulses a CPM ring laser pumped by an Ar-ion laser is used. Single pulse selection and amplification is made by a four-stage amplifier pumped by an excimer

Springer Proceedings in Physics, Vol. 49 Ultrafast Phenomena in Spectroscopy
Editors: E. Klose · B. Wilhelmi

laser [7]. The amplified pulses are chirp compensated by use of a double path arrangement through two glass prisms. The pulse width is measured by a second-order correlation technique to be 115 fs. For the excite-and-probe measurements the incoming light is divided into two parts, one for pumping and the other, attenuated, for probing. The delay of the probing pulse relative to the pump is controlled by a computer, which also monitors the pulse energy measured by a reference diode. Only pulses with an energy fluctuation <5% are accepted. In addition to the transmitted probe light the first-order diffraction signal of the pump is detected via a third, properly placed diode. Thus far, the device makes available only a limited number of wavelengths between 610 and 618 nm. However, as demonstrated below, this is already sufficient to get some useful information on various physical processes.

The samples studied are platelet-like single or mixed crystals grown from the vapour phase. They are of very good optical quality and the c-axis lies in the platelet plane. All measurements are made at room-temperature. The polarization of the pump pulse is always along the c-axis. The excitation spot is about 250 μm in diameter and that of the probe, placed in the centre of the pump, 200 μm.

2. Above-Gap Excitation of CdSe: Carrier Relaxation

The band-edge of CdSe at room-temperature is at E_g =1.770 eV. Thus, the excitation is here clearly above-gap with an excess energy of 235 meV relative to the low-density edge divided by 6:1 to electrons and holes. The concept of the experiment rests on the specific polarization selection rules for the uniaxial II-VI compounds as sketched in Fig. 1. As the polarization of the pump beam is chosen along the crystal c-axis, it creates only B-holes. For probe polarization likewise along c, we test the occupation of these initially populated states. However, for probing perpendicular to c we check in addition the occupation of 20 meV higher-lying electron states and holes in the A-band, respectively. The occupation numbers $F_B = f_{e1} + f_{hB}$ and $F_A = f_{e2} + f_{hA}$ of these different levels are related to the absorption $\alpha_{\parallel}$ and $\alpha_{\perp}$ of the sample

$$\alpha_{\parallel} = \alpha_{B,\parallel}(1-F_B), \quad \alpha_{\perp} = \alpha_{B,\perp}(1-F_B) + \alpha_A(1-F_A) \qquad (1)$$

with $\alpha_{B,\parallel}$: α_A= 2.60:2.25 and $\alpha_{B,\perp}$: α_A = 0.65:2.25 for CdSe. This allows for a direct calculation of the mean values of F_A and F_B across the sample from the experimental absorption change [8] the result of which is shown in Fig. 2. The increase of F_A occurs clearly delayed with respect to F_B, which rises together with the pump. This delay of ≈70 fs is a measure of the

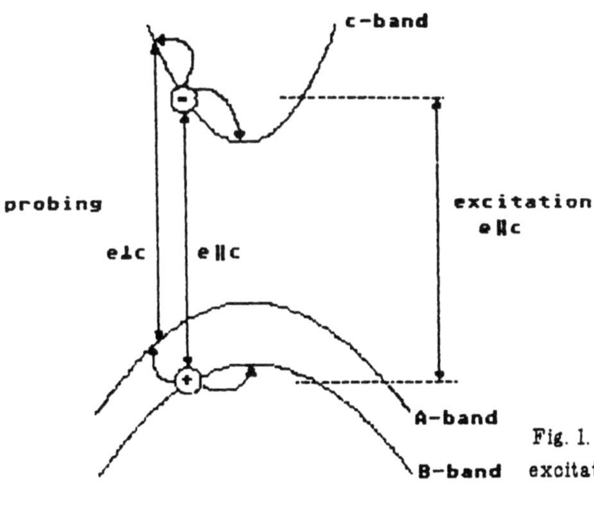

Fig. 1. Schematic for above-gap excitation of CdSe

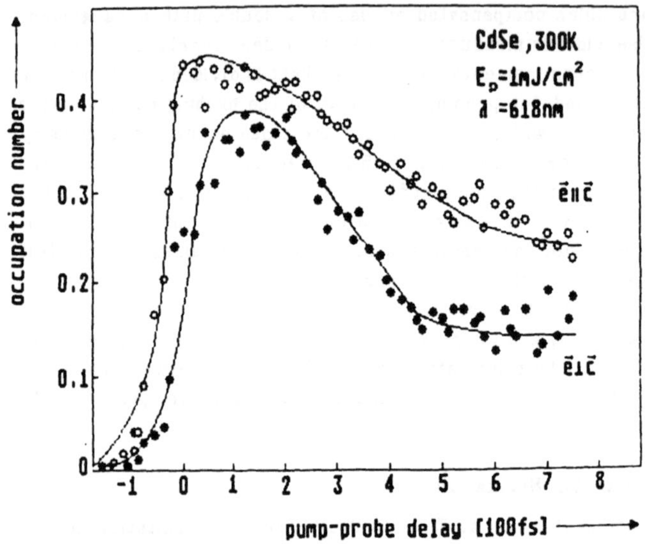

Fig. 2. Occupation numbers F_A and F_B for CdSe derived from excite-and-probe measurements with probe polarization perpendicular and parallel to the c-axis, respectively. ($\hbar\omega_{pump}$ = 2.0 eV)

arrival of the carriers at the formerly empty states. We believe that the main contribution stems from the B- to A-hole conversion, which involves spin change, while the electrons spread out much faster. At later delays both occupations behave quite similarly and recover down to a certain value, which is caused by thermalization of the carrier gas at an effective temperature. For the mean carrier density injected by the pump of 2×10^{19} cm^{-3} and reasonable carrier temperatures the Fermi-levels (gap-shrinking included) of both electrons and holes are clearly below the probe energies. From the occupation numbers at latest delays we estimate a carrier temperature of about 400 K.

Summarizing, the data of CdSe evidence a very fast redistribution of the carriers over a relatively large energy range within a few 10 fs after above-gap excitation and formation of an early stage of a quasi-equilibrium within 1 ps.

3. Below-Gap Excitation of CdS_xSe_{1-x} Mixed Crystals : Gap-Shrinkage

Using CdS_xSe_{1-x} mixed crystals one can tune the band-gap relative to the laser wavelength by composition. Previously [8], we have studied below-gap, exciton-resonant excitation on $CdS_{0.32}Se_{0.68}$. These measurements have revealed that the initially created excitons dissociate very fast (≈ 30 fs) into free carriers by LO-phonon absorption. Fig. 3 shows a study for a similar situation but at higher pump flux. The peak around zero-delay is a coherent artifact: diffraction of the pump light in the direction of the probe at the grating produced by pump and probe together. We have proved this by measuring the first-order diffraction signal of the pump which is also shown for comparison in Fig. 3. The agreement is obvious. Subtracting the artifact we find a considerably delayed switch-on of the transmission followed by partial recovery down to a plateau. Apparently, some of the carriers are scattered back to the energy of their creation and, subsequently, leave these states again. Because we excite below the gap, the carriers are created by LO-phonon absorption and are initially distributed somewhere above the photon energy . These carriers cause the gap to shrink. The gap-shrinkage is related to the total numbers of carriers and is almost independent of their actual distribution. Thus, after the pump pulse the band-edge lies below the photon energy and the carriers start to relax.

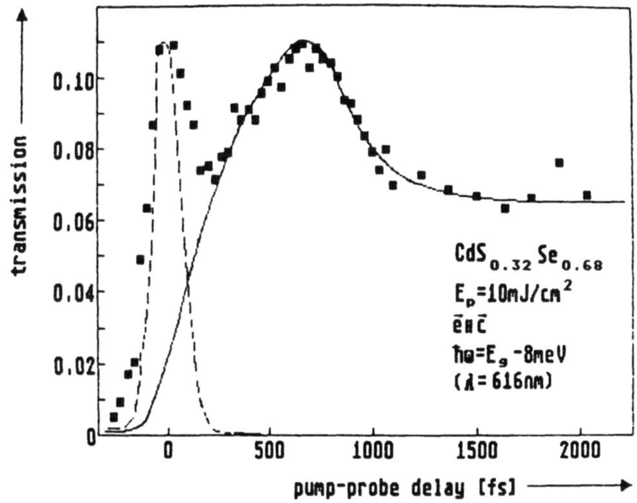

Fig. 3. Results of excite-and-probe measurements on a $CdS_{0.32}Se_{0.68}$ single crystal at excitation 8 meV below the gap, near the exciton resonances, (d = 3 μm). Dashed lines: First-order diffraction signal in arbitrary units

On the way down to the band bottoms the carriers from above the photon energy cause a transient transmission increase and, finally, in accordance with data for pure CdSe, a quasi-equilibrium with some temperature and chemical potential is established within 1 ps.

In the above experiment the excitation is only slightly below the gap and, thus, the band-edge crosses the laser-frequency already during the rise of the pump. That is, a large number of carriers are directly created from the absorbed photons without LO-phonon assistance. This can be changed by exciting further below the gap, so that the band-edge lies only slightly below the laser frequency after pumping. Then, initially, the states tested by the probe are practically not occupied and one should directly see the gap-shift by an increased density-of-states and the subsequent filling of the states by carriers. We have performed such an experiment on $CdS_{0.43}Se_{0.57}$ with 70 meV below-gap excitation. The result is presented in Fig. 4 . Subtracting the artifact again we indeed observe, in contrast to the former measurements, at first increasing absorption and then cross-over to absorption bleaching. Consistent with the data for slightly below-gap excitation a plateau appears around 1 ps indicating quasi-relaxation again. Therefore, we have directly measured and resolved in time the gap-shrinkage in an electron-hole plasma.

A very recent calculation [9] has treated theoretically below-gap excitation of pure CdS at 300K with 60fs pulses accounting for the coupling to LO-phonons on a microscropic level. Rescaling some parameters this theory can also be applied to the experiment presented above. However, the calculation predicts a predominant response via virtual carriers for excitation below the 1-LO threshold. Indeed , we observe a clear leading contribution from long-living pairs , although we excite more than two LO-phonon energies below the gap. Apparently, many-phonon processes are of importance, which are neglected thus far in the theory.

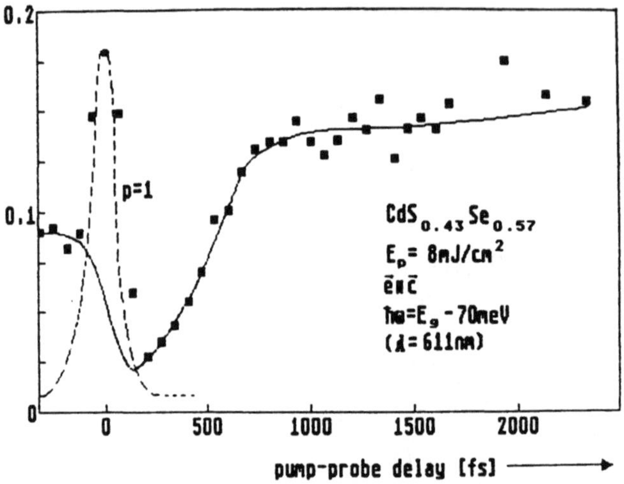

Fig. 4. Transmission vs delay-time of a $CdS_{0.43}Se_{0.57}$ crystal from excite-and-probe measurements at excitation 70meV below the gap. (d = 12 μm). Dashed line: First-order diffraction signal in arbitrary units.

4. Fs Light-Induced Grating: Nonlinear Refraction and Carrier Diffusion

In order to get information on nonlinear refraction and carrier diffusion we have studied a light-induced grating on the fs-time scale. The grating was created by two pump pulses coincident in time and space on the $CdS_{0.43}Se_{0.57}$ sample at an angle resulting in a grating spacing of $\Lambda = 3$ μm . The grating was tested by a third, weak probe pulse. We observe very large diffraction signals both in first- and in second order as seen in Fig. 5. However, the various signals behave very puzzlingly. Utilizing a recursion formula of the Bessel-functions one can easily compute from the zeroth-, first-, and second-order efficiency the change of refraction Δn present in the grating

$$\Delta n = \frac{\lambda}{2\pi d} \; \frac{\sqrt{I_{p=1}}}{\sqrt{I_{p=0}} + \sqrt{I_{p=2}}} \quad . \qquad (2)$$

Knowledge of Δn allows for evaluation of the absorption change by

$$\Delta\alpha d = \ln\,[(T/T_0)\,J_0^2(2\pi\Delta n d/\lambda)] \quad . \qquad (3)$$

These quantities are plotted versus pump-probe delay in Fig. 6 and behave very clearly. No essential difference is seen in the absorption change compared to the excite-and-probe measurements in Fig. 5 : After increasing absorption right after the pump pulses a plateau of bleached absorption occurs at longer delays. This is consistent with the 200 ps carrier recombination time discussed above. In contrast, we find recovery of the refraction change at later delays. This recovery is related to wash-out of the grating by diffusion. Thus, the application of fs-pulses enables us to resolve this very fast, but macroscopic process in time. From the recovery time of 4760 fs we obtain for the diffusion constant of the electron-hole plasma the very precise value $D = \Lambda^2/4\pi^2\tau_D$ = 480 cm^2/s. This value is considerably larger than that found in similar studies with ns-pulses on CdS by varying the grating period [10]. However, in these experiments the electron-hole plasma was Boltzmann-like, whereas in the

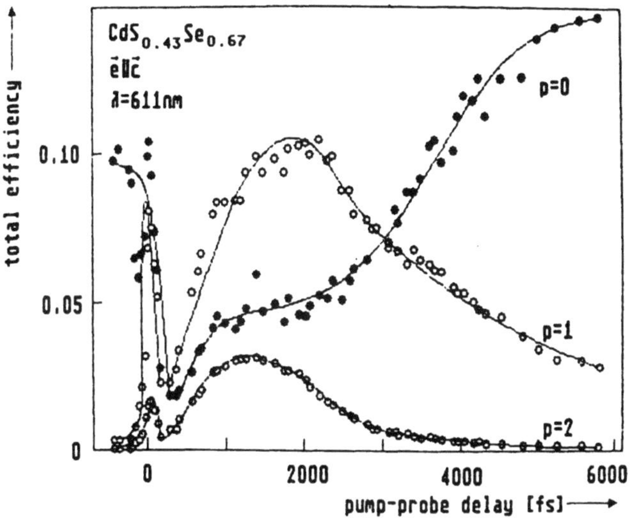

Fig. 5. Intensity of zeroth-, first-, and second-order diffraction signals vs pump-probe delay for the sample in Fig. 4

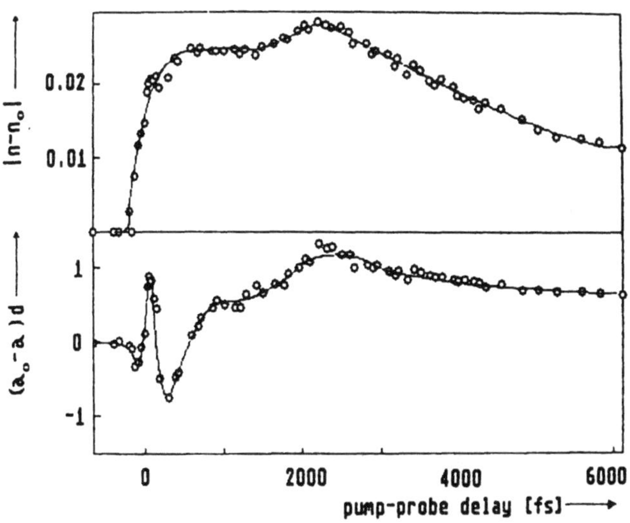

Fig.6. Change of refraction Δn and absorption $\Delta\alpha d$ vs pump-probe delay derived from the data in Fig. 5

present investigation the carrier gas is highly degenerate. As known from low-temperature measurements [10], the diffusion is strongly enhanced in this situation.

The refraction change seen in Fig. 6 - at maximum about 1% - is fairly large and suggests various applications, e.g. in fs optical phase conjugation.

6. Conclusion

We have observed very complex switch-on dynamics of the nonlinearities in the II-VI's on the fs time scale which strongly depend on the kind of excitation. We have directly resolved in time various basic processes as the relaxation of the carriers, the gap-shrinkage, and the carrier diffusion.

References

1. R. Zimmermann: Phys. Stat. Solidi (b) **146**, 371 (1988)
   H. Haug, S. Schmitt-Rink: Progr. Quant.Electron **9**,3 (1984)

2. F.Henneberger, J.Puls, H.Rossmann: J.Luminescence **30**,204 (1985)
   F.Henneberger, J.Puls , Ch.Spiegelberg, in: *Optical Bistability III*, ed. by H.M.Gibbs, P.Mandel, N.Peyghambarian , S.D.Smith, Springer Proc. Phys. 8 (Springer,Berlin,1986)
   F.Henneberger, J.Puls, H.Rossmann, Ch.Spiegelberg, M.Kretzschmar, I.Haddad:Physica Scripta **T13**, 195 (1986)

3. F.Henneberger, J.Puls, H.Rossmann, M.Kretzschmar, Ch. Spiegelberg, A. Schülzgen: J. Physique **C2**, 49 (1988)
   F .Henneberger, U.Woggon, J.Puls, Ch.Spiegelberg: Appl.Phys.B **46** ,535 (1988)

4. F. Henneberger, J. Puls, H. Rossmann,U . Woggon, S. Freundt, Ch. Spiegelberg, A. Schülzgen, Proc. VIth Int'l Conf. II-VI Compounds , Berlin (West) 1989, to be published in J. Crystal Growth

5. Ch.Spiegelberg, M.Kretzschmar, J.Puls, F.Henneberger:Phys.Stat. Solidi (b) **150**, 769 (1988)

6. J.Puls, F.Henneberger, F.Fink, in: Proc.Int'l.Symp. Ultrafast Phenomena in Spectroscopy, Minsk (1983) p. 339
   J.Puls, F.Henneberger, F.Fink, J. Jüpner, in: Proc.Int'l.Symp. Ultrafast Phenomena in Spectroscopy, Reinhardsbrunn (1985) p. 108

7. W. Dietl, E. Döpel, G. Marovski, V. Petrov, C. Rempel, W. Rudolph, F.P. Schäfer, B. Wilhelmi: Appl. Phys. B **46**,183(1988)

8. J.Puls, W.Rudolph, F.Henneberger, D.Lab:Phys. Stat. Solidi (b)**150** ,419 (1988)

9. R. Zimmermann and M. Hartmann:Proc. VIth Int'l Conf. II-VI Compounds, Berlin (West) 1989, to be published in J. Crystal Growth

10. C. Weber, U. Becker, R. Renner, C. Klingshirn, Z. Phys. B **72**, 379(1988)

Ultrafast IR Probing of Highly Excited Semiconductors

E. Kuokštis[1], *J. Cruz*[2], *and H.M. van Driel*[2]

[1]Laboratory for Nonlinear Spectroscopy of Condensed Media,
Vilnius University, Naugarduko 24, 232006 Vilnius, Lithuania
[2]Ontario Laser and Lightwave Research Center, University of Toronto,
Toronto, Ontario, Canada, M5S 1A7

Abstract. IR picosecond time-resolved reflectivity and transmission of highly photoexcited semi-insulating Si, GaSe, CdSe, $CdS_{1-x}Se_x$, and ZnTe single crystals have been analyzed taking into account free nonequilibrium carriers, interband transitions and interference effects in the theoretical model.

1. Introduction

Picosecond laser pulses can generate very high concentrations ($>10^{20}cm^{-3}$) of nonequilibrium carriers (NC) in semiconductors. This high density electron-hole plasma is of great interest due to a number of new fundamental effects. The study of time-resolved IR optical properties of highly excited semiconductors provides wide information about fast relaxation processes of such high electron-hole plasma.

NC change the complex dielectric constant, which is responsible for optical properties of the material and which, so far, has started to be analyzed in Ge, Si and GaAs [1,2]. Generally, the real and imaginary parts of the dielectric constant $\tilde{\varepsilon}=\varepsilon_1+i\varepsilon_2$ are the functions of probe frequency ω, as well as of a number of free NC and band structure parameters. They can be expressed as [3]

$$\varepsilon_1 = \varepsilon_\infty - 4\pi Ne^2\left[\frac{1}{m^*}\,\frac{\tau_e^2}{1+\omega^2\tau_e^2} + \frac{1}{m^*}\,\frac{\tau_h^2}{1+\omega^2\tau_h^2}\right] + 4\pi\chi\ , \qquad (1)$$

$$\varepsilon_2 = \frac{4\pi Ne^2}{\omega}\left[\frac{1}{m_e^*}\,\frac{\tau_e}{1+\omega^2\tau_e^2} + \frac{1}{m_h^*}\,\frac{\tau_h}{1+\omega^2\tau_h^2}\right] + \frac{4\pi\sigma}{\omega}\ . \qquad (2)$$

Note that the second term in (1) and the first one in (2) give the usual Drude expression result, while the last terms in (1,2) are responsible for interband transitions. Reflectivity and transmission of the crystal can be calculated once $\tilde{\varepsilon}$ is known [4].

2. Experimental

The bulk Si, CdSe, $CdS_{1-x}Se_x$ and ZnTe single crystalline samples were prepared as 0.5-2 mm thick plates with plane polished

Springer Proceedings in Physics, Vol. 49 **Ultrafast Phenomena in Spectroscopy**
Editors: E. Klose · B. Wilhelmi

surfaces, while those of GaSe were freshly cleaved as 10-20 μm Fabry-Perot resonators. An actively-passively mode-locked Nd:YAG laser (λ=1.06 μm) combined with a four-pass amplifier was used as a source of picosecond light pulses. The crystals were excited by the 2nd harmonic of the main laser frequency pulse (λ=0.53 μm, τ_p= 40 ps) with maximum energy per pulse of 1 mJ. Probing was carried out by IR radiation (τ_{pr}<40 ps), generated in a stimulated Raman scattering cell with Hydrogen (λ_{pr}=1.9 μm) or Methane (λ_{pr}=2.8 μm) by a portion of the 1.06 μm laser pulse.

3. Results and Discussion

Typical experimentally measured reflectivity kinetics of the crystals under investigation at different excitations are illustrated in Figs. 1-4. In all cases it is possible to resolve fast and slow components. The complicated dependence of the reflectivity in Si and GaSe can be explained by the free NC Drude model with carrier concentration exceeding that of the plasmon resonance for the probe radiation. The first minimum of the kinetics under high excitation is due to the passing of the plasmon resonance density of the NC during excitation whereas the second one corresponds to the same density during the relaxation process. The maximum evaluated NC density in Si and GaSe could be $\sim 10^{21}$ cm^{-3}, whereas the plasmon resonance density for the probe frequency was 2.6×10^{20} and 6.1×10^{20} cm^{-3}, correspondingly. The solid lines in Fig.2 illustrate theoretical kinetics obtained from the model based on ideas of [5] which included linear and Auger recombination, as well as interference effects in thin films and convolution of the probe pulse.

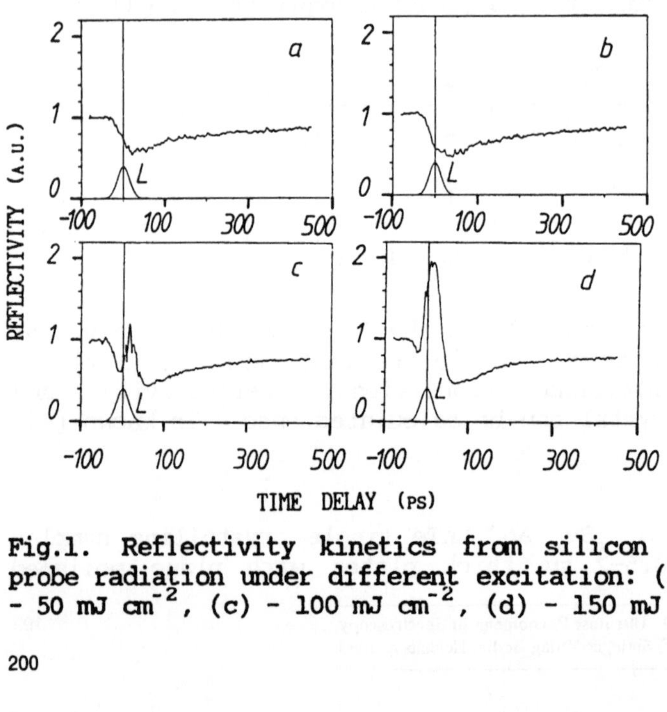

Fig.1. Reflectivity kinetics from silicon crystal for 2.8 μm probe radiation under different excitation: (a) - 30 mJ cm^{-2}, (b) - 50 mJ cm^{-2}, (c) - 100 mJ cm^{-2}, (d) - 150 mJ cm^{-2}.

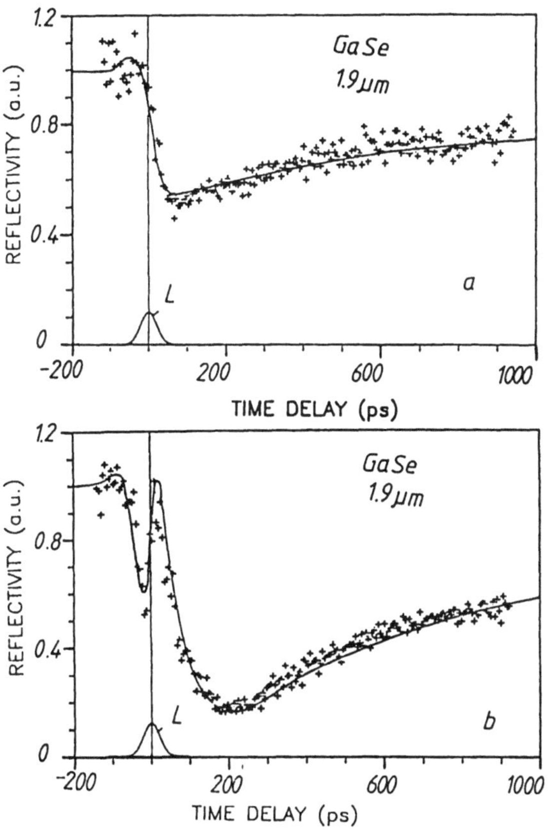

Fig. 2. Experimental (crosses) and theoretical (solid lines) kinetics of 1.9 μm probe beam reflectivity from 10 μm thick GaSe sample at different excitation: (a) - 120 $mJcm^{-2}$, (b) - 300 $mJcm^{-2}$.

The plasmon frequency wasn't reached in A^2B^6 crystals, however, maximum NC concentration was close enough to it. Consequently, the fast decrease of reflectivity in $CdS_{0.6}Se_{0.4}$, ZnTe (Fig.3,4) and CdSe (not shown) is caused also by free NC according to the Drude model. The slow reflectivity recovery which is clearly observed in $CdS_{1-x}Se_x$ crystals containing defects can be explained by taking into account defect/impurity levels, which change basically the imaginary part of optical constants due to the change of the population of these defects. The model explains the positive change of reflectivity following the excitation pulse (Fig.3,a). The participation of the defect/impurity center in the process of probe radiation absorption is improved by slow parts of transmission measured simultaneously in highly excited crystals (Fig. 3b,4b). The possible schema of the quantum transitions taking into account defect/impurity donor levels is shown in the

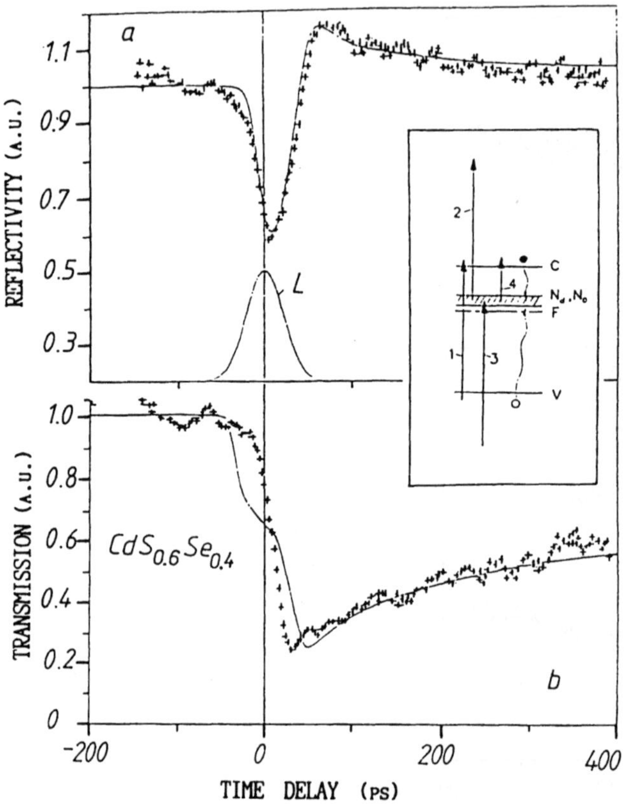

Fig. 3. Reflectivity (a) and transmission (b) kinetics for 2.8 μm probe beam in $CdS_{0.6}Se_{0.4}$ crystals under excitation of 120 $mJcm^{-2}$. Crosses correspond to experimental data, solid lines are theoretical curves (for parameters see text). The possible scheme of quantum transitions is shown in the inset.

inset in Fig.3. The theoretical solid lines in Fig.3 illustrate calculation for $CdS_{0.6}Se_{0.4}$ using the following parameters: NC generation rate $10^{31}cm^{-3}s^{-1}$; center capture coefficient for electrons $4x10^{-9}cm^3s^{-1}$ and holes $8x10^{-10}cm^3s^{-1}$; center capture cross-section for exciting radiation $6x10^{-14}cm^2$ (transition 3 in Fig.3) and $4x10^{-13}cm^2$ (transition 2), and for probe radiation $1.4x10^{-14}cm^2$ (transition 4); defect/impurity center density $3x10^{18}cm^{-3}$.

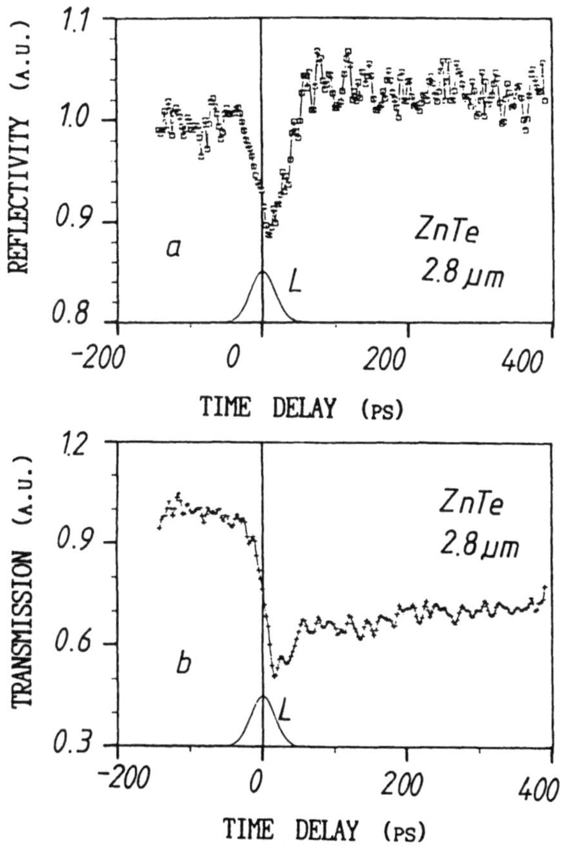

Fig. 4. Experimental data of time-resolved reflectivity (a) and transmission (b) for 2.8 µm probe beam in ZnTe crystals under excitation of 100 mJ cm^{-2} .

References

[1] H.M. van Driel, L.-A.Lompre, and N.Bloembergen, Appl.Phys. Lett. **44**, 285 (1984).
[2] G.N.Galkin, L.M.Blinov, V.S.Vavilov, and A.G.Solomatin, Pisma Zh.Eksp.Teor.Fiz. 7, 69 (1968).
[3] H.M. van Driel, Phys.Rev. 35, 8166 (1987).
[4] M.Born and E.Wolf, Principles of Optics (Pergamon Press, 1984), p. 630.
[5] J.C.Cruz, H.M. van Driel, and E.Kuokštis, Phys.Rev. (1989), to be published.

Picosecond Spectroscopy of Nonthermalised Electron–Hole Plasma in Highly Excited $CdS_{1-x}Se_x$ Crystals

R. Baltramiejūnas [1], *S. Juršėnas* [2], *E. Kuokštis* [2], *and A. Žukauskas* [2]

[1]Institute of Physics, Academy of Sciences of Lithuania, K. Požėlos 54, 232600 Vilnius, Lithuania

[2]Laboratory for Nonlinear Spectroscopy of Condensed Media, Vilnius University, Naugarduko 24, 232006 Vilnius, Lithuania

Abstract. The dynamics of nonthermalised nonequilibrium electron-hole plasma in highly photoexcited $CdS_{1-x}Se_x$ crystals is studied from time-resolved luminescence measurements.

1. Introduction

The application of mixed semiconductor crystals for development of quantum electronics devices requires investigation of the processes occurring at high excitation densities, particularly under conditions of formation of electron-hole plasma (EHP). Dense EHP in direct gap semiconductors is quite well investigated [1-3]. Luminescence properties of EHP are directly related to the temperature determined by energy relaxation of nonequilibrium carriers (NC). However, this fact is very often not taken into consideration. It should be noted that in highly excited crystals of direct gap semiconductors the values of the relaxation time of excess temperature and that of NC concentration are of the same order [4,5]. Under such conditions, the investigation of dynamics of radiative recombination requires a complex approach [6,7]. In this work such an approach has been applied to solid solutions of the A^2B^6 type, where the radiative characteristics of dense EHP have not been studied enough.

2. Experiment

Single crystals of $CdS_{1-x}Se_x$ grown by a vapor-phase technique have been investigated. The sample surface was excited by the third harmonic pulses of a $YAG{:}Nd^{3+}$ laser with passive mode-locking ($h\nu$=3.5 eV, pulse duration 25 ps) for x=0.23, and the second harmonic ($h\nu$=2.33 eV) for x=0.63. The temporal resolution of the set-up (~30 ps) was provided by means of an optical Kerr shutter. The experiment has been carried out at room temperature.

3. Results and Discussion

The observed luminescence band (Fig.1) should be treated as radiation of the dense EHP, since the experimental conditions (high density of excitation and high temperature) make the excitonic and localization effects unlikely [1-3]. The spectral

Springer Proceedings in Physics, Vol. 49 Ultrafast Phenomena in Spectroscopy
Editors: E. Klose · B. Wilhelmi

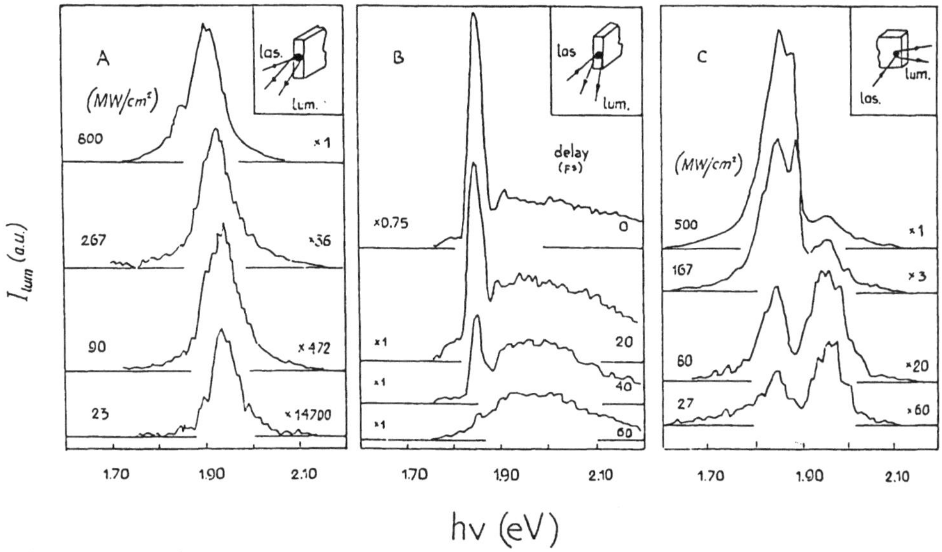

Fig. 1. Luminescence spectra of $CdS_{0.37}Se_{0.63}$ with manifestation of stimulated emission: A - time-integrated spectra for different excitation intensities, B - time-resolved luminescence spectra for I=800 $MWcm^{-2}$, C - time-integrated spectra for different excitation intensities with another excitation geometry.

and temporal characteristics of the luminescence were used to follow simultaneously the kinetics of the luminescence intensity and of the EHP effective temperature. The most interesting fact is that the recombination process takes place in the absence of thermodynamic equilibrium between the EHP and the crystal lattice. The EHP cooling is hindered by the accumulation of nonequilibrium phonons of a few generations and the corresponding energy relaxation time is (120±10) ps. The cooling rate substantially influences the kinetics of the recombination. This fact should be taken into account when determining the coefficient of the bimolecular recombination γ and the lifetime τ of the NC from the temporal evolution of the luminescence intensity. In the present work the values of γ and τ were obtained from the kinetic equation for a quantity of electron-hole pairs taking into account the experimental dependence of the EHP effective temperature (see Fig. 2).

Fig. 3 illustrates the manifestation of the stimulated emission under certain experimental conditions. The stimulation is seen to occur only synchronously with the excitation pulse. In this case only the start value of the EHP density is changed; the following recombination process can be treated in the same manner as above.

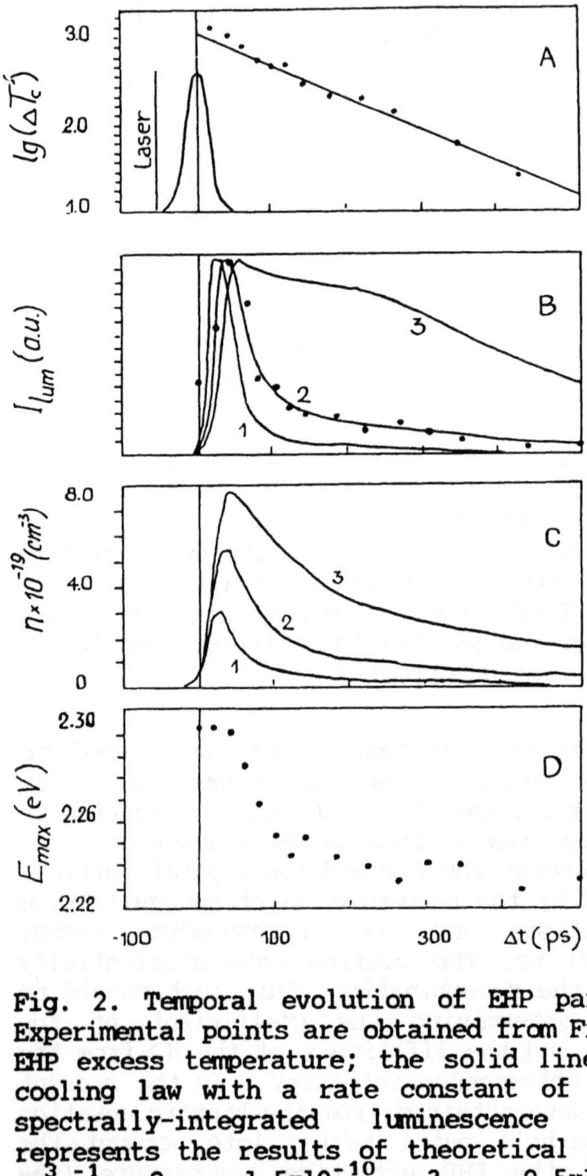

Fig. 2. Temporal evolution of EHP parameters in $CdS_{0.77}Se_{0.23}$. Experimental points are obtained from Fig. 1 B, C. A - kinetics of EHP excess temperature; the solid line exhibits the exponential cooling law with a rate constant of 120 ps. B - evolution of spectrally-integrated luminescence intensity; solid line represents the results of theoretical calculation: 1 - γ=5x10^{-10} cm^{3}s^{-1}, 2 - 1.5x10^{-10}, 3 - 0.5x10^{-10}. C - EHP density corresponding to the luminescence intensity depicted in Fig. 2B. D - kinetics of EHP luminescence spectra peak position.

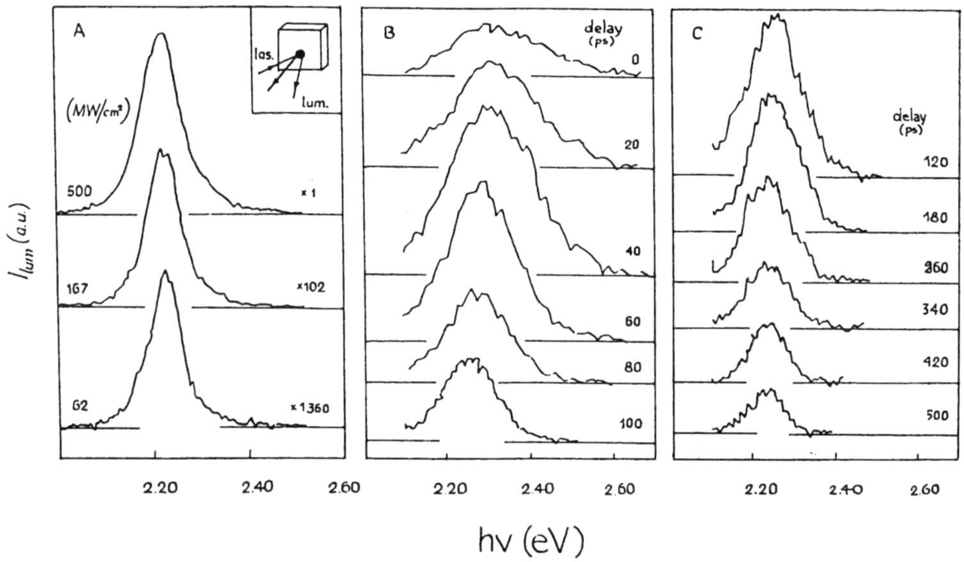

Fig. 3. Spontaneous luminescence spectra of $CdS_{0.77}Se_{0.23}$ single crystal: A - time-integrated spectra for different excitation intensities; B, C - time-resolved luminescence spectra for I=500 $MWcm^{-2}$, sensitivity factors are 1 (B) and 2.5 (C).

References

[1] V.G.Lysenko, V.I.Revenko, T.G.Tratas, and V.B.Timofeev, Zh. Eksp.Teor.Fiz. 68, 335 (1984).
[2] T.Amand, J.Collet, J.Phys.Chem.Sol.46, 1053 (1985).
[3] H.Saito, E.O.Gobel, Phys.Rev. 31, 2360 (1985).
[4] R.Baltramiejūnas, A.Žukauskas, V.Latinis, S.Juršėnas, Pisma Zh.Eksp.Teor.Fiz. 46, 67 (1987).
[5] R.Baltramiejūnas,A.Žukauskas, Phys.Stat.Sol.(b) 49, 337(1988).
[6] R.Baltramiejūnas, S.Juršėnas, A.Žukauskas, E.Kuokštis, V.Latinis, Fiz.Techn.Polupr. 23, 565 (1989).
[7] R.Baltramiejūnas, S.Juršėnas, E.Kuokštis, A.Žukauskas, O.M.Makienko, Zh.Prikl.Spektroskopii, 51, 443 (1989).

Transient Energy Transfer in ps Two-Photon Absorption Experiments in CdTe at Room Temperature

I. Rückmann [1], *J. Kornack* [1], *J. Kolenda* [2], *and M. Petrauskas* [2]

[1]Humboldt University, Department of Physics, Invalidenstr. 110, DDR-1040 Berlin, German Dem. Rep.

[2]Vilnius State University, Department of Physics, Sauletekio al. 9, 232054 Vilnius, Lithuania

Abstract. One-beam transmission experiments and two-beam coupling experiments with 35 ps pulses at 1.06 μm have been performed in CdTe and CdTe:Ge to investigate (i) the nonlinear (two-photon, two-step, and free-carrier) absorption processes, (ii) their kinetics, and (iii) the transient energy transfer (TEF) from stronger pump to probe due to coherent coupling, which is observed around zero delay. The competition between two-photon absorption and TEF is demonstrated by the intensity dependences of the probe-beam transmittivity measured and calculated at small delays. The influence of pump-to-probe intensity ratio on TEF is discussed.

1. Introduction

Time-resolved measurements of bleaching or of induced absorption require an experimental setup providing pump and probe beams with variable delay. Usually, if these beams are noncollinear and are split off the same laser beam the experimental result is not free of coherent interaction [1], occurring if both pulses overlap in time. This mostly undesired effect results from a diffraction of the stronger pump beam into the direction of the transmitted probe beam due to a light-induced grating with a small (but time dependent) contrast ratio generated by pump and probe if both overlap in time. On the other hand, the coherent coupling of the two beams – which has been called transient energy transfer (TEF) [2] – can be used to realize nonlinear optical amplifiers [3,4].

In our paper we present measurements of the nonlinear absorption and TEF in CdTe and CdTe:Ge at room temperature using 1.06 μm single pulses (35 ps FWHM) from a passive $m-l$ YAG laser exciting the volume of the sample via two-photon absorption (TPA) and additionally in CdTe:Ge via one-photon two-step absorption (TSA). Ge represents a deep level in the forbidden gap of CdTe with an activation energy of about 0.7 eV [5]. From the fitting of the one-beam transmission experiment the parameters of nonlinear absorption are obtained as presented in Sect. 2. In Sect. 3 results are given of a probe-and-pump experiment carried out to investigate the time behaviour of the laser-induced (TPA, TSA, and free-carrier) absorption and the TEF. Starting from the parameters obtained in Sect. 2, taking into account the free-carrier lifetime measured in DFWM experiments (see e.g. [6]), and including a transient free-carrier mixed grating generated by probe and pump, we calculate the probe-beam transmittivity versus intensity and time and compare it with the experiment. Using numerical results the influence of pump-to-probe intensity ratio on TEF is discussed.

Springer Proceedings in Physics, Vol. 49 **Ultrafast Phenomena in Spectroscopy**
Editors: E. Klose · B. Wilhelmi

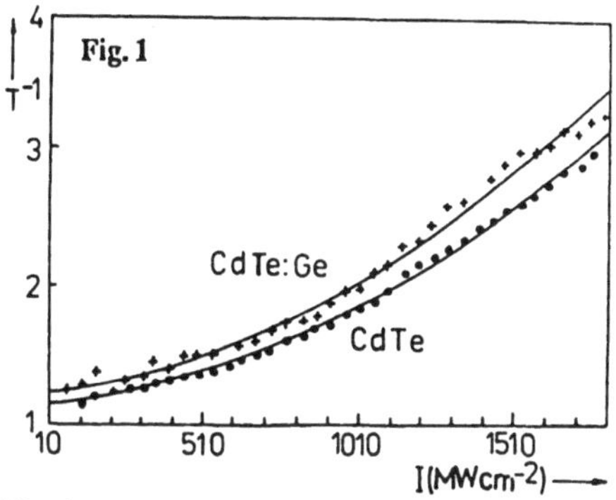

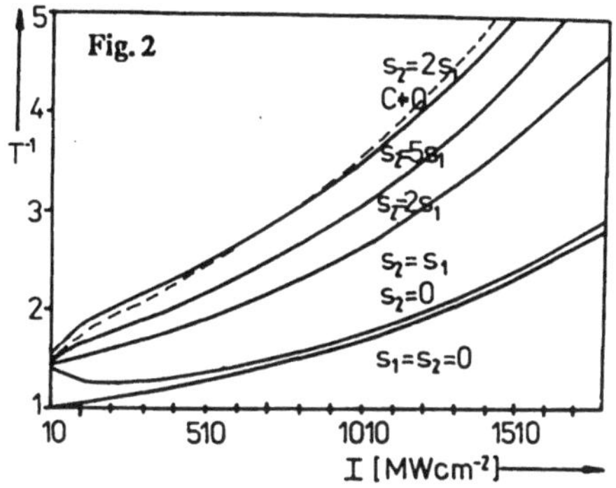

Fig. 1. Measured points of nonlinear absorption (T^{-1}) versus I of CdTe (circles) and CdTe:Ge (crosses) and fitted curves (lines)

Fig. 2. Calculated curves of T^{-1} vs I taking into account TPA, free-carrier absorption and demonstrating the influence of the TSA process at different cross-sections s_1 and s_2 (see text)

2. One-Beam Transmission Experiment

Measurements were carried out on 0.6 mm thick CdTe and CdTe:Ge ($N_{Ge} = 7.9 \times 10^{16}\,cm^{-3}$) single crystals grown from Te-rich melt by THM and having initial transmission values at 1.06 μm of $T_0 = 0.87$ and 0.8, respectively. In Fig. 1 intensity dependences of the measured nonlinear absorption (T^{-1}) are shown for CdTe and CdTe:Ge. The slightly higher values obtained in the case of CdTe:Ge are due to the contribution of the TSA process via the impurity level. To fit the experimental data, the coupled equation system

$$\frac{dI}{dz} = -a(I, f(I), N_{ch}(I))I\,,$$

$$\frac{dN_{ch}}{dt} = \frac{\beta I^2}{2E_L} + \frac{f s_2 N_{Ge} I}{E_L} - C N_{ch} N_{Ge}(1-f) - \frac{N_{ch}}{\tau_{ch}}\,,$$

$$\frac{df}{dt} = \frac{(1-f)s_1 I}{E_L} + C N_{ch}(1-f) - f\left(\frac{s_2 I}{E_L} + \frac{1}{\tau}\right)\,,$$

$$a(I, f(I), N_{ch}(I)) = (1-f)N_{Ge}s_1 + f N_{Ge}s_2 + \beta I + g N_{ch}$$

is solved – where N_{ch} is the free-carrier density generated, f the occupation number of the Ge level, s_1 and s_2 the TSA cross-sections, $CN_{ch}N_{Ge}(1-f)$ the trap rate, a the absorption coefficient, and z the light propagation direction through the sample – to calculate profiles $N_{ch}(z,t)$, $f(z,t)$, and $a(z,t)$ inside the sample after excitation with I at $t = 0$ using a 35 ps $\sin^2$ pulse. The inverse transmission is obtained by spatial and time integration. In Fig. 2 curves of nonlinear absorption are shown calculated at fixed values of $\beta = 0.01$ cm/MW and $g = 1 \times 10^{-17}\,cm^2$, demonstrating the contribution of the TSA process. Only for $s_1 \gg s_2$ does the TSA process become saturated with increasing intensity. The influence of nonzero trapping rate is shown in the case $s_2 = 2s_1$ ($1.6 \times 10^{-16}\,cm^2$). Experimental data were fitted very well, as shown by lines in Fig. 1, using parameters $\beta = 0.008$ cm/MW, $g = 1.2 \times 10^{-17}\,cm^2$ and, additionally in the case of the Ge-doped sample, $s_2 = 0.5s_1$. All other parameters (d, s_1, τ_{ch}) are obtained experimentally.

3. Two-Beam Coupling Experiment

Using noncollinear beams at 1.06 μm for probe and pump we have measured the pump induced change of the probe-beam transmittivity $\Delta T = T(0) - T(I_p)$ versus delay time as shown in Fig. 3 for pump intensities $I_p = 0.5$ and $1.2\,\mathrm{GW/cm^2}$. Positive values of ΔT correspond to an induced absorption, as observed at positive delays. At small negative delays an induced transmission (gain) is found due to TEF from pump to probe. Curve 3 represents the time behaviour of the induced absorption measured in CdTe:Ge. The higher ΔT values at $\Delta t > 50\,$ps result from the contribution of the TSA process. In the region $\Delta t < 50\,$ps curves 2 and 3 are nearly identical (not shown in Fig. 3).

To investigate the observed features in more detail, we have calculated the time behaviour of probe-beam transmission including the coherent coupling between pump and probe. The superposition of the two beams in the sample results in a spatially modulated intensity distribution. In the case of TPA we must consider

$$\begin{aligned} I^2(x,t,\Delta t) &= (I_p + I_z)^2 + 2I_pI_z + 4(I_p + I_z)\sqrt{I_pI_z}\cos\frac{2\pi x}{L} + 2I_pI_z\cos\left(\frac{2\pi x}{L/2}\right) \\ &= I_p^2(t) + I_{\mathrm{mod}}^2(x,t,\Delta t) \\ &= I_p^2(t) + 4(I_p(t) + I_z(t+\Delta t))\sqrt{I_p(t)I_z(t+\Delta t)}\cos(2\pi x/L)\,, \end{aligned}$$

where I_z is the probe-beam intensity and L the grating constant. The third term has cancelled, because only the first diffraction order of the pump beam due to the first grating with period L coincides with the direction of the transmitted probe beam. From I_p^2 an unmodulated free-carrier density is generated, from which the profile of the absorption coefficient and, hence, the transmission signals of probe and pump were calculated in accordance with the calculations described in Sect. 2. The second term is responsible for the generation of the modulated part of the free-carrier density

$$N_{\mathrm{ch,mod}}(x,t,\Delta t) = \beta/2E_L \int_{-\infty}^{t} I_{\mathrm{mod}}^2(x,t',\Delta t)dt'\,.$$

To calculate the first-order (self-)diffraction signal of the pump from the transient mixed grating, we have considered the Drude change of the refractive index, the absorption

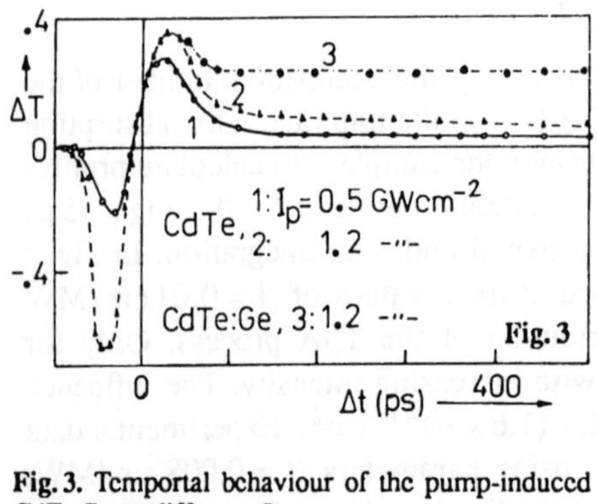

Fig. 3

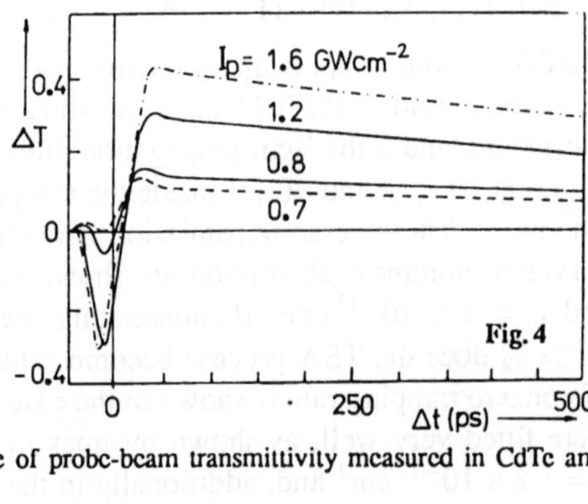

Fig. 4

Fig. 3. Temporal behaviour of the pump-induced change of probe-beam transmittivity measured in CdTe and CdTe:Ge at different I_p

Fig. 4. Calculated curves of ΔT vs delay time at different I_p including coherent coupling effects

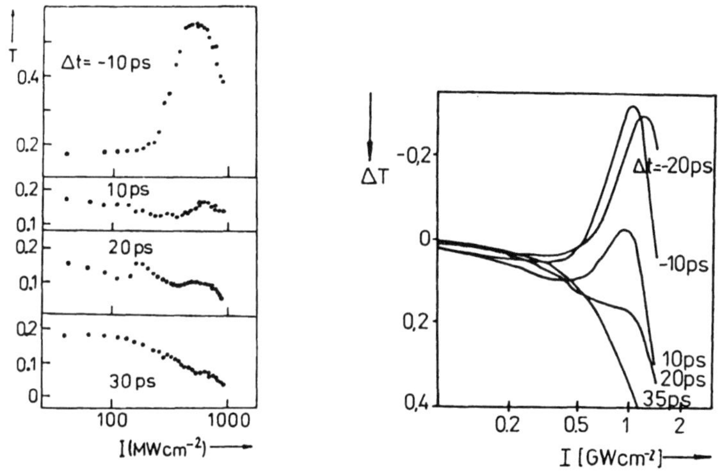

Fig. 5. Probe-beam transmittivity T vs pump intensity measured in CdTe at fixed delays (on the left) and ΔT vs pump intensity calculated using CdTe parameters

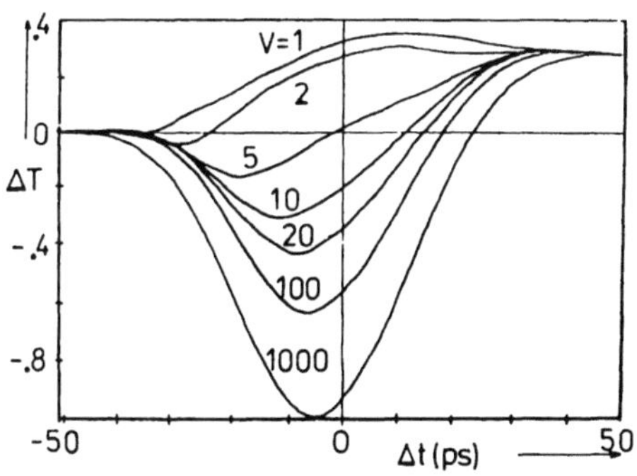

Fig. 6. Time behaviour of ΔT calculated at different intensity ratios $V = I_p/I_z$ at fixed I_p ($\beta = 0.008$ cm/MW, $g = 1.2 \times 10^{-17}$ cm^2)

index change due to free carrier absorption (both are caused by $N_{ch,mod}$) and due to TPA (caused by I_{mod}). We have used CdTe parameters obtained from Sect. 2, $\sin^2$ pulses, an intensity ratio $V = I_p/I_z = 10$, and $\tau_{ch} = 1$ ns. The grating erasure time due to diffusion and lifetime is long (some nanoseconds) in comparison with the diffracted pulse duration. In Fig. 4, ΔT vs Δt is shown calculated at different I_p. With increasing I_p the TEF leads to a strong nonlinear increase of T or $-\Delta T$ at small negative delays. With further growth of I_p the maximum of $-\Delta T$ is saturated and shifts to negative delays due to the stronger increase of the absorptive index change of the grating caused by TPA, in good agreement with the experiment. In Fig. 5 intensity dependences of T and ΔT measured and calculated at small delays are in good agreement, demonstrating the competition between TPA and TEF. At larger delays (see Figs. 3 and 4) ΔT is mainly due to free-carrier absorption. With decreasing probe-beam intensity (increasing V) the first-order pump-beam diffraction signal also decreases ($I^2_{mod} \propto I^2_p/\sqrt{V}$), but its contribution grows with respect to the probe-beam intensity as shown in Fig. 6.

References

1 A. von Jena, H.E. Lessing: Appl. Phys. **19**, 131 (1979)
2 V.L. Vinitski, N.V. Kukhatarev, S.G. Odulov, M.S. Soskin: Sov. Phys. – Usp. **22**, 742 (1979)
3 H.J. Eichler, M. Glotz, A. Kummrow, K. Richter, X. Yang: Phys. Rev. A **35**, 4673 (1987)
4 J. Dubard, A.L. Smirl, A.G. Cui, G.C. Valley, T.F. Bogges: Phys. Status Solidi (b) **150**, 913 (1988)
5 C. Scharager, P. Siffert, P. Höschl, P. Moravec, M. Vanecek: Phys. Status Solidi (a) **66**, 87 (1981)
6 I. Rückmann, M. Petrauskas, V. Netiksis, G. Tamulaitis, J. Halfpa: Phys. Status Solidi (b) **142**, 629 (1987)

Femtosecond Dynamics of Highly Excited CuBr Microcrystallites

U. Stamm[1], *M. Müller*[1], *M. Taiji*[2], *M. Yoshizawa*[2], *T. Kobayashi*[2], *and M. Kaschke*[3]

[1]Sektion Physik, Friedrich-Schiller-Universität Jena, Max-Wien-Platz 1, DDR-6900 Jena, German Dem. Rep.
[2]Department of Physics, University of Tokyo, 7-3-1 Hongo, Bunkyo-ku, Tokyo 113, Japan
[3]Zentralinstitut für Optik und Spektroskopie, Akademie der Wissenschaften, Rudower Chaussee 6, DDR-1199 Berlin, German Dem. Rep.

1. Introduction

Materials with large third order optical nonlinearities and ultrafast response are receiving intensive investigation currently. This is motivated by both fundamental physics and the search for effective basic materials for the fabrication of miniature all-optical switches.

Recently, a new class of material, namely, semiconductor colloids embedded in a host dielectric (usually glass or solvent), have moved into the center of interest. They have been shown to possess third order susceptibilities $\chi^{(3)}$ of about 10^{-8} - 10^{-9} esu /1/. Widely investigated have been $CdSe_xS_{1-x}$ microcrystallite doped-glasses. The carrier dynamics of these glasses were investigated from the femtosecond to the nanosecond range /2,3/ yielding an ultrafast response and an almost complete recovery of the excitation in 20 ps. Obviously it is of importance to examine other semiconductor colloids as well.

Here we report about the femtosecond dynamics of the elementary excitations in CuBr microcrystallites embedded in glass. The processes involved will be discussed in terms of the action of an optically generated dense electron-hole plasma.

2. Experimental

The sample is an 0.2mm thick NaAlB-silicate glass containing CuBr microcrystallites of an average diameter of 18 nm in a concentration of $1.5x10^{14}$ cm^{-3}. Because of the rather large crystallite size in comparison to the diameter of the bulk exciton (3.6 nm, /4/) the absorption spectrum of the sample given in Fig.1 is almost identical

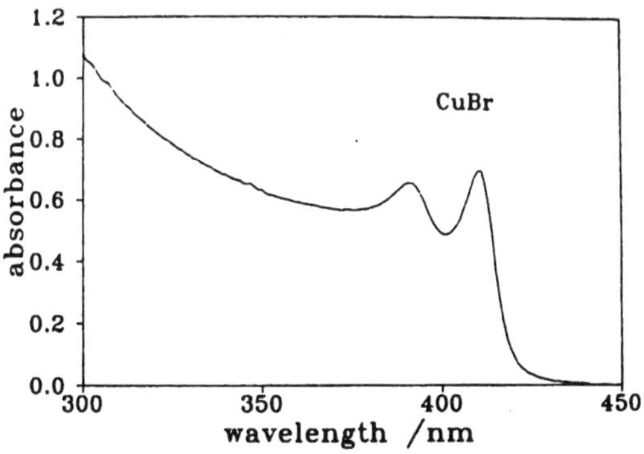

Fig.1. Absorption spectrum of 18 nm CuBr microcrystallites in glass.

Springer Proceedings in Physics, Vol. 49 Ultrafast Phenomena in Spectroscopy
Editors: E. Klose · B. Wilhelmi

to that of the bulk material. The two pronounced peaks at the low energy side of the spectrum correspond to the Z_{12} and Z_3 exciton of the bulk semiconductor.

The femtosecond experiments have been performed using a standard pump/probe technique with fs white light continuum. At 307 - 310 nm the sample is excited by the frequency-doubled amplified pulses of a colliding pulse mode-locked ring laser with a pulse width of less than 200 fs. The change in absorption after excitation is probed near the exciton energy in a range between 360 and 450 nm.

3. Results and Discussion

In the carrier dynamics of the CuBr glasses, nonthermal carrier populations at very early times, instantaneous band gap shrinking, exciton screening, and state filling effects were observed.

This is shown in Fig.2, where we have plotted the change in the optical density versus wavelength at various time delays between pump and probe pulses. At early time delays (-0.3 and 0 ps) we find an instantaneous rise following the excitation pulse of an absorption below the band gap (>420 nm) which indicates the energy gap shrinking resulting from the generation of a dense electron-hole plasma by the excitation pulse. The thermalization time of the originally nonthermal carrier population we determine to be 1.0 ps from the observation of gain at the shifted band edge (see the part at >425 nm in the spectrum at 1.8 ps delay time in Fig.2). This gain, or in other words the corresponding blue shift of the absorption, is equivalent to band filling effects.

The dominating feature in all spectra is the screening of the excitons visible as a full bleaching of the exciton absorption of both excitonic lines in Fig.2. In contrast to the below-gap absorption the rise of the exciton screening doesn´t follow the pump pulse instantaneously. To illustrate this, we have depicted in Fig.3 the change in the optical density as function of the delay time between pump and probe pulses at various wavelengths. The delay of the exciton bleaching (418nm and 397nm) with respect to the band gap renormalization is 500 fs at 10 K. In our opinion this can be explained by the temperature dependence of the Mott carrier density n_M at which the lowest exciton energy is reached by the downshifting energy gap, which is proportional to the temperature /5/. The very hot electron-hole plasma just after generation allows, after cooling, a more efficient screening of the excitons and the cooling rate of the plasma can be determined from the delay in the exciton bleaching /6/.

The decay of the bleaching of the exciton lines is found to be 3.1 ps and results from the very fast relaxation of the generated carriers into traps or to the valence band. A small part of the excitations still remains after 10 ps time delay and should have lifetimes of the order of the usual carrier lifetimes in CuBr.

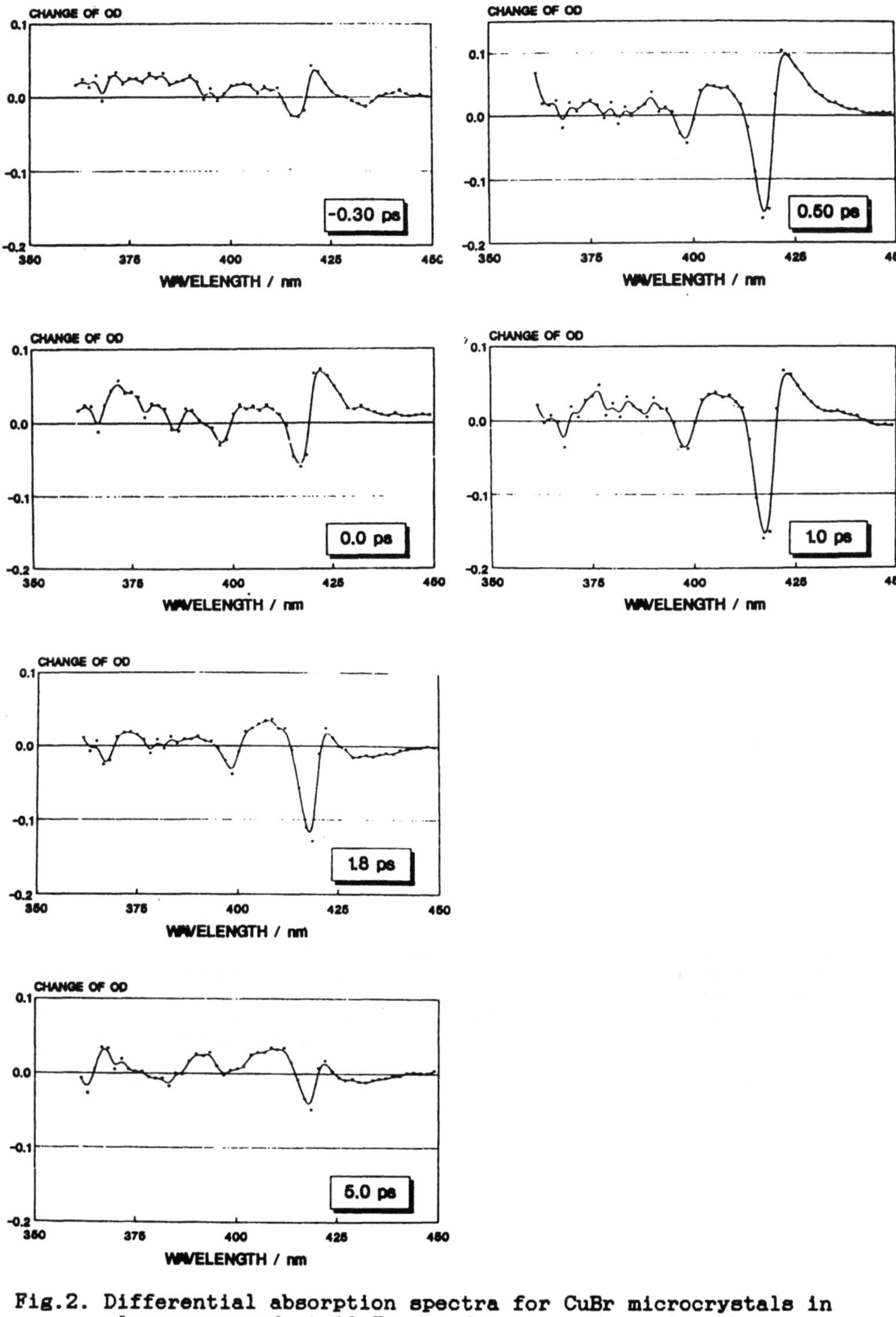

Fig.2. Differential absorption spectra for CuBr microcrystals in glass measured at 10 K as a function of wavelength for various time delays between pump and probe pulses.

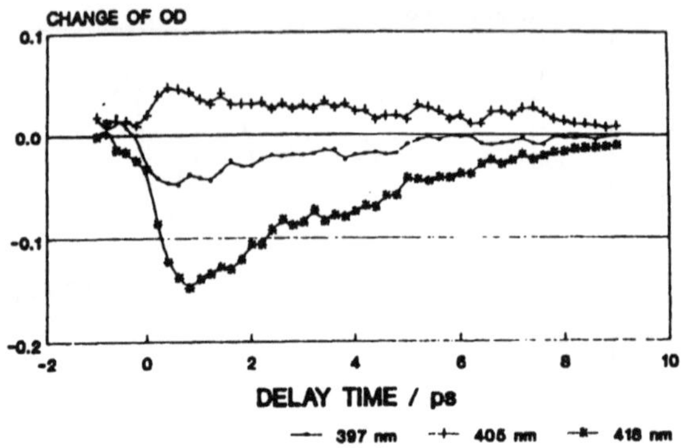

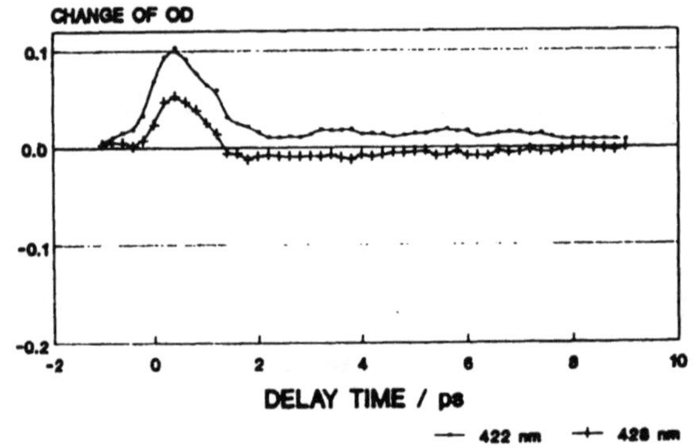

Fig.3. Differential absorption at 10 K as a function of the delay time between pump and probe pulses at various wavelengths.

References

1. R. K. Jain and R. C. Lind, J. Opt. Soc. Am. 73, 647 (1983).
2. N. Peyghambarian, SPIE Vol. 793, 147 (1987).
3. R. Roussignol, D. Ricard, J. Lukasik, C. Flytzanis, J. Opt. Soc. Am. B 4, 5 (1987).
4. A. I. Ekimov, Al. L. Efros, Proc. of the Third Binational USA-USSR Symposium on Laser Optics of Condensed Matter, Plenum Press, New York, 199 (1988).
5. R. Zimmermann, Many particle theory of highly excited semiconductors, BSB B.G. Teubner Verlagsgesellschaft, Leipzig (1988).
6. U. Stamm, M. Müller, M. Taiji, M. Yoshizawa, T. Kobayashi, M. Kaschke, to be published.

Study of Some Optical and Electronic Properties of GaAs Through Picosecond Laser Induced Transient Gratings

A. Agnesi[1], *G.P. Banfi*[1], *M. Costa*[1], *G. Reali*[1], *and V. Smid*[2]

[1]Università di Pavia, Dipartimento di Elettronica, Via Abbiategrasso 209, I-27100 Pavia, Italy
[2]Czech. Academy of Sciences, Institute of Physics, Prague, CSFR

Abstract. Laser induced transient grating measurements of $\chi^{(3)}$ and two-photon absorption coefficient β at 1 µm in bulk semi-insulating GaAs are presented. Preliminary results on ambipolar diffusivity are also reported.

1. Introduction

Laser induced transient grating (LITG) is a powerful technique for characterizing, both optically and electronically, semiconductor materials [1]. By this means it is possible to extract nonlinear optical susceptibilities of third and higher orders, and to estimate recombination times and carrier mobilities in a nondestructive way, scanning through the sample in such a way that a map of these characteristics is produced.

2. Description of the LITG technique

A carrier density grating is induced inside a GaAs crystal by the fringe pattern formed by two writing laser pulses WP_1 and WP_2, which simultaneously overlap in the sample. This grating induces a spatial modulation of the refractive index, which will scatter a reading pulse RP (impinging upon the sample along minus the direction of WP_1, and time-delayed by t_D with respect to WP_1 and WP_2) into the signal pulse SP in a direction opposite to WP_2. By setting $t_D=0$ and choosing all the frequencies of the beams to be the same, the LITG reflectivity $R=I_{SP}/I_{WP_2}$ (I is a pulse intensity) is proportional to $|\chi^{(3)}|^2 I_{RP} I_{WP_1} L^2$, and an estimate of $|\chi^{(3)}|$ can be obtained.

From a simple diffusive model of the photoexcited carrier dynamics, one expects an exponential time decay of SP by varying t_D. The time constant of the carrier grating is given by $1/\tau=1/\tau_R+1/\tau_D$, where $\tau_D=\Lambda^2/4\pi^2 D$ is the diffusion time and depends on the fringe spacing $\Lambda=\lambda_0/2n\sin(\theta/2)$, while the recombination time τ_R does not. So we are able to evaluate both τ_R and D by measuring τ for different angles θ between the writing beams [2].

3. Results and discussions

We used an active-passive mode-locked Nd-YAG laser operating at $\lambda=1.064$ µm, which provided TEM_{00} pulses, of time duration ≈ 30 ps and peak intensity of some GW/cm^2 on the sample. All the beams were sent through the sample with same polarization.

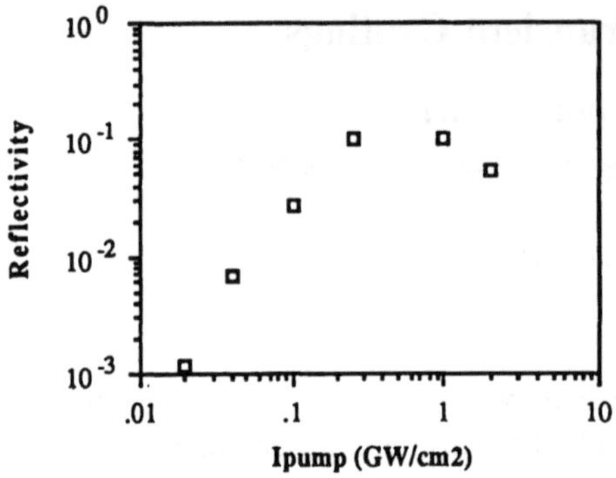

Fig.1. LITG reflectivity vs pump intensity. Experimental.

We carried out LITG reflectivity measurements (fig. 1) on a semi-insulating GaAs sample of dimensions of 5, 5 and 3 mm in the (001), (110), and ($\bar{1}10$) directions, respectively, with the beams impinging on the (001), (110) face. The reflectivity curve from GaAs were compared with those obtained by a CS_2 reference standard.

From the low intensity measurements of the R curve ($R \propto I_{pump}^2$, with $I_{pump} = I_{RP} = I_{WP_1}$) we found $|\chi^{(3)}| \approx 3 * 10^{-10}$ esu, which is due to plasma effect, i.e., index variation is proportional to free carrier concentration.

It is possible to consider two different regimes for the reflectivity curve. Below 100 MW/cm^2 in our reflectivity diagram, free electron plasma is generated by ionizing EL2 defects. Above 100 MW/cm^2 two photon absorption (TPA) dominates, giving rise to the saturation and eventually to the decreasing of $R(I_{pump})$. Anyway, plasma effect still determines $\chi^{(3)}$, though free carriers are now both electrons and holes (band-to-band excitation occurs through some virtual level).

We have considered a numerical model to explain saturation in fig.1. We solved the problem of two counterpropagating beams in a medium with TPA coefficient β,

$$dx/dz=-\beta(x+2y)x$$
$$dy/dz= \beta(2x+y)y$$
$$x(0)=y(L)=I_{pump},$$

and then perturbed the incoming beam x by δx, accounting for the beating between WP_1 and WP_2. In this way we have been able to calculate the amplitude of the index grating and the intensity of SP diffracted by such a grating taking into account its TPA, which leads to a direct evaluation of $R(I_{pump})$. It is seen that the experimental data fit very well with $\beta = \beta^* = 26$ cm/GW [3]. By setting $\beta = \beta^*$, $\beta = 3\beta^*$ or $\beta = \beta^*/3$ (a smaller variation than the spread usually found in the published TPA coefficient data), the R curve retains its shape but changes drastically in amplitude. LITG thus provides a sensible technique for a more direct estimation of β . The results are shown in the fig.2.

From the LITG experiments performed at the two angles $\theta \approx 5^o$ and $\theta \approx 19^o$ (fig.3), we found respectively $\tau \approx 740$ ps and $\tau \approx 400$ ps, from which we estimated $\tau_R \approx 1$ ns and an ambipolar diffusion coefficient $D \approx 4$ cm^2/s, corresponding to an ambipolar mobility $\mu \approx 150$ cm^2/Vs. These results are explained by TPA dominance due to the strong crystal excitation, being $I_{pump} \approx 200$ MW/cm^2, resulting in both electrons and holes photogeneration.

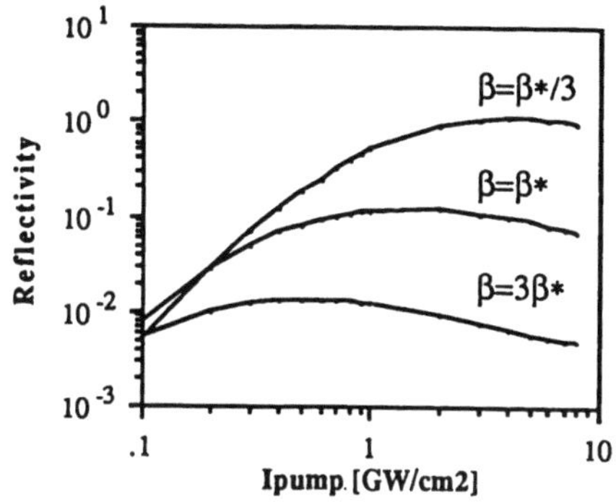

Fig.2. LITG reflectivity vs pump intensity. Numerical simulation.

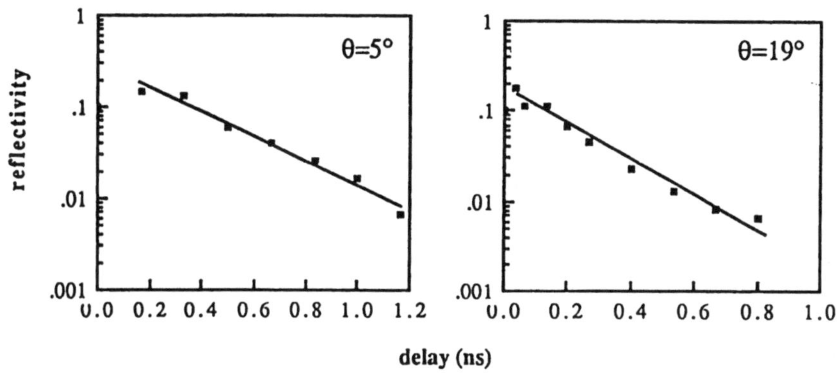

Fig.3. LITG signal decay at $\theta \approx 5^o$ and $\theta \approx 19^o$.

4. Conclusions

The optical LITG technique has been shown to be a powerful method in determining important optical ($\chi^{(3)}$, β) and electronic (D, τ_R) semiconductor properties. Low intensity measurements are under way for investigating other mechanisms contributing to the nonlinear susceptibility and carrier transport.

References

[1] H.J.Eichler, P.Gunter, D.W.Pohl, Laser-Induced Dynamic Gratings (Springer-Verlag, New York, 1986).

[2] A.L.Smirl, 'Dynamics of High-Density Electron-Hole Plasmas in Germanium', in Semiconductors Probed by Ultrafast Laser Spectroscopy R.R.Alfano ed. (Academic Press, London, 1984).

[3] T.F.Boggess, A.L.Smirl, S.C.Moss, I.W.Boyd, E.V.Stryland, IEEE J. Quantum Electron., 21 (1985) 5.

Investigation of the Gunn Effect in GaAs by Picosecond Photoconductivity

H. Bergner, V. Brückner, M. Lenzner, and R. Strobel

Sektion Physik, Friedrich-Schiller-Universität Jena, Max-Wien-Platz 1, DDR-6900 Jena, German Dem. Rep.

Introduction

In this paper we will discuss one example of the investigation of dynamic semiconductor parameters by transient photoconductivity measurements. The carrier drift velocity in semiconductors shows a nonlinear dependence on the applied field strength. Especially in GaAs the linear dependence vanishes due to electron transfer at high field strength from the central Γ-valley into the L-valley of the conduction band, where the electrons have lower mobilities because of the lower band bending. Moreover, a negative differential mobility can be found at certain field strengths, which causes the formation of charge domains. Therefore an oscillation of the current through the semiconductor is induced, which is known as the Gunn effect /1/. In this case the dependence of the electron drift velocity on the field strength cannot be measured in a stationary treatment.

Experimental method and results

We have investigated Cr-compensated GaAs samples ($n_{Cr}=2*10^{16}$ cm^{-3}) with polished and etched surfaces, respectively. Aluminium electrodes with a gap of the order of 1 mm were evaporated onto the surfaces. Substrate and electrodes form a microstrip electrical waveguide with an impedance of 50 Ω, which is adapted to the coaxial cables for input voltage (U_B) and output pulse. The gap region is illuminated with frequency-doubled pulses of a mode-locked Nd:YAG laser with 25 ps pulse duration. The pulse energy can be varied by neutral filters to generate different carrier concentrations in the GaAs samples. The photon energy of 2.34 eV exceeds the band gap of the semiconductor for 1 eV.

The electrical pulses generated in the switch will be detected by a boxcar integrator with a time resolution of 220 ps. The time dependent gap resistance is determined by the transient carrier concentration, the carrier drift velocity and the geometrical dimensions. So, at a given bias voltage the temporal behaviour of the output pulse represents the carrier dynamics.

In the Cr-compensated GaAs the main relaxation process of excess carriers takes place by trapping of the carriers at Cr-centers, causing an exponential temporal decay. In our case we found τ = 700ps, independent of the applied field strength (0.3-33 kV/cm) and of the exciting laser energy density (5nJ/cm² -10μJ/cm²). Also the surface preparation has no influence on the recombination time within the measuring accuracy (Fig. 1).

Measuring the maximum switched voltage U_o as a function of the applied voltage U_B the carrier drift velocity can be determined using the relation /2/

$$v = U_o / n_o e Z b d,$$

Springer Proceedings in Physics, Vol. 49 Ultrafast Phenomena in Spectroscopy
Editors: E. Klose · B. Wilhelmi

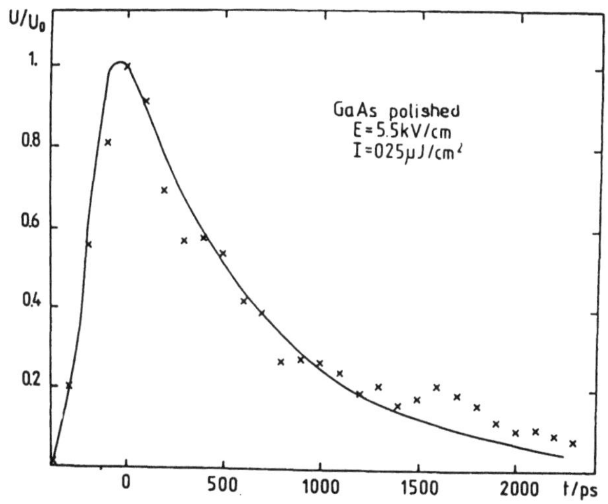

Fig. 1 Temporal behaviour of the transient photocurrent

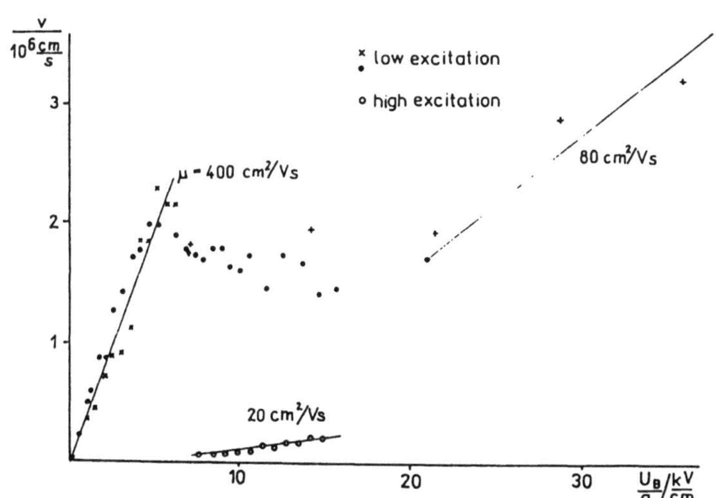

Fig. 2 Drift velocity in the polished sample

where Z is the impedance, b the stripline width, d the excited thickness of the semiconductor and n_o the induced carrier concentration. The electrical field strength in the main region of the gap with rectangular electrodes can be approximated by

$$E = U_B/g,$$

where g is the gap width. Figures 2 and 3 show the drift velocity v versus the electrical field strength E at different exciting laser energies in the polished sample (Fig. 2) and in the etched sample (Fig. 3), respectively.

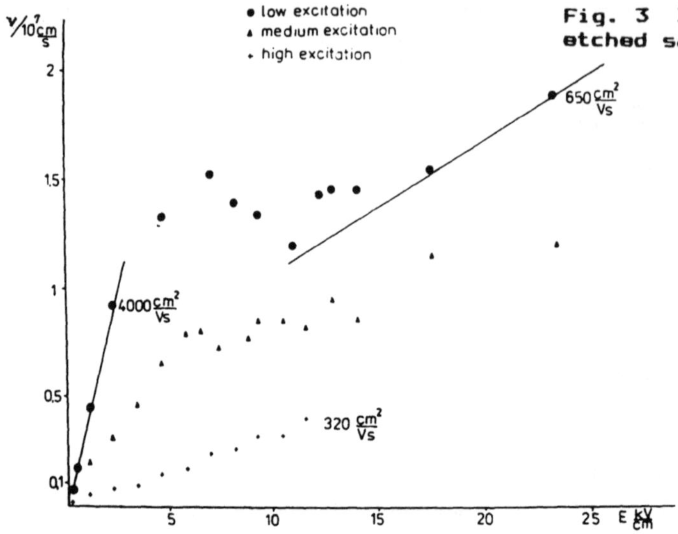

Fig. 3 Drift velocity in the etched sample

Discussion

First let us consider the case of low excitation. Up to a field strength of 5.3 kV/cm the carrier drift velocity increases linearly with the applied field strength. The slope of the curve v versus E provides a carrier mobility of $\mu \approx 400\ cm^2/Vs$ in the polished sample and $\mu \approx 4000\ cm^2/Vs$ in the etched sample, which is the mobility of the carriers in the Γ-valley of the conduction band. At higher field strength it is possible for electrons to overcome the potential barrier to the upper valley in the conduction band, which is characterized by a low mobility. The mobility of these carriers and thus the conductivity of the GaAs is drastically reduced. The distribution of the electrons in these two valleys depends on the actual value of the electric field in the bulk. Consequently we measured a decreasing drift velocity of the carriers in the range from 5.3 kV/cm up to ≈ 20 kV/cm. This decreasing drift velocity with increasing field strength corresponds to a negative differential mobility. The maximum slope provides $dv/dE = -300\ cm^2/Vs$ (polished) and $dv/dE = -1000\ cm^2/Vs$ (etched), respectively. Increasing the applied field strength again all electrons will be in the L-valley of the conduction band. We find once more a linear dependence v vs E but with a significantly lower slope. The corresponding electron mobilities are $\mu \approx 80\ cm^2/Vs$ and $\mu \approx 650\ cm^2/Vs$. So we can conclude that the effective mass of the electrons in the low-mobility upper valley is 5 times greater than the mass of electrons in the valley at the center of the zone. This is in good agreement with the measurements by other authors /3/.

The carrier mobilities found in the Γ-valley of $400\ cm^2/Vs$ (polished) and $4000\ cm^2/Vs$ (etched) clearly show the influence of the surface states as scattering centers on the photoconductivity because in our experiments photocarriers will only be excited near the surface. From the relaxation behaviour of both samples we can conclude that surface recombination is not the dominant process. Also we found that the recombination time for carriers in the L-valley (high electrical field strength) is the same as for carriers in the Γ-valley.

Now let us discuss the case of high excitation, where a decrease in drift velocity with increasing field strength is no longer found due to the formation of charge domains in the semiconductor. These domains move with an effective velocity through the material. The electron transfer from the Γ-valley to the L-valley cannot be measured if such charge domains exist. The build-up time T_D of the domains is inversely proportional to the square root of the carrier density

$$T_D = \sqrt{2}\ T_C\ \ T_R \quad ,$$

where T_C is the crossing time of the electrons through the gap

$$T_C = g/v$$

and T_R is the relaxation time of carrier fluctuations which will be negative in the case of a negative differential mobility

$$T_R = \epsilon_r\ \epsilon_o\ /(e\ n_o\ dv/dE).$$

In Table 1 we have listed the laser-induced carrier concentration in each experiment and the calculated build-up time of domains.

Table 1

sample	excitation	n_o (cm^{-3})	T_D (ps)
polished	low	$4.0 * 10^{16}$	260
	high	$4.0 * 10^{17}$	60
etched	low	$4.2 * 10^{14}$	290
	medium	$7.0 * 10^{16}$	20
	high	$1.8 * 10^{18}$	4

The temporal resolution of the electrical measuring equipment is 220 ps, so we can conclude that in the low excitation case charge domains were not built up within the measuring time. That gives us the possibility of measuring the real drift velocity dependence on the applied field strength, whereas in the high excitation case charge domains occur and an effective carrier drift velocity will be measured.

References

/1/ J.B. Gunn; Sol. State Commun. 88 (1963) 1

/2/ H. Bergner, V. Brückner, F. Kerstan, M. Lenzner; Scientific Instrumentation 3 (1988) 41

/3/ M.A. Osman, D.K. Ferry; Phys. Rev. B 36 (1987) 6018

Time-Resolved Luminescence Studies of Vertical Transport in Short-Period Superlattices with Enlarged Well

J. Puls [1], *F. Henneberger* [1], *I.N. Uraltsev* [2], *A.M. Vasiliev* [2], *U. Woggon* [1], *and M. Luhn* [1]

[1]Humboldt-Universität zu Berlin, Sektion Physik, Invalidenstr. 110, DDR-1040 Berlin, German Dem. Rep.

[2]A.F. Ioffe Physico-Technical Institute, Academy of Sciences, 194021 Leningrad, USSR

Abstract: Studies of time-resolved photoluminescence of three different types of GaAs-AlGaAs microstructures are reported: A short period superlattice, a second one, where an enlarged well is introduced into the SL and, a third one, where the well is separated from the superlattice by enlarged barriers. Very high sensitivity is achieved by an electro-optical cross-correlation technique using a gain modulated Si avalanche photodiode as a detector. From the data obtained we are able to determine the relevant parameters of vertical transport and nonradiative recombination in the structures investigated.

1. Introduction

The transport of carriers perpendicular to the layers of semiconductor superlattices (SL) is of both fundamental and applicational interest. Time-resolved photoluminescence experiments are a powerful tool to study this process on the ps-time scale [1]. Usually, an enlarged well (EW) is incorporated into the SL to detect the arrival of the photo-excited carriers at the position of the EW. Here, we report on first comparative studies of such GaAs-$Al_x Ga_{1-x}As$ structures and those modified by thick barriers between the EW and the SL regions. Additionally, a single SL grown under the same conditions is used to study the contribution of nonradiative recombination in the SL regions of the above-mentioned structures [2].

2. Experimental

All samples investigated are nominally undoped SL grown by molecular-beam epitaxy. The thickness of the GaAs wells (l_W =3 nm) is chosen equal to that of the $Al_xGa_{1-x}As$ barriers l_B (x=0.35). The single SL consists of about 50 periods and is sandwiched between 200 nm $Al_xGa_{1-x}As$ cladding layers to avoid carrier leakage. In the second sample a single GaAs well in the middle of the SL is enlarged to l_{EW} =13.5 nm, whereas in the third sample the central GaAs well *and* the neighbour barriers are enlarged (l_{EW} =l_{EB} =13.5 nm).

The kinetics of the SL- and EW-luminescence are measured in a temperature range from 80 to 300 K using excitation by a mode-locked Ar^+ laser (514.5 nm). The luminescence is spectrally decomposed by interference filters and detected by a Si avalanche photodiode (APD: SP 114). Time resolution is achieved by modulating the gain of the avalanche diode with very short electrical pulses (200 ps) [3]. These pulses are formed by a reference APD from a small part of the laser output. The delay of the electrical pulses with respect to the excitation pulses can be scanned by a computer controlled optical delay line. Due to the high quantum efficiency of Si photodiodes the sensitivity of this electro-optical cross-correlation technique is comparable to that of time correlated single photon counting using photomultipliers with S1 or S25 cathode in the spectral range under consideration. By means of standard deconvulation

Springer Proceedings in Physics, Vol. 49 Ultrafast Phenomena in Spectroscopy
Editors: E. Klose · B. Wilhelmi

methods [4] a time resolution of some ten ps is achieved. The excitation intensity is varied between 10 and about 300 W/cm^2 yielding carrier densities clearly above the residual doping of the samples.

3. Results

3.1 Short-Period Superlattice

In Fig. 1 the decay curves measured on the short-period SL are given for various values of the mean excitation intensity $\bar{I}_0$. For comparison, the cross-correlation of the excitation pulse is drawn by a dashed line. The halfwidth of the pulse (FWHM) is about 300 ps. The oscillations found especially for negative delay are due to the electrical connections between reference diode and detecting diode. In principle, they can be depressed by an improved electrical arrangement [3], but, on the other hand, these oscillations make a more precise deconvolution possible.

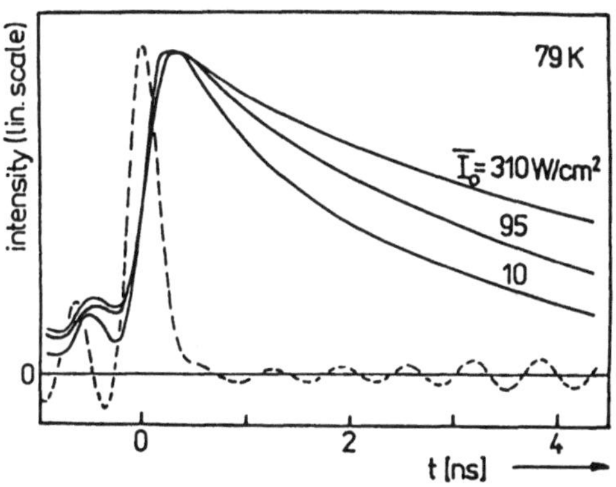

Fig. 1 Decay curves of the photoluminescence measured at the peak position (λ=750 nm) for different mean excitation intensities $\bar{I}_0$. Dashed line: Measurement of the excitation pulse

For the SL the decay is not simply exponential. For simplicity, we have defined the respective time constant at the 1/e-point, which increases from 2.2 ns up to 5.1 ns increasing the excitation intensity by a factor of 30. Furthermore the time-integrated luminescence intensity shows a superlinear dependence on $\bar{I}_0$ with an exponent of 1.6. These results can be well understood if we assume a dominating nonradiative recombination in the SL, where the nonradiative parallel channel tends to saturate.

3.2 Superlattice with Enlarged Well

In Figs. 2 and 3 the decay curves of SL- and EW-photoluminescence of the sample with enlarged well are given for 79 and 298 K, respectively. For the luminescence at the spectral position of the enlarged well practically no intensity dependence of the decay time is found, whereas the decay kinetics of the very weak SL-photoluminescence could be measured only at the highest excitation intensity used. Here, at 79 K a very fast decay of the luminescence is observed with a time constant of about 20 ps. This has to be attributed to the capture of photo-excited electrons and holes into the enlarged well, where the barrier thickness of 3 nm allows for Bloch transport of electrons *and* holes in the SL regions [1]. Neglecting recombination of carriers during the first tens of ps we can estimate an ambipolar diffusion constant of about 5 cm^2/s in the short period superlattice [5]. At room temperature (Fig. 3) a much slower transport of the carriers due to phonon scattering results in a decay time of the SL luminescence of 210 ps.

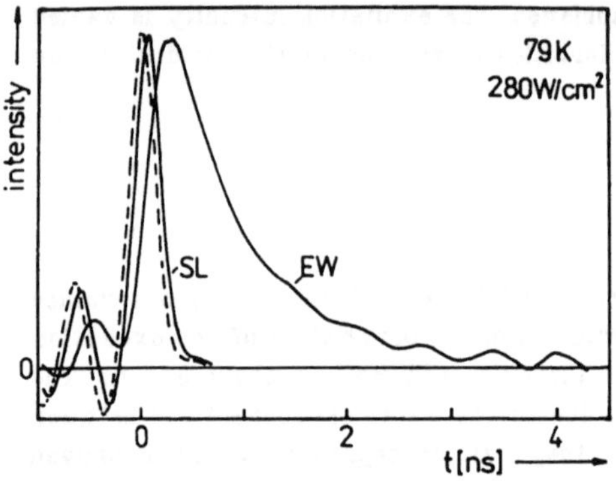

Fig. 2 Decay curves of the EW- (811 nm) and SL-luminescence (745 nm) at 79 K

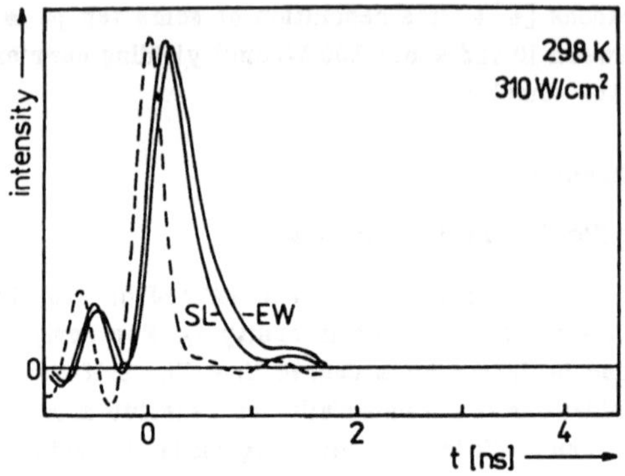

Fig. 3 Decay curves of the EW- (854 nm) and SL-luminescence (774 nm) at 298 K

The decay curves of the luminescence from the enlarged well can be well fitted assuming single exponential decay. At 79 K a time constant of 700 ps is found, which increases at higher temperatures (950 ps at 120 K) and then drops down to 330 ps at room temperature (Fig. 3). The latter is accompanied by a drastic reduction of the quantum yield. The temperature dependence of the carrier lifetime observed is also known from time-resolved photoluminescence studies on multiple-quantum-well structures [6].

3.3 Superlattice with Enlarged Well and Enlarged Barriers

The results obtained on the sample with an enlarged well separated by thick barriers from the SL are plotted in Figs.4 and 5 for 79 K and 298 K, respectively. Comparing Figs. 2 and 4 the influence of the thick barriers on the capture of carriers into the EW is clearly seen. Here, the decay time of SL-luminescence (150 ps) is determined by the tunneling of the carriers, especially the heavier holes, through the enlarged barriers. At room temperature this time is further increased (200 ps) due to the additional action of phonon scattering.

At 79 K the decay time of the photoluminescence from the EW is comparable to that of the sample without thick barriers, but no essential change is observed on increasing the temperature up to 300K and the quantum yield decreases only by a factor of two. The reason for this somewhat surprising behaviour is not clear at present and needs further investigation.

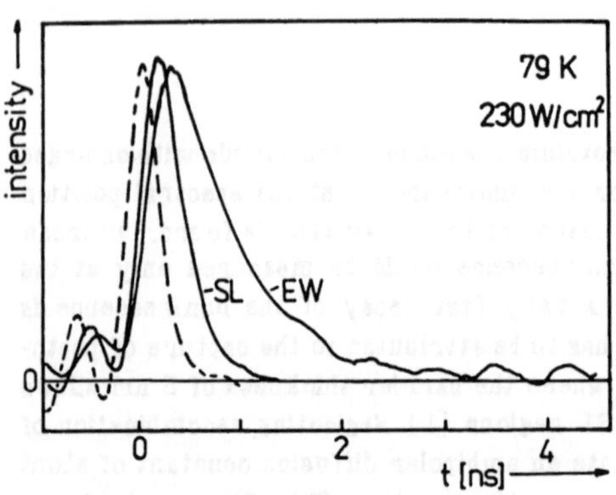

Fig.4 Decay curves of the EW- (808 nm) and SL-luminescence (740 nm) at 79 K

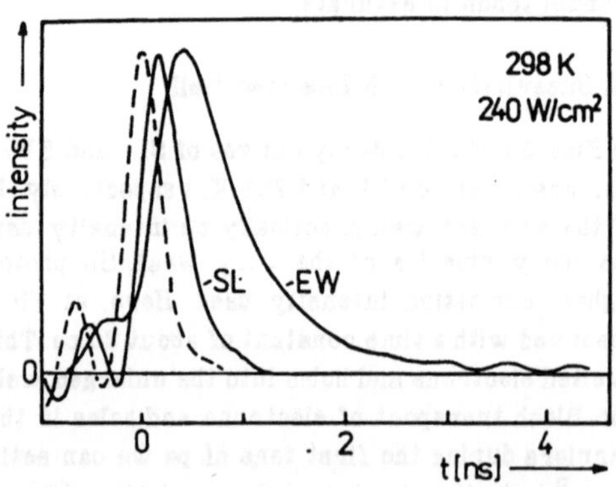

Fig. 5 Decay curves of the EW- (858 nm) and SL-luminescence (773 nm) at 298 K

4. Conclusion

We have observed the transfer of photo-excited carriers from the SL to EW by a dramatic shortening of photoluminescence decay. The data suggest diffusion-like transport in this situation. Separation of the EW from the SL by enlarged barriers leads indeed to a reduced capture of carriers by the EW. An additional effect of the enlarged barriers is a drastic suppression of the nonradiative recombination rate at room temperature.

References

1. see e. g.: B. Deveaud, J. Shah, T. C. Damen, B. Lambert, A. Chomette, A. Regreny: IEEE J. Quant. Electronics **24**, 1641(1988)
2. B. Lambert, B. Sermage, B. Deveaud, I. Clerot, A. Chomette, and A. Regreny: Abstracts MSS-4, Ann Arbor (USA) 1989 (p. 321)
3. K. Berndt, Optics Commun. **61**, 33(1987)
4. D. V. O'Connor and D. Phillips: Time-Correlated Single Photon Counting, Chap. 6 (Academic, London 1984)
5. J. Puls, F. Henneberger, I. N. Uraltsev, A. M. Vasiliev: to be published
6. T. Miyoshi, Y. Aoyagi, Y. Segawa, S. Namba: Japan. J. Appl. Phys. **24**, 369(1985)

Femtosecond Spectroscopy of Semiconductor Colloids

M. Kaschke[1], *N.P. Ernsting*[2], *and H. Weller*[3]

[1]Zentralinstitut für Optik und Spektroskopie, Akademie der Wissenschaften, Rudower Chaussee 6, DDR-1199 Berlin, German Dem. Rep.
[2]Max-Planck-Institut für biophysikalische Chemie, Abteilung Laserphysik, Postfach 2481, D-3400 Göttingen, Fed. Rep. of Germany
[3]Hahn-Meitner-Institut, Bereich Strahlenchemie, D-1000 Berlin 39, Germany

Abstract. Quantum-size semiconductor microcrystallites of ZnCdS and CdS in aqueous and isopropanol solution were optically excited using intense sub-picosecond UV laser pump pulses. Broad band optical probing monitored the yield of electron emission. With typical pump fluences of femtosecond light pulses, multiple photons are absorbed by one quantum dot. The electron yield curves, as a function of pump fluence, are described by a simple kinetic model which involves the trapping rate constant and a maximum number of electrons which can be emitted per particle.

1. Introduction

The non-linear optical properties of direct-gap semiconductors have been examined intensively over the past few years, both with bulk materials /1-3/ as well as microcrystals /4-6/ embedded in a glass host. A relatively young class of materials consists of very small crystallites, with diameters ranging up to 100 Å. These "microcrystallites" are usually prepared by chemical reaction, either in a polymer film or as a colloid from solution /7, 8/. If the particle radius is less than the excitonic radius in the bulk material we speak of quantum-size microcrystallites or quantum dots. For the description of such tiny particles a molecular approach is more appropriate than a semiconductor one, however, standard pictures of semiconductor theory, such as excitons, band filling, are frequently applied. In quantum dots the electronic wavefunctions of an electron and hole always overlap appreciably due to the confinement. In this sense, the lowest states of an electron and hole may be called "exciton" /7/.

Recently, we have investigated the bleaching kinetics after absorption of multiple photons from an intense ultrashort light pulse at 308 nm /9/. It was shown that recombination of charge carriers takes place on a picosecond time scale and is fluence dependent. Further, the occurrence of rapid surface trapping of charge carriers, as well as the ejection of photons out of the quantum dot after light pulse excitation, was established.
Here, we will distinguish experimentally between the original "exciton" and trapped electron-hole pairs. We examine quantum-size CdS and $Zn_{1-x}Cd_xS$ as a colloid in water and in isopropanol.

2. Experimental

The pump/probe - absorption experiment used a laser system combining an excimer laser with a colliding pulse mode (CPM) dye laser yielding an amplified UV pulse at 308 nm of high energy. The probe continuum was generated by focusing a 100 fs pulse at 616 nm (60 µJ) in 1 mm o-phosphoric acid in a 5 mm fused silica cell. The continuum pulse

Springer Proceedings in Physics, Vol. 49 **Ultrafast Phenomena in Spectroscopy**
Editors: E. Klose · B. Wilhelmi

contains about 4 μJ of energy covering a spectral range from 350 to 750 nm. The temporal resolution was 250 fs as inferred from a cross-correlation measurement /10/.

3. Results

The absorption spectrum of aqueous colloidal $Zn_{0.95}Cd_{0.05}S$ is shown in fig. 1. The corresponding bulk bandgap (estimated from the bulk bandgaps of ZnS and of CdS) is also indicated in the figure. The red-edge absorption of the colloidal solution is clearly shifted to higher energies from the bulk bandgap, indicating quantum confinement in both cases.

Also given in the figure is the particle diameter as function of the wavelength of the first "excitonic" transition, according to simple theory /11/. By comparison of the spectrum with published data /7/, we estimate the mean particle diameter for the CdS colloid to be ca. 25 Å and for the ZnCdS to be ca. 40 Å. These values have to be compared with the excitonic radius of the bulk material (53 Å for CdS and 40 Å for ZnCdS).

Photophysical and photochemical changes after short UV pulse excitation were monitored by broad band transient absorption measurements in the wavelength range 370 .. 750 nm. The rise of transient absorption occurred within the time resolution of this measurement. Fig. 2

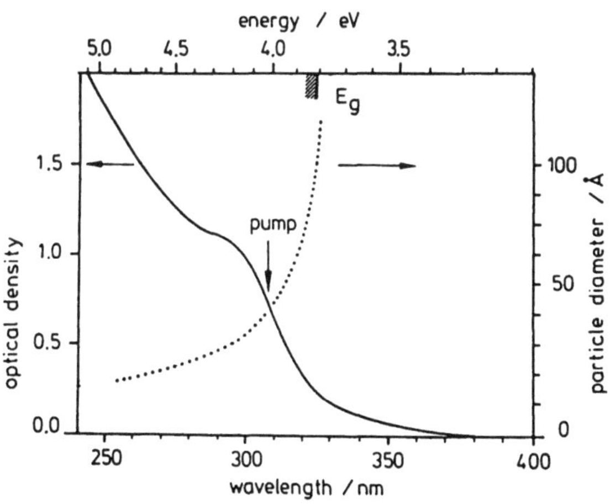

Fig. 1 Solid line: absorption spectrum of aqueous colloidal $Zn_{0.95}Cd_{0.05}S$.
Dotted line: particle diameter estimated from the spectral location of the "exciton" band.

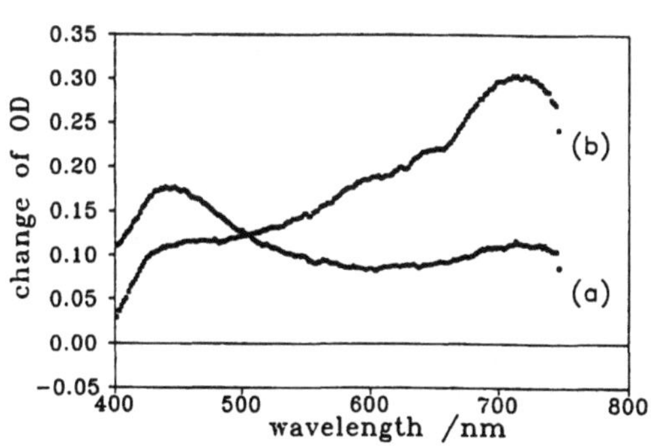

Fig. 2 Transient absorption spectrum of CdS in isopropanol (a) and water (b) at 1.67 ps after excitation by a 300 fs UV pulse.

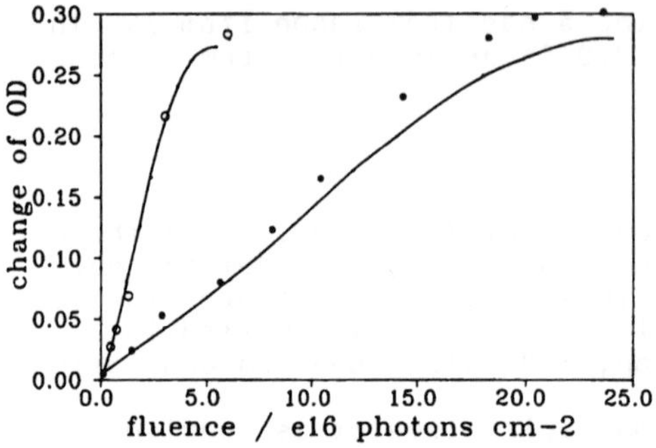

Fig. 3 Change of optical density of aqueous CdS (solid circles, at 720 nm) and $Zn_{0.95}Cd_{0.05}S$ colloids (hollow circles, at 616 nm) at 1 ps after excitation by a 300 fs UV pulse. Lines are the result of model calculations (see text).

shows a transient spectra at 1 ps after passage of the pump through the sample for CdS colloid in water and in isopropanol. Evidently, a strong additional absorption can be observed in the red spectral range. This is in agreement with findings in ns and ps measurements /12/. By comparison with the spectrum of a hydrated electron we can conclude that in aqueous colloidal solution after intense UV pulse excitation we observe an "instantaneous" (within our temporal resolution) ejection of electrons. The almost structureless transient absorption in the range 440 .. 720 nm in the case of the colloidal solution in isopropanol can be assigned to photochemical changes due to trapped charge carrier.

Fig. 3 shows the maximum changes of optical density at 720 nm (CdS) and at 616 nm (ZnCdS) for 2 mm of the colloidal solution, which are proportional to the number of ejected electrons. Evidently, in the case of the CdS colloid saturation of the ejection occurs at considerably higher pump fluences. At the top of the figure, the pump fluence has been translated into number of photons absorbed per particle.

4. Discussion

Considering the fact that at highest pump fluences there are more than 50 photons available for every particle we have to take absorption of multiple photons into account. We therefore propose the following model to explain the ejection of electrons (Fig. 4).

Absorption of a photon with absorption cross section σ_0 leads to the first excited state a, which may be called "excitonic" in the sense of confined charge carriers. This is followed by trapping, presumably at the surface, with a rate constant k_T. In the case of strong pumping this "excitonic" state alternatively may absorb another photon with a reduced absorption cross section σ_0^* leading to a doubly excited state b. Besides trapping with rate constant $2\,k_T$ a second decay channel for state b is given by the emission of an electron (which can only occur from this state) with rate constant k_A, leading to a deexcited particle with an excess hole. Since recombination of trapped carriers proceeds on a ps timescale /9/ the effect of an intense fs pump pulse is to drive population up along the main diagonal in the scheme of fig. 4, creating a typical particle which has absorbed an average number n_{abs} of photons and emitted an average number of electrons n_{el}. This suggests a simpler energy level scheme shown on the l. h. s. of fig. 4. "Screening" by the first

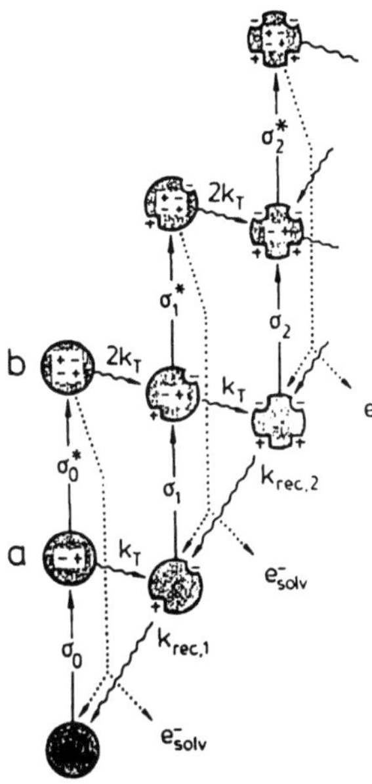

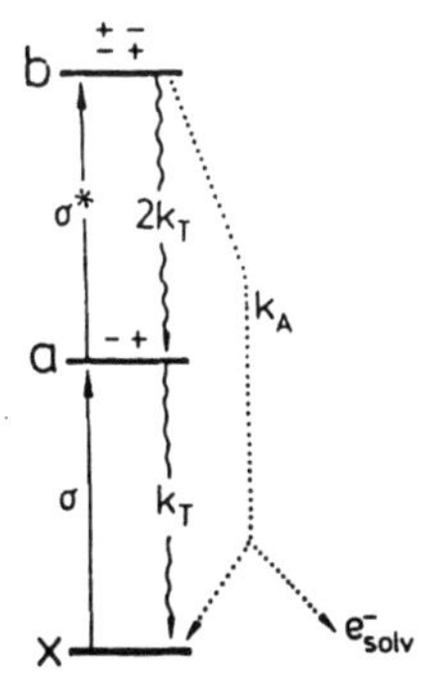

Fig. 4 Full and simplified formal rate model for multiphoton absorption by semiconductor microcrystallites.

"exciton" is included through

$$\sigma^* = \alpha \cdot \sigma \ . \qquad (1)$$

Screening by trapped charge carriers is accounted for by letting σ depend on $n_{abs}-n_{el}$, for example

$$\sigma = \sigma_0 \cdot [\beta + (1-\beta) \cdot \exp\{-(n_{abs}-n_{el})/n_\sigma\}] \quad \text{with } \alpha<\beta<1 \ . \qquad (2)$$

Electron emission should also proceed more slowly as positive charge accumulates on the particle. For every emitted electron, the ionization potential is increased by an amount $e^2/(\varepsilon \cdot R)$. Eventually, the electron affinity of the solvent will be exceeded and electron emission must cease. Therefore we set

$$k_A = k_{A0} \cdot (1-n_{el}/n_A) \text{ for } n_{el}<n_A, \text{ and } k_A = 0 \text{ otherwise} \ . \qquad (3)$$

The rate equations for the simplified model of fig. 4 are then

$$n_a = I \cdot \sigma \cdot (1-n_a-n_b) - [I \cdot \sigma^* + k_T] \cdot n_a + 2 \cdot k_T \cdot n_b \qquad (4)$$

$$n_b = I \cdot \sigma^* \cdot n_a - [2 \cdot k_T + k_A] \cdot n_b \qquad (5)$$

$$n_{el} = k_A \cdot n_b \qquad (6)$$

$$n_{abs} = I \cdot \sigma \cdot (1-n_a-n_b) + I \cdot \sigma^* \cdot n_a \ . \qquad (7)$$

The rate equations were integrated numerically and thus saturation curves like those in fig. 3 were obtained. The free parameters for best overall agreement with experiment (the number of ejected e^- is obtained from the change of optical density in the red spectral range) are the trapping rate constant k_T and the maximum number of ejected electrons n_A. For CdS we find $k_T = 300\ ps^{-1}$, and $n_A = 7$, whereas for the larger ZnCdS $k_T = 28\ ps^{-1}$, and $n_A = 20$ have been obtained. k_T corresponds to an "exciton" lifetime of 3 fs for the CdS particle and 36 fs for the ZnCdS colloid, respectively. Calculating the tran-

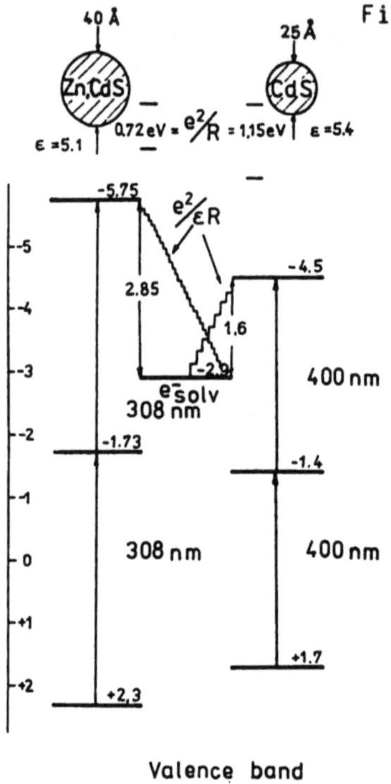

Fig. 5 Electrochemical potentials of bulk CdS and ZnS in water and potential energy of solvated electron.

sit time of an electron through such a small semiconductor particle, one finds characteristic times in the order of 10 fs /9/. Thus, such very fast trapping times are indeed reasonable. However, the calculated values can only be considered as estimates, since they are already far beyond the limits where the rate equation description is applicable. We can take it therefore as a mere indication of an almost instantaneous photochemical reaction on the surface.

Let us discuss the maximum number of electrons which can be ejected from such a particle which differs by a factor of three for the two colloids. To this end we have depicted in fig. 5 the electrochemical potential of the bulk semiconductor materials (CdS, ZnS) and the energy of an solvated electron /12/. It is evident, that from the first excited ("excitonic") state there can not occur an electron ejection. With every electron ejected from the doubly excited state, the remaining excess hole reduces the photochemical potential by $e^2/\varepsilon R$, where ε is the dielectric constant. Since in the ZnCdS particle the potential energy decrease with every ejected e^- is smaller (due to the larger R) and since the energy difference between the initial potential of the doubly excited state and the solvated electron is larger more electrons can be emitted by this particle in comparison to the smaller CdS microcrystallite. Thus, our experimental findings and model calculations are in agreement with thermodynamical considerations.

Acknowledgement

We are grateful to the Deutsche Forschungsgemeinschaft for support through the Leibniz Prize Program. M. K. thanks the Max-Planck-Gesellschaft for a grant.

References

/1/ I. Broser, J. Gutowski; Appl. Phys. B **46** (1988) 1

/2/ Ch. Spiegelberg, M. Kretzschmar, J. Puls, F. Henneberger; phys. stat. sol. (b) **150** (1988) 769

/3/ V. D. Egorov, P. Floegel, H. X. Nguyen, M. Kaschke; phys. stat. sol. (b) **146** (1988) 351

/4/ J. Warnock, D. D. Awschalom; Appl. Phys. Lett. **48** (1986) 425

/5/ M. C. Nuss, W. Zinth, W. Kaiser; Appl. Phys. Lett. **49** (1986) 1717

/6/ G. R. Olbright, N. Peyghambrian, S. W. Koch, L. Banyai; Opt. Lett. **12** (1987) 413

/7/ R. Rosetti, J. L. Ellison, J. M. Gibson, L. E. Brus; J. Chem. Phys. **80**, 4464 (1984)
N. Chestnoy, R. Hull, L. E. Brus; J. Chem. Phys. **85** (1986(2237
L. E. Brus; IEEE J. QE-**22** (1986) 1909

/8/ A. Henglein, A. Kunar, E. Janata, H. Weller; Chem. Phys. Lett. **132** (1986) 133

/9/ N. P. Ernsting, M. Kaschke, H. Weller, L. Katsikas; submitted for Journ. Opt. Soc. Amer. b (1989)

/10/ N. P. Ernsting, M. Kaschke; submitted for Appl. Phys. B (1989)

/11/ L. E. Brus; Journ. Chem. Phys. **80** (1984) 4403

/12/ M. Haase, H. Weller, A. Henglein; J. Phys. Chem. **92** (1988) 4706

Pump Energy Dependent Relaxation in Semiconductor-Doped Glasses

D.V. Lap, Th. Peschel, P. Heist, W. Rudolph, and H.-E. Ponath

Sektion Physik, Friedrich-Schiller-Universität Jena, Max-Wien-Platz 1,
DDR-6900 Jena, German Dem. Rep.

Abstract: Absorption transients of commercial semiconductor-doped glasses are investigated for different excitation energies. Above a certain threshold energy a fast relaxation process is observed additionally to the slow interband one. The theoretical model presented includes inhomogeneous broadening due to the size distribution of the semiconductor microcrystals, thermalization and saturation of absorption.

1. Introduction

Semiconductor-doped glasses consist of nanometer-sized semiconductor crystals embedded in a glass matrix. Due to this structure quantum size effects occur that are responsible for the large optical nonlinearities of those materials. Their relaxation behaviour is of great importance for any future application. It has been the subject of several papers in the past /1-3/. Different mechanisms like Auger /1/ or thermalization /2/ were proposed to explain the measured relaxation curves. In the present paper absorption transients of semiconductor-doped glasses after excitation with femtosecond pulses (τ_L=100fs) are investigated for different excitation pulse energies of 2-20 mJ/cm^2. We investigated the appearence of a fast relaxation process /1,3/ (3-25ps depending on the excitation energy). In particular we observed fast relaxation for excitation of higher levels. We found the fast relaxation to be more pronounced at high pulse energies and the slow (several 100ps) interband relaxation to be dominant at low excitation.

A theoretical model is presented that explains the experimental findings as resulting from the interplay of

i) inhomogeneous broadening of the discrete absorption lines /4/ of the single semiconductor microcrystals due to their size distribution /5/,

ii) saturation of absorption and thermalization after excitation of higher levels.

2. Experimental results

Pulses at λ=610nm with a duration of 100fs and a spectral width of 5nm are gained from a CPM-ring laser. After four-stage amplification they reach a pulse energy of 200μJ /6/. Pulses at 530nm were selected from a continuum generated in CCl_4. After two-stage amplification and time compression we obtained pulses of 20 μJ with a duration of 140fs. The incoming pulse is divided into a pump pulse and a test one, the latter of which may be delayed. The widths of the pump and probe beams were 250μm and 150μm respectively. The excitation energy was controlled with an accuracy of 5%.

The specimens investigated are commercial filter glasses (RG5, RG8, OG2, VEB Jenaer Glaswerke) with thicknesses of 300-500μm. Their transmission spectra are given in fig.1. The excitation wavelengths are above the absorption edges of the respective filters (λ_{ex}=610nm for RG5 and RG8, λ_{ex}=530nm for OG2).

The absorption-change-versus-time curves obtained for the different materials are plotted in figs.2-4. For higher excitation a fast (τ≈5-25ps) relaxation component appears while slow relaxation (τ≈100ps) dominates at lower excitation levels.

Springer Proceedings in Physics, Vol. 49 Ultrafast Phenomena in Spectroscopy
Editors: E. Klose · B. Wilhelmi

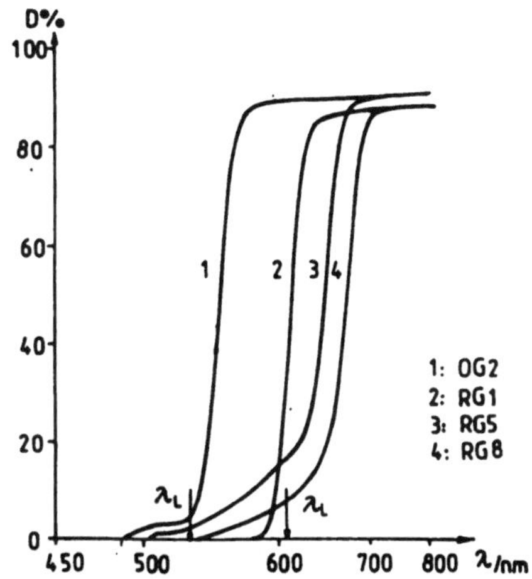

Fig.1 Transmission spectra of the specimen investigated

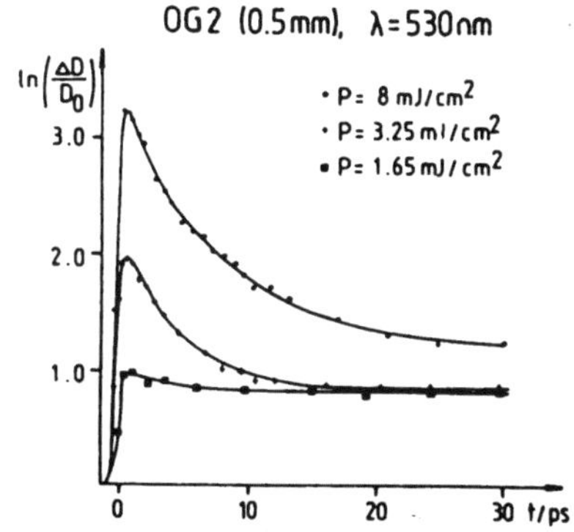

Fig.2 Relaxation curves for OG2

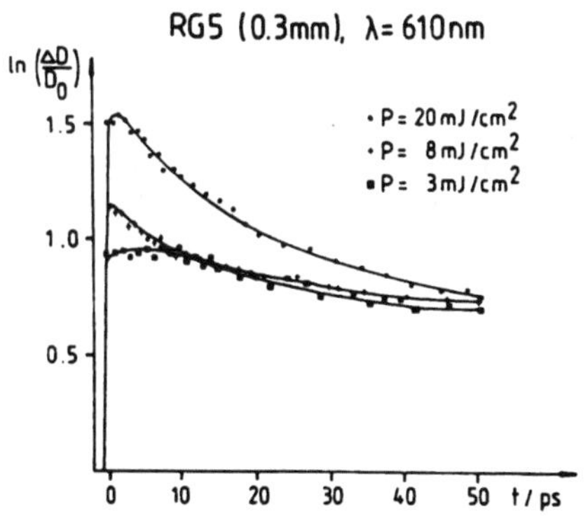

Fig.3 Experimental relaxation curves obtained for RG5

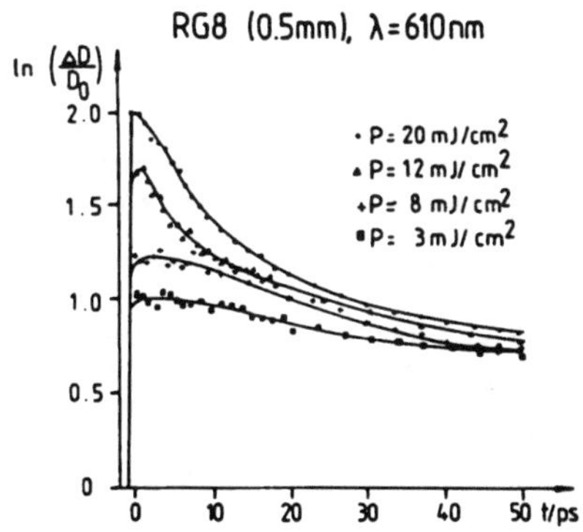

Fig.4 Experimental relaxation curves obtained for RG8

3. Theoretical model

Our theoretical considerations are based on the following assumptions: The absorption lines are inhomogeneously broadened due to the size distribution of the semiconductor microcrystals in the glass. Hence at an excitation wavelength a fair way above the absorption edge of the material small crystallites are still excited into the lowest (1s) electron-hole-pair level while bigger ones are excited into higher levels (1p or still higher ones). This situation is sketched in fig.5. The small and the big microcrystals are labeled systems I and II respectively.

The behaviour of system I can easily be described as a two-level system. From density matrix theory there results an equation of motion for the excitation probability ρ_s of the 1s level /7/ (neglecting its excitation probability in thermal equilibrium):

$$\frac{d\rho_s}{dt} + \frac{\rho_s}{T_s} = \left(\frac{1}{2} - \rho_s \right) a_s I \,. \tag{1}$$

The constant a_s labels the linear absorption of the microcrystal while I represents the incident intensity. The generation term on the right-hand side of eq.(1) contains saturation of absorption due to state filling. The total absorption change turns out to be proportional to $a_s\rho_s$ and the number of system I microcrystals in the sample.

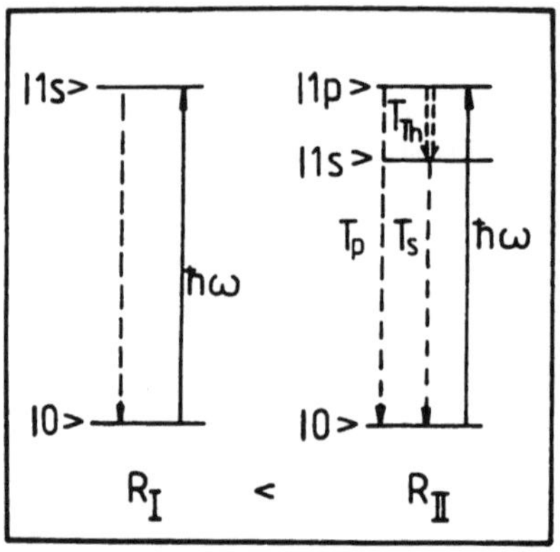

Fig.5 Sketch of the energy spectrum taking into account different microcrystals

To describe the behaviour of larger microcrystals a modified formalism is needed that includes thermalization. We assume that the generation term should have the same form for excitation of any level. These assumptions lead us to the following ansatz:

$$\frac{d\rho_s}{dt} + \frac{\rho_s}{T_s} + \frac{\rho_s - \rho_s^*}{T_{Th}} = 0$$
$$\frac{d\rho_p}{dt} + \frac{\rho_p}{T_p} + \frac{\rho_p - \rho_p^*}{T_{Th}} = \left(\frac{1}{2} - \rho_p \right) a_p \, I \, . \qquad (2)$$

The thermalization terms on the left-hand side contain quasi-equilibrium probabilities ρ^* determined by a fermi distribution. Since thermalization does not change the number of excited particles N we may calculate the fermi level from the equation

$$2\,\rho_s + 6\,\rho_p = 2\,\rho_s^* + 6\,\rho_p^* = N \qquad (3)$$

taking into account the degeneracies of the s and p levels /8/. If we may neglect thermal energy kT in comparison with the level separation $\Delta E = E_p - E_s$ the system of equations (2) has a very simple form:

$$\frac{d\rho_s}{dt} + \frac{\rho_s}{T_s} - \frac{3}{T_{Th}} \min\left(\rho_p, \frac{1-\rho_s}{3} \right) = 0$$
$$\frac{d\rho_p}{dt} + \frac{\rho_p}{T_p} + \frac{1}{T_{Th}} \min\left(\rho_p, \frac{1-\rho_s}{3} \right) = \left(\frac{1}{2} - \rho_p \right) a_p \, I \, . \qquad (4)$$

The thermalization term flips when all states below the excitation level may be filled. Consequently we have two different relaxation regimes:

i) $N<2$, $\rho_p<(1-\rho_s)/3$
Since the absorption change is determined by ρ_p its relaxation time is identical with the thermalization time T_{Th} as described in /2/.

ii) $N>2$, $\rho_p>(1-\rho_s)/3$
As in case i) the system will reach quasi-equilibrium after a time T_{Th}. A stationary value of $\rho_s \approx 1$ is reached. Inserting it into (4) we get for ρ_p and hence for the absorption change:

$$\rho_p(t) = \rho_p(0)\, \exp\left(-\frac{t}{T_p}\right) + \frac{T_p}{3(T_s+T_{Th})} \left(\exp\left(-\frac{t}{T_p}\right) - 1\right) . \qquad (5)$$

Equation (5) is valid until N becomes less than two. This restriction determines the duration of the fast process under consideration.

Case ii) demands a certain minimum pump energy since at least two electron-hole pairs have to be excited in each system II crystallite. Consequently the model predicts femtosecond relaxation for low pump intensities, as reported in /2/, and a picosecond one above a certain threshold that may be derived by solving (4) and keeping in mind that $N=2$ at the end of the pump pulse has to be fulfilled.

To compare our model with the experimental findings we calculated relaxation curves for RG5 and RG8. The results given in figs.7 and 8 are in good agreement with

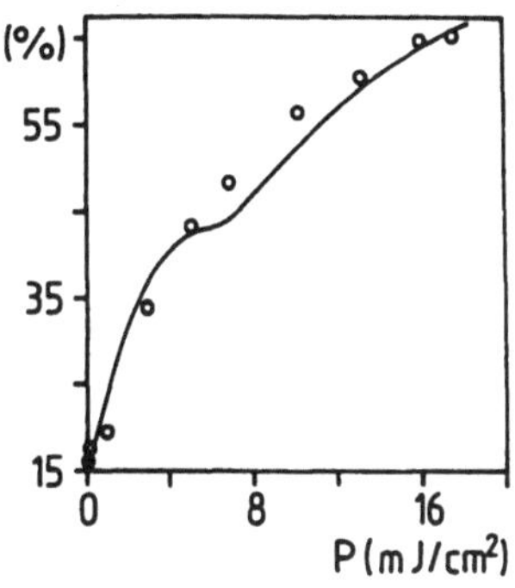

Fig.6 Maximum transmission versus pulse energy for RG5. The full line represents results of our model. The respective experimental values are indicated by circles.

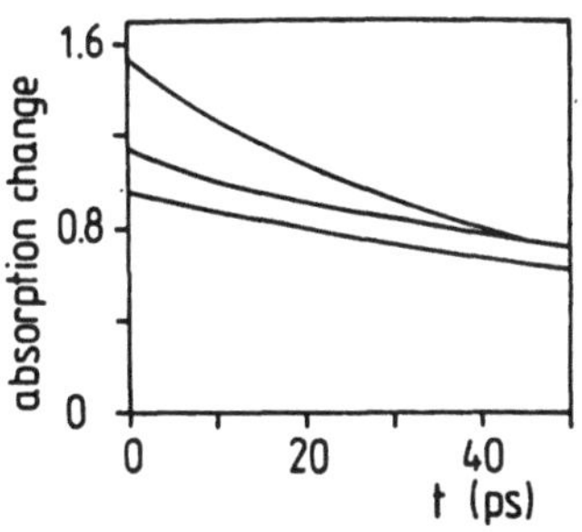

Fig.7 Model relaxation curve for RG5

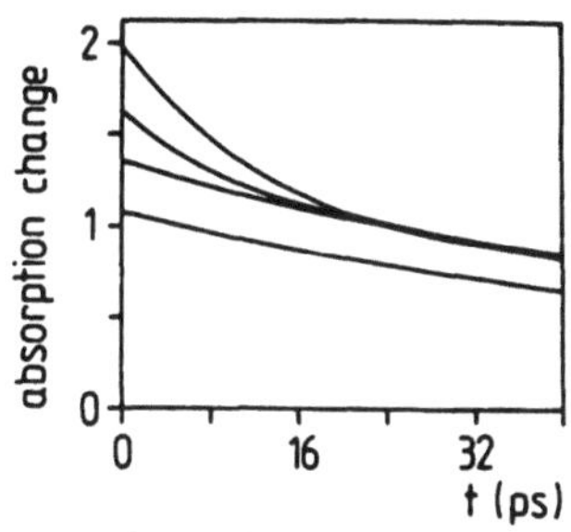

Fig.8 Model relaxation curve for RG8

the experimental ones (compare figs.3 and 4). Additionally the saturation of absorption has been calculated for RG5 (see fig.6).

The relaxation times were fitted to the experiments. The respective parameters are:

for RG5: T_s=120ps, T_r=25ps

for RG8: T_s=80ps, T_r=10ps

4. Discussion

Our theoretical model gives an explanation for the power threshold occurring in the fast relaxation process in semiconductor-doped glasses. The respective relaxation times have to be fitted from the experiment.

The plateau in the theoretical curve in fig.6 may result from the lack of accuracy of the fitting procedure, since it is very sensitive to the choice of the absorption constants for systems I and II respectively.

For the delay occurring in the experimental relaxation curves even at low energies (see the respective figures) another explanation is needed: Since CdS_xSe_{1-x} crystals are of hexagonal symmetry the excited states are split into two corresponding to light and heavy holes respectively. Using a theory analogous to that outlined above but with pumping into both excited levels one will get plateaulike structures in the relaxation curves if the oscillator strength of the lower state exceeds that of the higher one. The plateau length will be determined by the transition time from the light-hole to the heavy-hole state.

5. References

/1/ P.Roussignol, M.Kull, D.Ricard, F.de Rougemont, R.Frey, C.Flytzanis, Appl.Phys.Lett. 51,1882 (1987)

/2/ M.C.Nuss, W.Zinth, W.Kaiser, Appl.Phys.Lett. 49,1717 (1986)

/3/ M.Tomita, T.Matsumoto, M.Matsuoka, JOSA B6,165 (1988)

/4/ S.Schmitt-Rink, D.A.B.Miller, D.S.Chemla, Phys.Rev. B35,8113 (1987)
/5/ I.M.Lifshiz, V.V.Slezov, Zh.E.T.F. 35,479 (1958)
/6/ V.Dietel, V.Rudolph, B.Wilhelmi, J.C.Diels, J.J.Fontaine, Izw. Akad. Nauk SSSR 48,480 (1984), JOSA B2,681 (1985)
V.Dietel, F.Döpel, G.Marovski, V.Petrov, C.Rempel, V.Rudolph, F.P.Schäger, B.Wilhelmi, Appl.Phys.B45,183 (1988)
/7/ Th.Peschel, Phys.Stat.Sol. B150,657 (1988)
/8/ Here we consider the case of a single 1s state only. If the level is split due to different hole masses or there are other states like traps in the vicinity of the 1s state we have to replace its degree of degeneracy by the total number of states. Additionally the relaxation time of the 1s state has to be replaced by an averaged one. The result will be higher threshold intensities for a saturation of the 1p state but the theory in general remains unchanged.

Part V

Applications to Molecular Spectroscopy

Molecular Photochemistry with Femtosecond Laser Pulses

T. Baumert, B. Bühler, R. Thalweiser, and G. Gerber

Fakultät für Physik, Universität Freiburg, D-7800 Freiburg, Fed. Rep. of Germany

Abstract. Femtosecond laser pulses generated from an amplified colliding pulse modelocked ring dye laser have been employed in molecular beam experiments to study the dynamics and the pathways of multiphoton induced ionization, autoionization and fragmentation of Na_2. Energy distributions of photoelectrons arising from these processes and the mass and released kinetic energy of the corresponding fragment ions are measured by time-of-flight spectroscopy.

Alkali dimers and small clusters have recently been the subject of a number of ionization and fragmentation studies (1). In particular the sodium molecules Na_2 and Na_3 have received increasing attention. Because of their low ionization potentials two or three visible photons are sufficient for photoionization. With the intense lasers applied in most of the studies multiphoton processes are unavoidable. While purely nonresonant processes play only a minor role for the alkali systems, the dominant contribution to the ionization process is due to resonantly enhanced multiphoton ionization (REMPI). However, despite the numerous studies carried out with sodium, the ionization and fragmentation processes are in many cases not yet well known. This is mainly because i) the final continuum states are usually not analyzed and ii) dissociative ionization and neutral fragmentation with subsequent photoionization of excited fragments cannot be distinguished when using nanosecond or even picosecond laser pulses, since both processes lead to the same fragment ions.

Only recently have experiments been reported on sodium dimers with clearly identified fragmentation and ionization pathways: fragmentation of neutral Na_2 into two excited $Na^*(3p)$ atoms — the complementary process to laser-induced associative ionization — applying Doppler spectroscopy (2) and multiphoton ionization and dissociation of Na_2 via the double minimum state (2) $^1\Sigma_u^+$ and via Rydberg states using time-of-flight spectroscopy (3).

Springer Proceedings in Physics, Vol. 49 **Ultrafast Phenomena in Spectroscopy**
Editors: E. Klose · B. Wilhelmi

We report on first results obtained with femtosecond laser pulses applied to molecular beam studies of the dynamics and the mechanisms of ionization, autoionization and fragmentation of highly excited molecular states of Na_2. In particular we have focussed our efforts on the understanding of the excitation and different decay processes of highly excited neutral states of Na_2 which are embedded in the molecular ion continua. Electronic autoionization of bound doubly excited molecular states Na_2^{**} (nl,n'l') and fragmentation of highly excited neutral and ionic states of Na_2 have hardly been investigated and are generally not well understood. The only two other studies of doubly excited molecular states we are aware of are concerned with the role of the repulsive $Q_1\,{}^1\Sigma_g^+(2p\sigma_u)^2$ state in multiphoton ionization processes of H_2 (4) and with the determination of spectroscopic constants for the ${}^1\Sigma_g^-$ (4p+3d) state of K_2 (5). Doubly excited states play a major role in the reaction dynamics of diatomic molecules, since these states directly couple the different continua of dissociation and ionization. Electronic autoionization and neutral dissociation are therefore competing processes, but information from both channels may be used to characterize the doubly excited molecular states.

To study the dynamics of multiphoton processes leading to excitation, autoionization and fragmentation we have applied ultrashort laser pulses to induce the transitions and time-of-flight (TOF) spectroscopy to determine the mass and initial kinetic energies of the fragments and the energy and angular distributions of ejected electrons. The experiments were carried out with a supersonic molecular beam. Because of the strong cooling in supersonic expansions we predominantly produce Na_2 in the lowest vibrational state $v''=0$ and in a very few low J states. The final continuum states can definitely be assigned from the measured electron and ion kinetic energy distributions. Application of femtosecond laser pulses considerably facilitates the interpretation of fragmentation processes, since the time duration of a laser pulse is much shorter than the fragment separation time. The terminal velocities of the separating fragments are typically 0.002 nm/fs for $W=3000\ cm^{-1}$ released kinetic energy. Therefore no further laser-induced excitation or ionization of the fragments can occur and all observed signals have to be related to processes which occur at small internuclear distances.

Femtosecond pulses are generated in a home-built colliding pulse modelocked ring dye laser (CPM) with 4 intracavity prisms to adjust for the group velocity dispersion. The emission peak has been shifted to 616 nm by adjusting the DODCI absorber concentration. The output

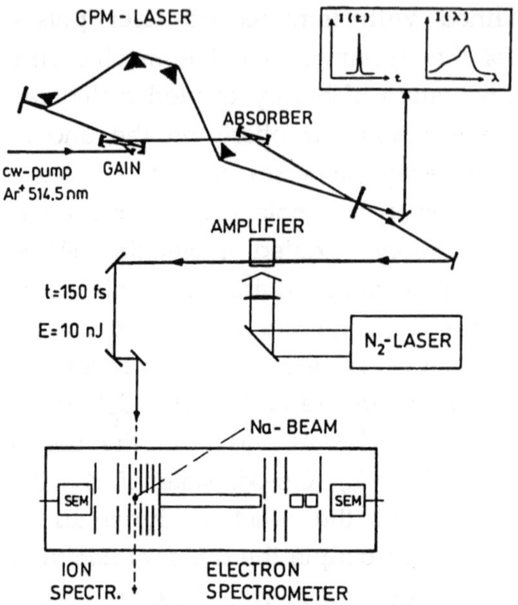

Fig.1. Femtosecond laser – molecular beam arrangement.

of the CPM dye laser was amplified at a rate of 100 Hz in a N_2-laser pumped dye amplifier to produce pulses of 10 nJ energy and t=150 fs time duration. The pulse length was measured using the autocorrelation by second harmonic generation (SHG) in a nonlinear crystal. The laser pulse energy was intentionally kept at this low level in order to be able to investigate the basic physical processes.

The laser system and the schematic experimental arrangement of the ion and electron time-of-flight spectrometers are shown in fig. 1. The electron TOF-spectrometer had been calibrated by a series of REMPI measurements of atomic Na where electrons of definite kinetic energy are produced when using the two-photon-excited even states 3d, 4s, ... as ionization enhancing intermediate levels. The laser-molecular beam interaction region is placed between parallel plates and since the ions are extracted perpendicular to the beam with a low electric field, parent ions can be distinguished from fragment ions having initial kinetic energy. The released kinetic energy leads to a broadening or a double peak structure in the TOF spectrum due to fragments recoiling parallel and antiparallel to the extracting field. Therefore, from the observed difference in flight time the kinetic energy of the fragments can be inferred. This experimental technique is well known and has been used by several groups (6). The TOF spectrum in fig.2 clearly shows the observation of Na_3^+, Na_2^+ and "slow" Na^+ as well as "fast" Na^+ fragment

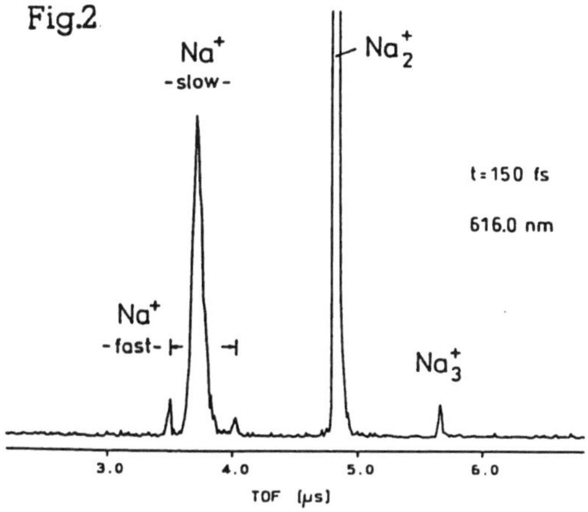

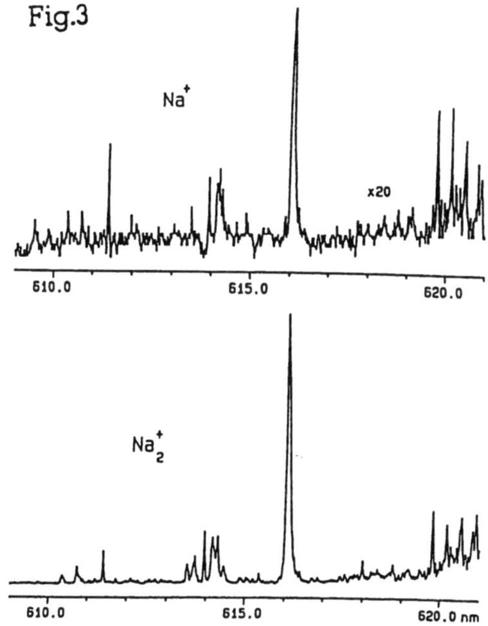

Fig.2. Time-of-flight spectrum of ions formed by the interaction of femto-second laser pulses with a sodium molecular beam.

Fig.3. One color REMPI spectra of Na and Na_2

ions resulting from the fs-laser excitation at λ= 616 nm. "Fast" and "slow" Na^+ ions originate from fragmentation processes occurring at small inter-nuclear distances of Na_2^*. Predissociation of Na_2^* and photoionization of Na^* as the origin of observed Na^+ ions can be ruled out considering the time duration of the fs-laser pulse. Based on this result, which is rather difficult to obtain from other experiments, and with the known molecular potential curves we completely determined for this model case the multiphoton excitation and fragmentation pathways.

With a pulsed tunable dye laser (τ=2.5 ns) we observed in the ionization spectra between 610 nm and 620 nm an isolated and very strong peak at 616.08 nm in both the Na_2^+ and the Na^+ channel. The wavelength dependent ionization spectra of atomic and molecular (Na_2) sodium are shown in fig.3. Based on the known spectroscopy of molecular Rydberg states of Na_2 (7) it is evident that this peak in the Na_2^+ spectrum is due to a resonance enhanced three-photon ionization of Na_2. This particular process is indicated in the Na_2 potential energy curve diagram given in fig. 4. The ionization enhancing intermediate electronic state

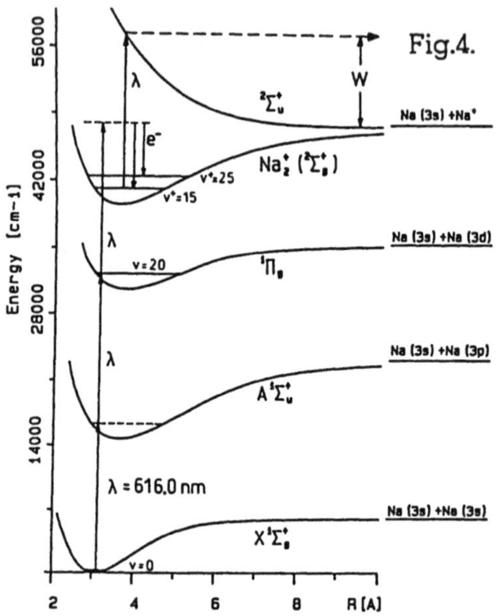

Fig.4. Potential energy diagram of Na_2 illustrating the process leading to "fast" Na^+ ions.

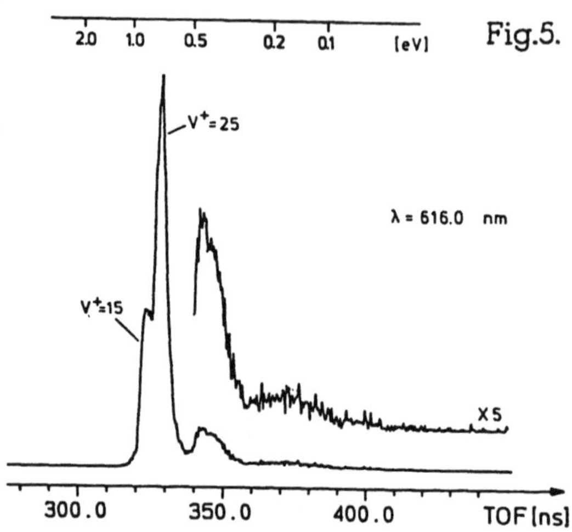

Fig.5. Time-of-flight electron spectrum.

$^1\Pi_g$ is populated by the two-photon process Na_2 ($X\ ^1\Sigma_g^+$, v"=0) + 2hν --> Na_2 (Ryd $^1\Pi_g$, v*=20). Because of the Na_2 molecules predominantly produced in the v"=0 level, the two-photon Franck-Condon factors and the upper $^1\Pi_g$ Rydberg state vibrational spacing ΔG, which is of the order of 100 cm^{-1}, approximately 90% of the excited Na_2 molecules are in the v*=20 level, despite the broad spectral distribution (≈150 cm^{-1}) of the femtosecond laser pulse. The Rydberg molecule is then photoionized by absorption of a third photon. The dimer ions are preferentially formed in the v^+= 24,25 and v^+= 14,15 vibrational states of the electronic ground state $X(^2\Sigma_g^+)$ due to favorable Franck-Condon factors. This direct ionization of the $^1\Pi_g$ Rydberg state leads to electrons having kinetic energies of E=810 +/- 10 meV and E=940 +/- 10 meV, which are actually observed in the TOF electron spectrum shown in fig.5. In addition to these strong peaks the electron spectrum shows less intense broad structures in the range between 300 meV and 500 meV and around 200 meV extending to very low energies. The created dimer ions may now undergo a bound-free transition by absorption of one more photon from still the same fs-laser pulse:

$$Na_2^+ (X\ ^2\Sigma_g^+, v^+) + h\nu \rightarrow Na_2^{+*} (^2\Sigma_u^+) \rightarrow Na^+ + Na(3s) + W$$

Taking into account the populated v^+-levels and the known potential curves for the ionic ground and first excited states (8) the corresponding bound-free transitions lead to recoil energies W between 9910 cm^{-1} and 11037 cm^{-1}. The energy W=10500 +/- 500 cm^{-1} obtained from the analysis of the TOF ion spectrum perfectly agrees with that. We therefore conclude that "fast" Na^+ ions are produced by the ionization and fragmentation process shown in fig.4.

The observation of "slow" Na^+ ions however cannot be explained within this framework since photoionization of the vibrational level v^*=20 of the $^1\Pi_g$ (3s + 3d) Rydberg state directly into the $^2\Sigma_u^+$ - continuum is energetically forbidden for the applied laser wavelength. Based on the measured electron energy distribution and the recoil energy W=900 +/- 500 cm^{-1} , obtained from the analysis of the "slow" Na^+ fragment ions, a consistent explanation is found by considering the excitation of doubly excited molecular states of Na_2 as is shown in fig.6. The doubly excited states Na_2^{**} (nl,n'l') form a Rydberg series converging versus the $^2\Pi_u$ state of Na_2^+, whose potential curve is known theoretically (8). Assuming that the shape of a $^1\Pi_u$ state potential curve, correlated to Na(3p)+Na(4s), is similar to the ionic $^2\Pi_u$ curve and that it has a potential barrier at large internuclear distances like the B $^1\Pi_u$ state built from Na(3s)+Na(3p), we conclude that in the resonance enhanced three-photon process vibronic levels close to the dissociation limit of the doubly excited $^1\Pi_u$ (3p+4s) state are excited. The wavefunctions of these vibronic levels extend from 3Å to approximately 10Å. These doubly excited levels may autoionize into the $X(^2\Sigma_g^+)$ ground state of Na_2^+ giving rise to electron energies between 260 meV and 500 meV which are actually observed. For internuclear distances greater than 6Å the vibronic levels cross into the continuum of the repulsive $^2\Sigma_u^+$ state of Na_2^+. Therefore for R ≥ 6Å there is a second open autoionization channel which is responsible for the observed electrons having energies in the range from 0 meV to 160 meV. This autoionization process

$$Na_2^{**}\ ^1\Pi_u (3p+4s) \rightarrow Na_2^{+*}(^2\Sigma_u^+) + e^-(E_{kin})$$

and the subsequent fragmentation

$$Na_2^{+*} (^2\Sigma_u^+) \rightarrow Na^+ + Na(3s) + W$$

produce "slow" Na^+ ions whose kinetic energies depend on the internuclear distance R where the autoionization takes place. A calculation

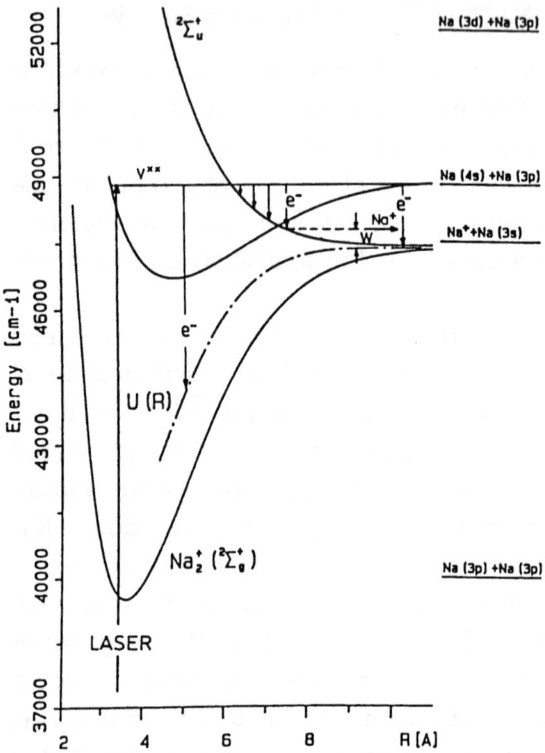

Fig.6. Excitation and autoionization processes of doubly excited Na_2^{**}

of the electron spectrum arising from bound-free transitions $^1\Pi_u$(4s+3p) ——> $^2\Sigma_u^+$, assuming a constant transition moment for the R-range under consideration, shows good agreement with the actually observed electron energy distribution.

In conclusion, this is the first reported experiment where in a molecular beam experiment a fs-laser has been used in combination with ion and electron spectroscopy to clarify the ionization, autoionization and fragmentation mechanisms of molecules which are excited by multiphoton processes into an energy range of singly and doubly excited electronic states.

REFERENCES:

1) S. Leutwyler, M. Hoffmann, H. Härri, E. Schumacher
Chem. Phys. Lett. 77, 257 (1981)
M. Broyer, J. Chevaleyre, G. Delacretaz, P. Fayet, L. Wöste
Surface Science 156, 342 (1985)

C. Brechignac, Ph. Cahuzac, J. Roux, D. Pavolini, F. Spiegelmann
J. Chem. Phys. 87, 5694 (1987)
C. Brechignac, Ph. Cahuzac, J. Leyguier, J. Weiner,
J. Chem. Phys. 90, 1492 (1989)
2) G. Gerber, R. Möller, Phys. Rev. Lett. 55, 814 (1985) and
Proceedings in Physics Vol. 16 (Springer 1987) p. 126
eds. N.K. Rahman, C. Guidotti, M. Allegrini
3) R. Haugstätter, A. Goerke, I.V. Hertel Z. Phys. D 9, 153 (1988)
and Phys. Rev. A 39, 5084 (1989)
4) J.W. Verschuur, H.B. van Linden van den Heuvell
Chem. Phys. 129, 1 (1989)
5) A. Katern, P. Kowalczyk, F. Engelke, Chem. Phys. Lett. 146, 325 (1988)
6) e.g. R. Ogorzalek Loo, G.E. Hall, H.P. Härri, P.L. Houston,
J. Phys. Chem. 92, 5 (1988)
7) N.W. Carlson, A.J. Taylor, K.M. Jones, A.L. Schawlow
Phys. Rev. A 24, 822 (1981)
C. Bordas, P. Labastie, J. Chevaleyre, M. Broyer
Chem. Phys. 129, 21 (1989)
8) W. Müller, W. Meyer, J. Chem. Phys. 80, 3311 (1984)
and Thesis I. Schmidt Univ. Kaiserslautern 1987

Femtosecond Spectroscopy of Polydiacetylene and Polythiophene

T. Kobayashi [1], *M. Yoshizawa* [1], *U. Stamm* [1,*], *M. Taiji* [1], *and K. Yoshino* [2]

[1] Department of Physics, Faculty of Science, University of Tokyo, Hongo 7-3-1, Bunkyo-ku, Tokyo 113, Japan

[2] Department of Electrical Engineering, Osaka University, Yamada-Oka 2-1, Suita, Osaka 565, Japan

* Permanent address: Sektion Physik, Friedrich-Schiller-Universität Jena, Max-Wien-Platz 1, DDR-6900 Jena, German Dem. Rep.

Abstract. Femtosecond time-resolved absorption spectroscopy was performed on polydiacetylene (PDA-3BCMU) and poly(3-methylthiophene) (P3MT) films. Several nonlinear optical processes, i.e. hole burning, Raman gain, and dynamic Stark shift, were observed at the same time. The relaxation of photoexcited free excitons to self-trapped excitons was observed in both PDA-3BCMU and P3MT. The formation and decay times of self-trapped excitons at 10 K were 150±50 fs and 2.0±0.1 ps in PDA-3BCMU and 70±50 fs and 800±100 fs in P3MT, respectively.

1. Introduction

Optical and electrical properties of conjugated polymers have been extensively investigated because of their large and fast optical nonlinearities and their role as model compounds for quasi-one-dimensional semiconductors. Recent progress in the field of high speed optoelectronics has encouraged the search for new materials with large third-order nonlinear optical susceptibilities [1]. Conjugated polymers seem to be very promising candidates for future practical applications in nonlinear optical devices.

Polydiacetylenes (PDAs) and polythiophenes (PTs) belong to the rather well studied conjugated polymers. PDAs show phase transitions with dramatic color changes and their properties can be modified by changing the side-groups R and R' in the backbone structure of $\{RC-C\equiv C-CR'\}_n$ [2]. PTs possess a simple backbone geometry resembling that of *cis*-polyacetylene. The *cis*-like structure is stabilized by the sulfur which is known to interact only weakly with the π-electron system of the backbone so that one can consider PTs as a kind of pseudopolyenes [3].

The elementary excitations in PDAs and PTs are excitons, polarons, and bipolarons. Solitons cannot be generated because of their nondegenerate ground states. In PDAs photogenerated singlet and triplet excitons have been observed [4,5]. Polarons and bipolarons are formed in PTs after photoexcitation or electrochemical doping [6,7]. The results of photoluminescence measurements have indicated that the decay time constants of PTs and red-phase PDAs are several picoseconds [8,9]. The fluorescence is expected to be due to either free charge carriers or excitons. In blue-phase PDAs photoluminescence has not been observed. This suggests that the nonradiative decay in blue-phase PDA is much faster than in red-phase PDA.

In the present study, we investigated the relaxation dynamics of photoexcitations in a blue-phase PDA film (PDA-3BCMU) and poly(3-methylthiophene) (P3MT) by femtosecond time-resolved absorption spectroscopy.

2. Experimental

The femtosecond spectroscopy system consists of four parts, a colliding-pulse mode-locked ring dye laser, a four-stage dye amplifier pumped by the second harmonic pulses of a Q-switched Nd:YAG laser, a high-index prism pulse compressor, and an optical system of pump-probe absorption spectroscopy [10]. The center

Springer Proceedings in Physics, Vol. 49 Ultrafast Phenomena in Spectroscopy
Editors: E. Klose · B. Wilhelmi

wavelength of the amplified pulses is 628 nm (1.97 eV). The duration and the energy per pulse are 70-100 fs and about 200 μJ, respectively. The probe white continuum generated by self-phase modulation was detected with a polychromator/multichannel detection system and analyzed by a microcomputer.

The sample of PDA-3BCMU (3-butoxycarbonylmethylurethane) was synthesized by a procedure given in the literature [11] and prepared by casting condensed solution in chloroform on a glass substrate. The thickness of the films was about 100 nm. The PDA-3BCMU film has an absorption peak of the lowest singlet (^1B_u) excitons at 1.93 eV at 10 K and 2.00 eV at 290 K. The 1.97-eV pump pulse excites the ^1B_u excitons selectively and efficiently.

About 1 μm-thick films of poly(3-methylthiophene) (P3MT) were prepared by electrochemical polymerization on glass substrates [12]. The absorption peak of P3MT is 2.4 eV and the pump photon energy lies very close above the band edge.

3. Experimental Results

3.1 Polydiacetylene

The transient photoinduced absorption spectra of the PDA-3BCMU film observed at 10 K are shown in Fig. 1. The polarizations of the pump and probe pulses are parallel to each other. At delay times between -0.1 ps and 0.1 ps, i.e. the pump and probe pulses temporally overlap each other at the sample, a sharp bleaching peak at 1.97 eV and two small minima at 1.79 and 1.71 eV are observed.

The sharp peak of bleaching at 1.97 eV is mainly due to a hole burning, but it has an asymmetric shape. The asymmetry can be explained in terms of the optical Stark effect. Since the pump photon energy in the present study is slightly higher than the absorption peak of the excitons, the transition energy shifts to lower energy and the absorbance change has a maximum at 1.95 eV just below the pump photon energy. However, the energy shift of the excitons could not be estimated from the observed absorbance change spectra. Further careful investigations of the absorbance change and effect of the inhomogeneous and homogeneous broadening of the spectrum of excitons are needed to confirm the explanation. The two small minima at 1.79 and 1.71 eV can be explained in terms of Raman gain. The corresponding Raman shifts are 1450 and 2100 cm^{-1}, which are assigned to the stretching

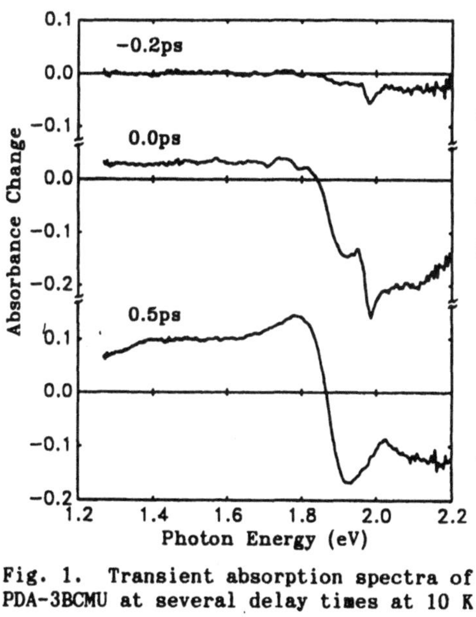

Fig. 1. Transient absorption spectra of PDA-3BCMU at several delay times at 10 K.

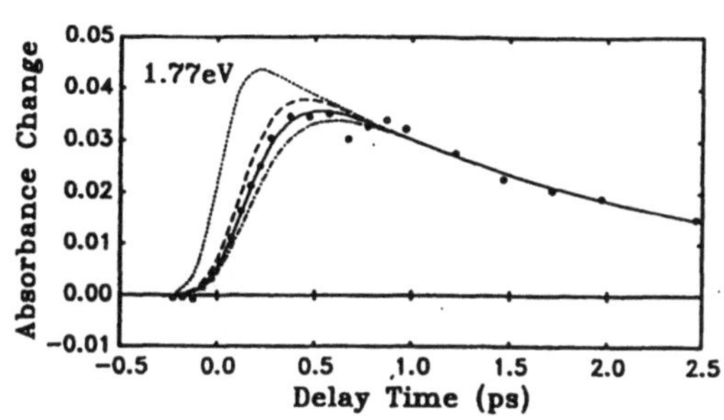

Fig. 2. The time dependence of the absorbance change of PDA-3BCMU at 1.77 eV. Curves are the calculated absorbance changes with the formation times of 0 fs (dotted curve), 100 fs (dashed curve), 150 fs (solid curve), and 200 fs (dash-dotted curve). The decay time is set at 2.0 ps in the calculation, using the pulse shape of sech^2 with 150 fs half-width for both pump and probe pulses.

vibrations of the C=C bond and C≡C bond, respectively. The optical Stark effect mediated by the same phonon modes was also observed in PDA-*p*TS [13].

The photoinduced bleaching around 1.92 eV is due to the absorption saturation of the ^1B_u excitons. The photoinduced absorption below 1.85 eV is not observed at -0.2 ps and appears more slowly than the bleaching. At 0.0 ps the absorption spectrum is very broad and featureless in the whole spectral region between 1.3 and 1.8 eV except for two small minima at 1.79 and 1.71 eV, while it has a maximum at 1.80 eV and drops below 1.35 eV at delay times longer than 0.5 ps. It suggests that the photoinduced absorption shifts to higher energy with time from 0.0 ps to 0.5 ps. The photoinduced absorption is assigned to the self-trapped excitons formed from the photoexcited free ^1B_u excitons. The free excitons relax to the self-trapped excitons and the transitions from the excitons to the conduction band shift to higher energy.

Figure 2 shows the time dependence of the absorbance change in PDA-3BCMU at 1.77 eV. The lifetime of the self-trapped excitons is estimated to be 2.0±0.1 ps from the absorbance change at delay times longer than 1 ps. The formation time of the self-trapped excitons can be determined by the convolution of the pump and probe pulses with several formation times, compared with the rise of the absorbance change at 1.77 eV and the early decay process of the bleaching. The calculated absorbance changes are plotted in Fig. 2. At 290 K, the formation and decay times of the self-trapped excitons are determined to be 150±50 fs and 1.5±0.1 ps, respectively.

3.2 Poly(3-methylthiophene)

The transient absorption spectra of P3MT at 10 K are shown in Fig. 3. At the delay times of -0.1 and 0.0 ps a negative peak due to Raman gain was again observed at 1.79 eV. The Raman shift of 1.97-1.79 = 0.18 eV is assigned to the C=C stretching mode. The strong bleaching around the excitation energy results from the generation of excitations near the band edge as well as from the ground-state depopulation. The sharp bleached line at zero delay time indicates nonthermalized free-exciton in the conduction and the valence band. The bleached part at later times after thermalization becomes much broader. On the lower energy side of the broad absorption band, the induced absorption arises during excitation caused by the onset of a strong excited-state absorption.

The time dependence of the absorbance changes in P3MT is similar to PDA-3BCMU. The formation and decay time constants of the self-trapped excitons at 10 K are

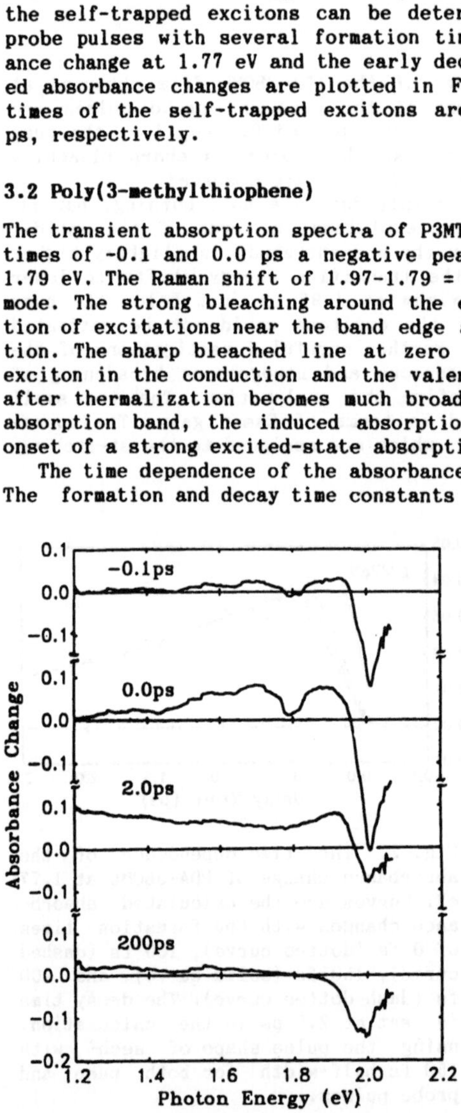

Fig.3. Transient absorption spectra of P3MT at several delay times at 10 K.

estimated to be 70±50 fs and 800±100 fs, respectively. The time dependence of absorbance change observed at 290 K does not show a significant difference from the low temperature data and the 800-fs decay time is also found. After the 800-fs relaxation, it was observed that the bleaching becomes larger again and the induced absorption decays slowly following the power law, $t^{-0.8}$. Both the remaining bleaching higher than 1.8 eV and the induced absorption with a peak around 1.4 eV have a decay time much longer than several hundred picoseconds. This long lifetime component is due to triplet excitons, polarons and/or bipolarons. The definite assignment of the component is to be made by an elongated delay time experiment together with the intensity dependence.

4. Discussion

Since the formation process of self-trapped excitons has no barrier in the one-dimensional system, the formation process is expected to take place within the coupled phonon cycle. The excitons in PDA-3BCMU and P3MT are strongly coupled with the C-C stretching modes. The frequencies of the stretching modes are 1500-2100 cm^{-1} and correspond to the oscillation periods of 15-25 fs, while the trapping time constants of excitons are 150 fs in PDA-3BCMU and 70 fs in P3MT and they are much longer than the phonon periods. It can be explained as follows. The binding energy remains as the kinetic energy of the lattice oscillation and the formation time of the self-trapped excitons is determined by the energy redistribution rate from the strongly coupled phonon to the other phonon modes. The formation of self-trapped excitons in PDA-3BCMU is slower than in P3MT. It may be due to the bulky side groups in PDA-3BCMU and its lower frequency phonon modes than in P3MT.

The fluorescence quantum yield in P3MT was reported to be 10^{-5} [9]. The low fluorescence quantum yield has been explained by very fast nonradiative decay processes. The temperature dependence of the decay time constants is very small in both PDA-3BCMU and P3MT. It means that the nonradiative decay processes have only low potential barriers. The lifetime of the self-trapped excitons in P3MT is shorter than that in PDA-3BCMU, while luminescence has not been observed in PDA-3BCMU. Our preliminary results shows that the lifetime of self-trapped excitons in PDA-4BCMU, which is a red-phase PDA and has photoluminescence, is about 3 ps at 10 K [14,15]. It is only one and a half times longer than in PDA-3BCMU. This means that the difference between red-phase PDA and blue-phase PDA is not only the rate of the nonradiative decay but also the mechanism of photoluminescence. Experiments on the photoluminescence are in progress.

References

[1] See, e.g., *Nonlinear Optics of Organics and Semiconductors*, edited by T. Kobayashi, (Springer-Verlag, Berlin, 1989).
[2] See, e.g., *Polydiacetylenes*, edited by D. Bloor and R. R. Chance, (Martinus Nijhoff Publishers, Dordrecht, Netherlands, 1985).
[3] Z. Vardeny, et al., Phys. Rev. Lett. **56**, 671 (1986).
[4] B. I. Greene, et al., Chem. Phys. Lett., **139**, 381 (1987).
[5] T. Kobayashi and H. Ikeda, Chem. Phys. Lett., **133**, 54 (1987).
[6] Y. H. Kim et al., Phys. Rev. B **36**, 7486 (1987).
[7] K. Kaneto et al., J. Phys. Soc. Jpn. **57**, 747 (1988).
[8] D. Bloor et al., in *Electronic properties of polymers and related compounds*, edited by H. Kuzmany et al., (Springer-Verlag, Berlin, 1985) p.253.
[9] K. S. Wong et al., J. Phys. C **18**, L843 (1985).
[10] M. Yoshizawa et al., IEEE J. Quantum Electron, **25**, (1989) 2532.
[11] H. Ohnuma et al., Macromolecules **17**, 1285 (1984).
[12] K. Kaneto et al., J. Phys. Soc. Jpn. **54**, 1146 (1985).
[13] G. J. Blanchard et al., Phys. Rev. Lett. **63**, 887 (1989).
[14] T. Kobayashi et al., J. Opt. Soc. Am. B (Special Issue on the Application of the Ultrafast Spectroscopy to Chemistry and Biology), to be published.
[15] M. Yoshizawa et al., to be submitted.

Investigations on the Dynamics of Hydrogen Bonds by Picosecond IR Spectroscopy

H. Graener, T.Q. Ye, and A. Laubereau

Physikalisches Institut, Universität Bayreuth, D-8580 Bayreuth, Fed. Rep. of Germany

During the past decades a variety of spectroscopic information on hydrogen bonded systems has been accumulated, giving detailed insight into the structure and chemical properties of hydrogen bridges /1/. Direct observations of the dynamics of this class of molecular systems are still lacking. We report on a novel experimental approach by the help of transient IR spectroscopy. After well-defined vibrational excitation by an infrared picosecond pulse, we observe the time evolution of the hydrogen bonds by time and frequency resolved IR transmission spectroscopy using the OH stretching vibration as a spectroscopic probe.

It is well established in the literature that the OH stretching absorption is strongly influenced by the hydrogen bonds /1/. The different situations are illustrated in Fig. 1. Similar to ethanol monomers and dimers, the OH end groups of larger open chains have to be distinguished from hydroxyl groups in other (internal) positions. The stretching vibration for acceptor ("non-bonded") end groups is situated at ν_β ~ 3620 cm^{-1}, close to the monomer value ν_α, whereas the donor (H-bonded) end group has an absorption frequency of ν_γ ~ 3500 cm^{-1}. The broad band around ν_δ ~ 3300 to 3400 cm^{-1} is explained by the absorption of internal OH groups in hydrogen-bonded open chains or cyclic oligomers. The integrated absorption coefficient of the OH stretching vibration of oligomeric ethanol (ν_δ) is a factor of 12 (3.5) larger than that for the monomeric (dimeric) molecules /2/. The binding energy of a H-bridge of ethanol was determined from temperature studies to be roughly ~ 6 kcal/mol corresponding to ~ 2000 cm^{-1} per bond /2,3/. In other words,

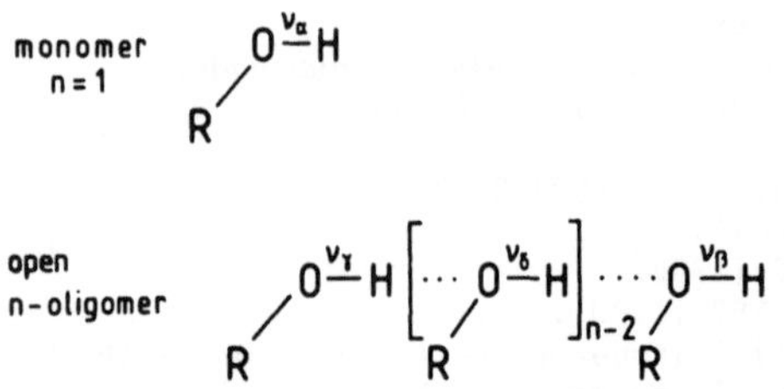

Fig. 1: Chemical structure of an open n-oligomer; the frequencies ν_α–ν_δ of the OH stretching mode are explained in the text; ethanol R = C_2H_5

Springer Proceedings in Physics, Vol. 49 **Ultrafast Phenomena in Spectroscopy**
Editors: E. Klose · B. Wilhelmi

the energy of the first excited states of OH and CH stretching vibrations is larger than the H-bond energy.

Our experimental system for ultrafast IR spectroscopy has been described recently /4/. The experiment starts with picosecond pulses of an actively-passively mode-locked Nd:YAG laser. A large fraction of the laser emission is used in a four-step parametric device to generate single tunable infrared excitation pulses (duration 10 ps, bandwidth 8-15 cm^{-1}, tuning range 2700 - 6700 cm^{-1}, energy 100 μJ). The smaller part of the original laser pulse is time delayed and directed into a separate double-step parametric system for the generation of the probe pulse (energy < 1 μJ, other parameters as for the pump pulse). The two infrared pulses are focused into the thin sample cell filled with ethanol:CCl_4 solutions. As a direct measure of population effects, the energy transmission change $\ln(T/T_0)$ of the interrogation pulse is measured as a function of probe frequency ν, delay time t_D and pump frequency ν_{pu}. Here $T(\nu)$ denotes the transient probe transmission: $T_0(\nu)$ represents the equilibrium transmission value (pump beam blocked). While previous IR spectroscopic studies of H-bridges in liquids were carried out close to thermal equilibrium, the rapid population changes generated by our excitation pulse represent a distinct non-equilibrium situation of the molecular system. Examples of our transient IR transmission spectra with picosecond time resolution are depicted in Fig. 2.

Two different situations are considered: excitation of the OH stretching mode of internal hydroxyl groups of (open and/or ring) oligomers with frequency ν_δ = 3320 cm^{-1} (Fig. 2a) and pumping of a CH stretching band of the ethyl groups at 2925 cm^{-1} (Fig. 2b). The delay time is kept constant at t_D = 10 ps shortly after the maximum of the excitation pulse. The measured transmission change $\ln(T/T_0)$ is plotted as a function of probe frequency. The ordinate scales of Figs. 2 a and b should not be compared because of slightly different excitation amplitudes. The striking similarity of the two spectra in spite of the different excitation conditions should be noted. They display the following common features /5/:

(i) bleaching ($\ln(T/T_0)>0$) of the broad oligomer band around 3300 cm^{-1},

(ii) induced absorption ($\ln(T/T_0)<0$) at the frequency positions 3500 - 3600 cm^{-1} of hydroxylic end groups,

(iii) bleaching in the CH stretching region, e.g. of the strongest CH band at 2980 cm^{-1}; spectra similar to Fig. 2b are obtained for ν_{pu} = 2980 cm^{-1},

(iv) induced absorption at 2960 cm^{-1} which is interpreted as excited state absorption of a CH stretching mode.

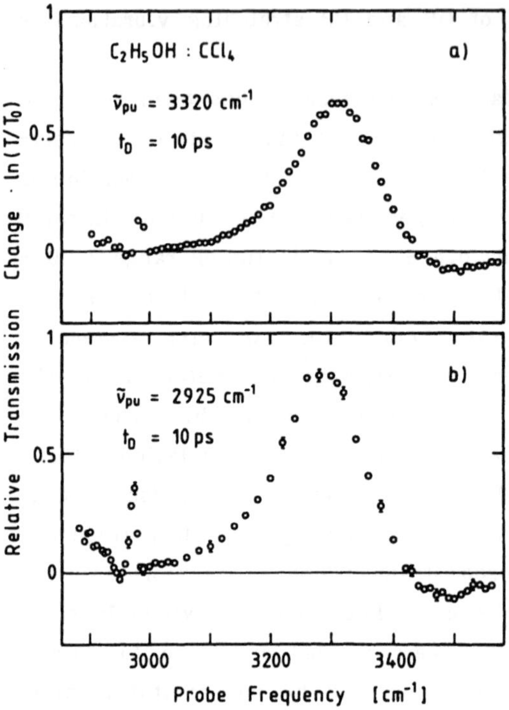

Fig. 2:
Picosecond IR probe spectra of ethanol:CCl_4 (1.5 mol/l, 293 K) shortly after an IR excitation pulse, t_D = 10 ps; a) excitation of internal hydroxylic groups (ν_δ) with ν_{pu} = 3320 cm^{-1}; b) excitation of a CH stretching mode with ν_{pu} = 2925 cm^{-1}

Finding (i) may be explained by the population of the first excited state of the OH mode and/or by predissociation of hydrogen bonds of internal OH groups. Observation (ii) is particularly important for the following since it provides clear evidence for the breaking of hydrogen bridges after CH excitation (Fig. 2b). Points (iii) and (iv) have to be explained by the population of the first excited states of CH stretching modes; these bands are only weakly affected by association or dissociation of oligomers. The observation of CH population changes is expected for CH excitation (Fig. 2b), of course, and gives experimental support to the rapid energy distribution reported previously /6/. For OH excitation (Fig. 2a) findings (iii) and (iv) indicate V-V energy transfer from OH to CH vibrations; in other words, an additional decay channel is seen for the primarily excited ν_δ vibrations competing with vibrational predissociation of H-bonds. It is mentioned that bleaching of the strong CH band at 2980 cm^{-1} is also observed after excitation of donor end groups (and dimers) with ν_{pu} = 3500 cm^{-1}.

Some time resolved data are presented in Fig. 3. The transmission change $\ln(T/T_0)$ is measured versus delay time for different pump and probe frequencies. Pumping and probing at ν_δ, the frequency of internal OH groups, (Fig. 3a), one observes a rapid transmission increase with a time constant of 20

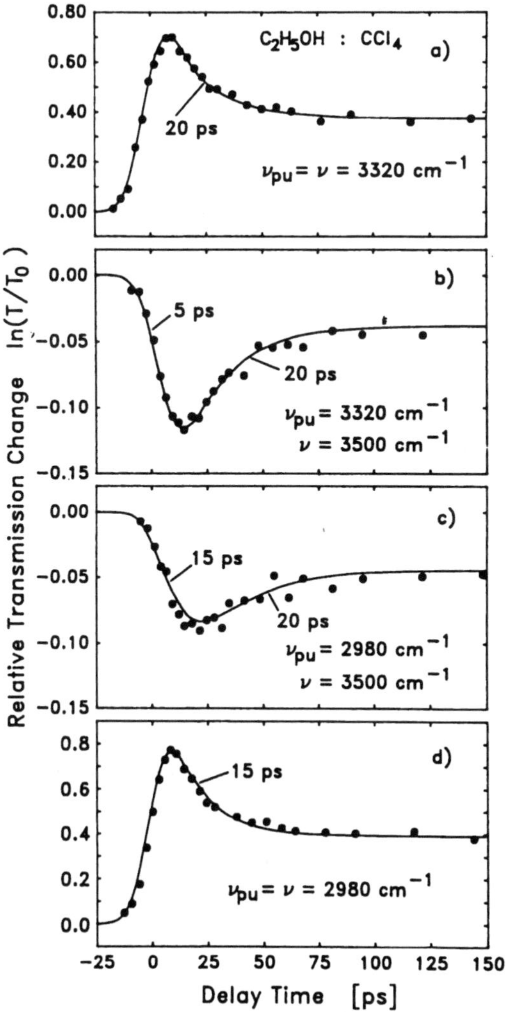

Fig. 3: Relative transmission changes of probe pulse versus delay time for different pump and probe frequencies. a) $\nu_{pu} = \nu = 3320\ cm^{-1}$, pumping and probing the internal OH groups (ν_δ). b) $\nu_{pu} = 3320\ cm^{-1}$, $\nu = 3500\ cm^{-1}$; pumping at ν_δ and probing the OH donor end groups; c) same as b) for CH excitation. d) Bleaching of the CH band $\nu_{pu} = \nu = 2980\ cm^{-1}$, for a measurement of the effective CH population lifetime; experimental points. calculated curves.

ps. For longer delay times (up to 4 ns, not shown in the Fig.) a nearly constant transmission increase persists. Changing the probe frequency to 3500 cm^{-1} (Fig. 3b) and monitoring the bonded end groups of the ethanol chains, one observes a transmission decrease which is slightly delayed compared to the transmission increase of Fig. 3a /7/. For later times the transmission recovers with a time constant of $\tau = 20$ ps. Exciting the strong CH band at 2980 cm^{-1} and probing at the same CH frequency, one observes the transmission changes depicted in Fig. 3d. A rapid increase occurs, followed by a decrease with a time constant of 15 ± 5 ps. From the experimental situation it can be concluded that the latter time constant is the effective lifetime of the CH stretching vibrations.

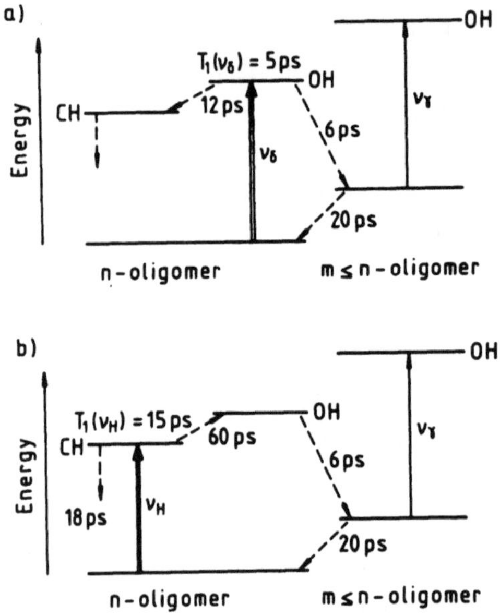

Fig. 4: Simplified energy level scheme for vibrational predissociation of ethanol oligomers; a) for OH excitation ν_δ; b) proposed mechanisms for CH excitation; double arrows: excitation process; single arrows: probing step; dashed arrows: relaxation process

Fig. 3c shows the result for pumping the CH vibrations and probing the OH end groups at 3500 cm^{-1}. The observed first absorption increase is clearly slower than that of Fig. 3b. The involved time constant is 15 ps, the effective lifetime of the CH stretching vibrations, and the subsequent absorption decrease is governed by the same time constant of 20 ps as in the case of OH excitation (Figs. 3 a,b). The experimental results are explained by the simplified energy level schemes of Fig. 4.

a) OH excitation (Fig. 4a)

After exciting the OH stretching vibration of internal hydrogen bonded molecules two decay channels exist for the vibrational energy: (i) the relaxation to the vibrational manifold of the ethanol molecules, dominantly to the CH stretching vibrations with a time constant of 12 ps, and (ii) the dissociation of the hydrogen bond, which is estimated to be a factor of two faster; the time constant of τ = 20 ps is explained by a partial reassociation of the broken bonds.

b) CH excitation (Fig. 4b)

After population of the first excited state of a CH stretching vibration the energy relaxes to other vibrational modes of ethanol. The energy transfer includes the OH stretching mode with an estimated time constant of 60 ps; a populated stretching vibration of internal OH groups may subsequently lead to bond-breaking as indicated above.

The long time transmission changes are due to an overall temperature change after equilibration of the vibrational energy.

In summary we point out that vibrational predissociation of hydrogen bonds was observed on the ps time scale for the system ethanol:CCl_4 after rapid excitation of OH or CH stretching modes. Our observations clearly indicate the IR photodissociation in the liquid at room temperature to be mode selective.

References

/1/ The Hydrogen Bond, eds. P. Schuster, G. Zundel, and C. Sandorfy (Elsevier, Amsterdam, 1976);
G. Geiseler and H. Seidel, Die Wasserstoffbrückenbindung (Vieweg, Braunschweig, 1977). and ref. cited therein

/2/ U. Liddel and E.D. Becker, Spectrochim. Acta 10, 70 (1957)

/3/ A. Biondi and D.J. Simkin, J. Chem. Phys. 25, 1073 (1956)

/4/ H. Graener, R. Dohlus and A. Laubereau, Chem. Phys. Lett. 146, 306 (1987)

/5/ H. Graener, T.Q. Ye and A. Laubereau, J. Chem. Phys. 91, 1043 (1989)

/6/ A. Laubereau, G. Kehl and W. Kaiser. Opt. Commun. 11, 74 (1974)

/7/ H. Graener, T.Q. Ye and A. Laubereau, J. Chem. Phys. 90, 3413 (1989)

Sub-picosecond Time-Resolved Intramolecular Electronic Energy Transfer in Bichromophoric Molecules

M. Kaschke[1] *and N.P. Ernsting*[2]

[1]Zentralinstitut für Optik und Spektroskopie, Akademie der Wissenschaften, Rudower Chaussee 6, DDR-1199 Berlin, German Dem. Rep.

[2]Max-Planck-Institut für biophysikalische Chemie, Abteilung Laserphysik, Postfach 2481, D-3400 Göttingen, Fed. Rep. of Germany

Abstract. The intramolecular electronic energy transfer in different bichromophoric molecules is resolved in time. Model calculations, including the intramolecular distributions of orientations and separations are presented. No hints for contributions of Dexter transfer were found.

Introduction

Over the past decade, there has been an increasing interest in the study and in applications of intramolecular electronic energy transfer (IMET) /1, 2/. The synthesis of bichromophoric compounds with variations of the number of spacers should elucidate the contributions from long-range Förster /3/ and short-range Dexter transfer /4/ for a number of molecular systems. On the one hand, such bichromophoric molecules are of importance as active media in laser technology because they provide broad spectral gain. On the other hand, they represent interesting examples where an insight into the molecular arrangement and its dynamics is gained as far as it influences the mutual separations and orientations of the chromophoric moieties. For small separations R (less than 15 Å) typical rates of Förster-like intramolecular electronic

$$K_{ET} = 1/T_D \cdot (R/R_o)^6 \qquad (1)$$

with R_o being the Förster radius (typically 30 .. 60 Å) and T_D being the flüorescence lifetime of the donor (typically a few nanoseconds) can be estimated to be of the order of $10^{12}\ s^{-1}$. Thus, the process proceeds on a sub-picosecond timescale and will be complete before changes in chromophoric distance or orientation have occurred. Hence, appropriate fs techniques record IMET kinetics of an essentially static ensemble of molecules. Such kinetics would in principle allow us to draw conclusions on the underlying intramolecular distance and orientational distributions.

We will discuss here fs investigations on rhodamine bichromophores /5/, in which the static distribution of orientations determines the transfer kinetics, and on coumarin bichromophores /6/, in which the distribution of separations determines the transfer rate.

Rhodamine Bichromophores

In the first case, Rhodamine dyes were linked to a UV dye (PPO). The absorption spectrum of the bichromophore is the sum of the individual constituents, indicating negligible interaction between the chromophores. To reveal the fast IMET the decay of the population density for the excited donor chromophore may be observed. Since the IMET transfer to the acceptor chromophore is accompanied by repopulation

Springer Proceedings in Physics, Vol. 49 Ultrafast Phenomena in Spectroscopy
Editors: E. Klose · B. Wilhelmi

Fig. 1

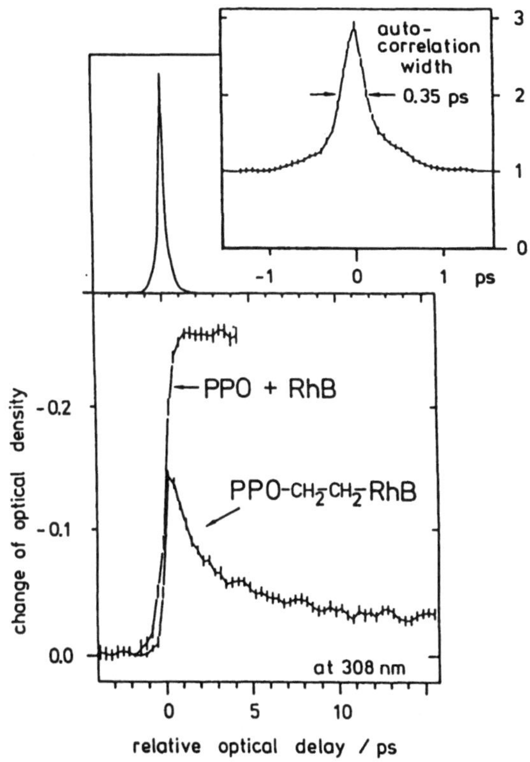

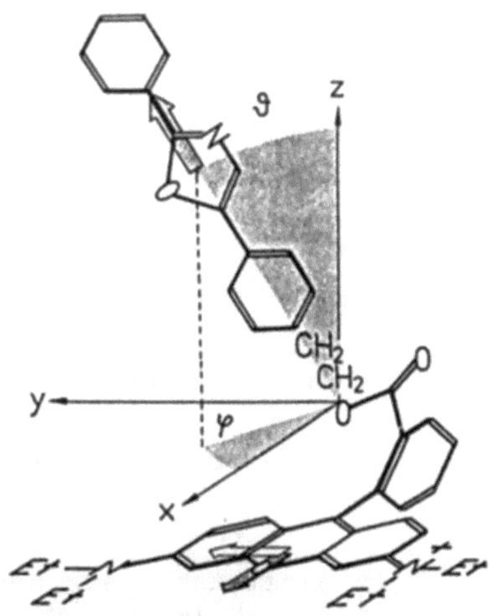

Fig. 2

Fig. 1 GSR at 308 nm for one bichromophore PPO-CH_2-CH_2-RhB after UV fs pulse excitation

Fig. 2 Geometrical model for Rhodamine bichromophores

of the donor ground state, measurements of the donor ground state recovery (GSR) should mainly reflect the IMET kinetics.
For the GSR measurements a hybrid excimer/dye laser system /7, 8/ was used providing high energy pulses at 308 nm with about 250 fs pulse duration. Fig. 1 shows the rapid IMET for one bichromophore, as well as the autocorrelation trace of the excitation (and probe) pulse on the same timescale. For comparison the kinetics of an equimolar mixture of the non-linked constituents (PPO, Rhodamine B) is also shown. Clearly, intermolecular transfer gives no significant contribution to the GSR signal on our timescale.
For a quantitative description of the observed kinetics of PPO-GSR the geometrical model of Fig. 2 is used. The main feature of the linkage between PPO and the rhodamine is a variation in mutual orientation angles and not so much a variability in length. In a numerical model calculation on the basis of Förster dipole-dipole interaction the kinetics associated with each orientation (and consequently with a slightly different intramolecular chromophore separation) are then averaged to yield the expected overall response. By comparison with experimental observations for the molecule PPO-CH_2-CH_2-RhB, we are then forced to assume an even distribution of PPO orientations relative to the rhodamine chromophore over the hemisphere $0 \leq \vartheta \leq \pi$ (polar angle), $0 \leq \varphi \leq 2\pi$ (azimuthal angle). The result (dotted curve) is shown for this bichromophore in Fig. 3, giving excellent agreement.

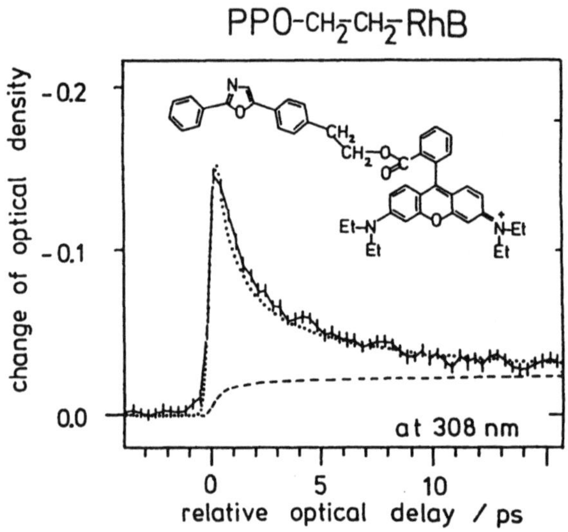

Fig. 3 Calculated and measured IMET kinetics for PPO-CH_2-CH_2-RhB

No free fit parameters were used in the calculations. In summary so far, we can attribute the observed kinetics of the IMET in such closely linked bichromophores primarily to the distribution of static mutual orientations, which results in a distribution of transfer rates. It can be also concluded that the energy transfer between two large chromophores linked by one or two methylene groups is still adequately described by the basic Förster theory, although the separation of the chromophores is comparable to their size, and contributions from the Dexter mechanism might be expected.

Coumarin Bichromophores

In a second example, bichromophoric coumarin was studied after UV donor excitation by broad band fs spectroscopy. 7-ethoxyumbel-liferone as a model for the donor was linked via a flexible polymethylene chain of various lengths (n=3,4,8,12 carbon atoms) to the acceptor chromophore coumarin 500 /7/ (Fig. 4). The bichromophoric was excited by a 250 fs UV pulse at 308 nm, at which the acceptor absorption is negligible but donor absorption is strong.
Radiationless IMET mainly by the Förster mechanism to the green emitting coumarin 500 acceptor is now viewed not by the GSR of the donor but by the rise of the excited acceptor population. The corresponding rise of the gain in the green spectral range is monitored by amplification of the probe pulse. To this end a femtosecond continuum was used for interrogation of the excited sample /8/. The recorded kinetics at the wavelength of the maximum of the acceptor gain (500 nm) are shown in Fig. 5 (points) for two linkage lengths (n = 4,8). For comparison the rise of acceptor gain in the case of direct acceptor excitation is plotted as full circles. Clearly, a delayed rise of the gain due to IMET as well as a significant dependence of this delay on the chain length is observed.
The distribution of chromophorie distances for such a flexible polymethylene chain has been calculated by Valeur et al. /9/. According to this calculation the intramolecular distances between the chromo-

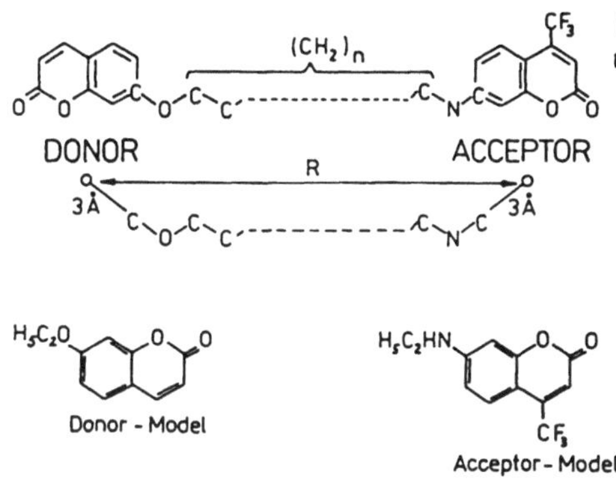

Fig. 4 Coumarin bichromophores and model compounds

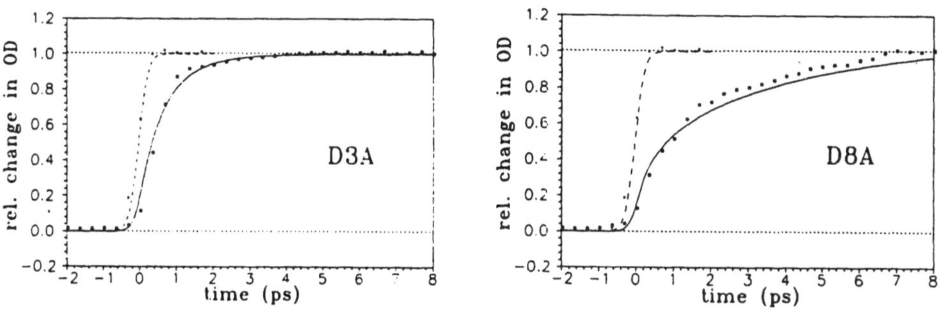

Fig. 5 Experimental and calculated rise of acceptor gain due to IMET after donor excitation with a fs UV pulse for two linkage lengths

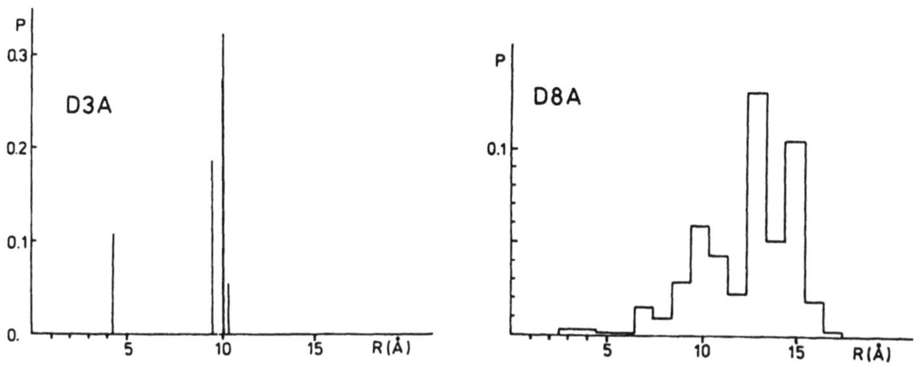

Fig. 6 Calculated distribution of intramolecular chromophore separations for two Coumarin bichromophores with different linkage lengths /9/

phores are distributed in the range 5 Å and 20 Å, depending on the chain length (Fig. 6). In a similar numerical model calculation as for the rhodamines, the kinetics associated with each intramolecular chromophoric distance are again averaged taking into account the calculated distributions of the intramolecular separations to yield the expected overall response. In this calculation for each particular intramolecular chromophore separation a random mutual orientation is assumed which is supported by independent measurements /6/. The results of these calculations are also shown in Fig. 5 as solid lines. Very good agreement has been obtained for all four bichromophores. Thus, we may conclude that the distributions of intramolecular separations as calculated by Valeur et al. which enter significantly in our calculations are very well confirmed by our time resolved measurements. Again there have been found no hints for a significant contribution of the Dexter mechanism to the IMET in such bichromophores. In conclusion, such femtosecond measurements have resolved the IMET between closely linked organic chromophores and allow us to obtain direct information on the statistical (static) distributions in distances and mutual orientations within the molecule.

Acknowledgement

We are grateful to the Deutsche Forschungsgesellschaft for support through the Leibniz Prize Program. M. K. thanks the Max-Planck-Gesellschaft for a grant.

References

/1/ F. C. DeSchryver, N. Boens, J. Put; Adv. Photochem. 10 (1977) 359
/2/ B. Valeur in Fluorescent Biomolecules; Eds. D. M. Jameson and G. D. Reinhardt (Eds.), Plenum Press, New York 1989, p. 296
/3/ T. Förster, Z. Naturforsch. 49 (1949) 321
/4/ D. L. Dexter, J. Chem. Phys. 21 (1952) 836
/5/ N. P. Ernsting, M. Kaschke, J. Kleinschmidt, K. H. Drexhage, V. Huth, Chem. Phys. 122 (1988) 431
/6/ M. Kaschke, N. P. Ernsting, B. Valeur, J. Bourson, Chem. Phys. Lett., submitted (1989)
/7/ J. Mugnier, J. Pouget, J. Bourson, B. Valeur, J. Luminescence 33 (1985) 273
/8/ N. P. Ernsting, M. Kaschke, Appl. Phys. B, submitted (1989)
/9/ B. Valeur, J. Mugnier, J. Pouget, J. Bourson, F. Santi, J. Chem. Phys. 93 (1989) 6073

Sub-picosecond Transient Optical Absorption Spectroscopy of the Spiropyran–Merocyanine Photochromism

N.P. Ernsting

Max-Planck-Institut für biophysikalische Chemie, Abteilung Laserphysik, Postfach 2481, D-3400 Göttingen, Fed. Rep. of Germany

"BIPS" (= 1',3',3'-trimethylspiro-[2H-1-benzopyran-2,2'-indoline]; structure I in fig. 1) is the parent compound of a large class of photochromic materials. We examine the development of the colored form of BIPS in methyl-cyclohexane after UV excitation, and with sub-ps time resolution. Its risetime is 0.7 ps. Band structure appears with a time constant of 5 ps. The latter process is interpreted as vibrational cooling.

Fig. 1 shows the likely structural evolution in the course of the photochromic reaction. Upon excitation of BIPS in the near UV, a ring-opening reaction leads to various merocyanine isomers (eg. structure V) which have characteristic absorption bands in the red region of the visible spectrum. The excitation pulse in our experiment, at 308 nm, had a pulse duration of 0.26 ps; the transient spectra were recorded in the range 370-700 nm with an overall time resolution better than 0.4 ps.

In ethanol, a broad transient absorption centered around 440 nm had been found to decay with lifetime of 1.3 ps. This decay is paralleled by the rise of merocyanine absorption [1].

Fig. 2 summarizes the assignments of the observed transient spectra. The process was described as ultrafast barrierless

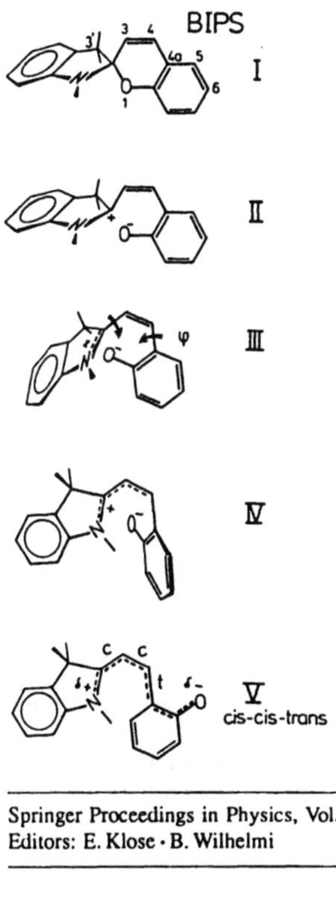

Fig. 1

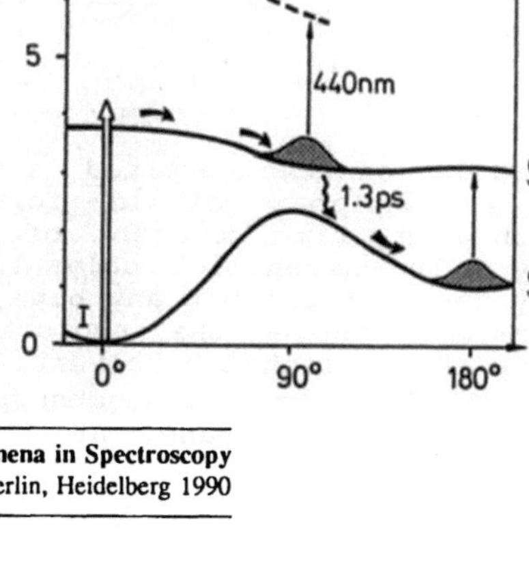

Fig. 2

Springer Proceedings in Physics, Vol. 49 **Ultrafast Phenomena in Spectroscopy**
Editors: E. Klose · B. Wilhelmi

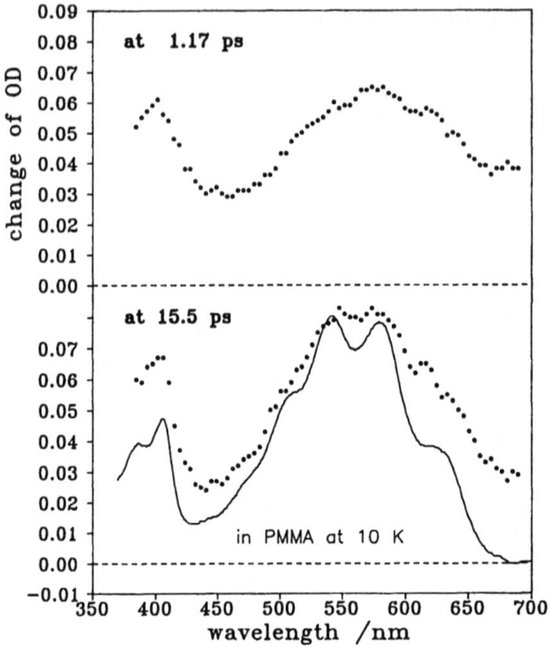

Fig. 3

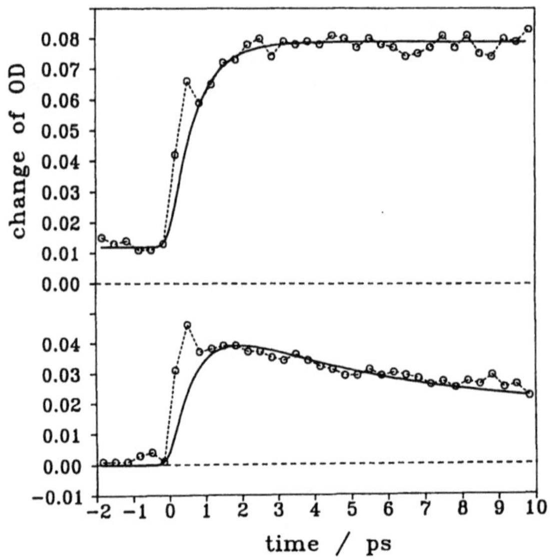

Fig. 4

isomerisation on the excited singlet potential energy surface towards a nonplanar cisoid form (IV), followed by internal conversion with a time constant of 1.3 ps, and fast attainment of planarity. In alcoholic solvents, the merocyanine absorption generally does not exhibit any band structure [2]. .

Here we examine the photochromic spiropyran - merocyanine conversion in an aliphatic solvent.

Fig. 3 shows transient absorption spectra for a solution of BIPS in methyl-cyclohexane. At 1.17 ps after UV excitation, the

Fig. 4 shows kinetic traces, at 572 nm (top) and at 690 nm (bottom). The initial spike which is present in both traces indicates broad, intense transient absorption which probably originates from a transient population on the excited singlet potential energy surface. The risetime of merocyanine absorption is ca. 0.7 ps - significantly faster than in ethanol. The absorption at 690 nm has a decay component with time constant of 5 ps. This decay is correlated with the appearance of structure in the central part of the merocyanine absorption band. The structure may be due to vibrations of the primary isomer, or else it may indicate the simultaneous existence of several isomers. For comparison, the absorption spectrum of the coloured compound - at 10 K in a PMMA host - is also shown. The appearance of structure, and the decay of red-edge absorption, may also be a result of cooling of the initially hot merocyanine isomer in the ground state.

Transient absorption spectra have so far not given any indication of a primary, stable, zwitterionic intermediate "X" having the parent geometry. In order to elucidate structural change in the sub-ps time range, the degree of polarisation should be studied in following work.

References

[1] N.P. Ernsting, Chem. Phys. Letters **159**, 526 (1989)
[2] T. Bercovici, R. Heiligman-Rim, E. Fischer, Mol. Photochem. **1**, 23 (1969)

Observation of Time-Resolved Absorption Anisotropy Decay Resulting from Rotation of Free Organic Molecules

N.A. Borisevich[1], *E.V. Khoroshilov*[1], *I.V. Kryukov*[1], *P.G. Kryukov*[1], *G.B. Tolstorozhev*[2], *M.Yu. Shakhbazyan*[1], *and A.V. Sharkov*[1]

[1]P.N. Lebedev Physics Institute, USSR Academy of Sciences, 117924 Moscow, USSR

[2]B.I. Stepanov Institute of Physics, Belorussian Academy of Sciences, 220603 Minsk, USSR

Abstract. A photoinduced orientational anisotropy of an ensemble of free molecules has a complex time dependence, and differs significantly from the relaxation in condensed media. We have studied the time-resolved $S_1^* \to S_n^*$ absorption anisotropy decay resulting from rotation of free POPOP molecules in vapour at 600 K after $S_0 \to S_1^*$ excitation by 70-femtosecond linearly polarized pulses at 308 nm. The maximum absorption anisotropy r=0.4, recorded at the moment of excitation, speaks in favour of parallel dipoles of $S_1^* \to S_n^*$ and $S_0 \to S_1^*$ transitions of POPOP molecules. Within 4.5 ps value r decays approximately to zero, then increases and gets a stationary level close to the calculated value r=0.05 which is due to a nonchaotic rotation of free molecules.

1. Introduction

A molecular absorption of linearly polarized light gives rise to an ensemble of anisotropically excited molecules. As a result of reorientation (rotation) of molecules, the anisotropy (absorption or luminescence) changes with time. The theoretical calculations performed classically by Blokhin et al. [1,2] and later quantum mechanically by Myers and Hochstrasser [3] and Felker and Zewail [4] showed that the relaxation of optically induced orientational anisotropy of the ensemble of free molecules in gas phase has a complex time dependence, and differs significantly from the relaxation in condensed media. If we consider vapours of complex organic molecules with pressure of several Torr and temperature 300-600 K the average periods of molecule rotation of 10^{-11}s would be significantly less than intervals between collisions (approximately 10^{-8}s). Under these conditions, according to the calculation made by Blokhin et al. [1,2], the luminescence or absorption anisotropy is,

Springer Proceedings in Physics, Vol. 49 Ultrafast Phenomena in Spectroscopy
Editors: E. Klose · B. Wilhelmi

at first, maximal. Then, as a result of rotation, it decreases abruptly and reaches a minimum within the time close to $\tau = \sqrt{I/kT}$, where I is the momentum of inertia of a molecule (mean value out of three values, which characterize the molecule). This time may be interpreted as an average time of molecule turning by 1 radian at temperature T. For dye molecules the relaxation time of orientational anisotropy lies in the region of several picoseconds or less at the temperature mentioned above.

At temperatures close to liquid helium temperature the anisotropy decay occurs within tens picoseconds, and may be observed by means of time-resolved picosecond technique. An experiment with stilbene was taken by Scherer et al. [5] at temperature 2K in a molecular beam. The anisotropy decay time was about 50 ps. But the application of a picosecond technique did not make it possible to measure in detail the kinetics of decay of luminescence anisotropy in the experiment with the stilbene vapour at 463 K (Myers et al. [6]).

2. Materials and Methods

We have studied decay of a photoinduced absorption anisotropy in vapour (of 1 Torr pressure) of 1,4-Di-[2- (5-phenyloxazolyl)]-benzene (POPOP) molecules at T= 600 K. The observation was made with the help of a pump-probe absorption spectroscopy with femtosecond time resolution. The pulses from colliding pulse mode-locked dye laser with duration of 70 femtoseconds were amplified in a multipass jet amplifier designed by Khoroshilov et al. [7]. The amplifier was pumped by pulses of a copper-vapour laser at 10 kHz-rate. The amplified radiation was transformed to the second harmonic (the wavelength 308 nm). This ultraviolet linearly polarized radiation served for the excitation of $S_0 \rightarrow S_1^*$ transition of POPOP molecules. The probing was performed by the pulses at the fundamental wavelength of 616 nm with the polarization parallel or perpendicular to that of an exciting pulse. This wavelength corresponds to $S_1^* \rightarrow S_n^*$ transition of the molecule. The values of photoinduced absorbance changes, $\Delta A_\parallel(t)$ and $\Delta A_\perp(t)$, were measured under the conditions of mutually parallel and perpendicular polarization of the exciting and probing radiation. From those data we calculated the value of anisotropy, $r(t)=(\Delta A_\parallel(t)-\Delta A_\perp(t)) \ / \ (\Delta A_\parallel(t)+2\Delta A_\perp(t))$.

3. Results and Discussion

Figure 1 illustrates the experimental dependences $\Delta A_\parallel(t)$ and $\Delta A_\perp(t)$ on the delay between probing and exciting pulses, and Fig. 2 shows

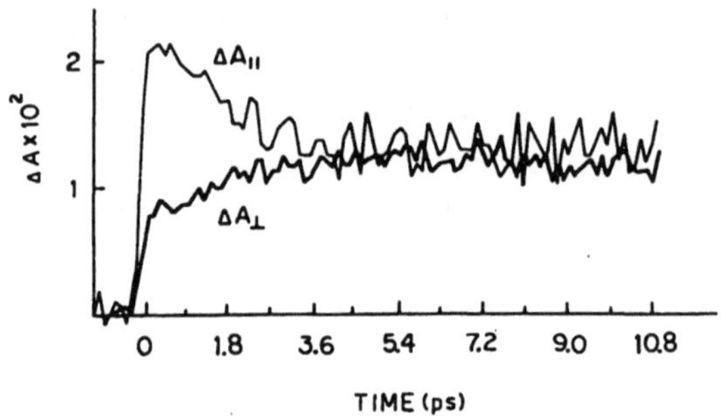

Fig.1.Photoinduced absorbance changes, $\Delta A_{\parallel}$ and $\Delta A_{\perp}$, vs delay time between exciting and probing pulses for POPOP molecules at T=600 K.

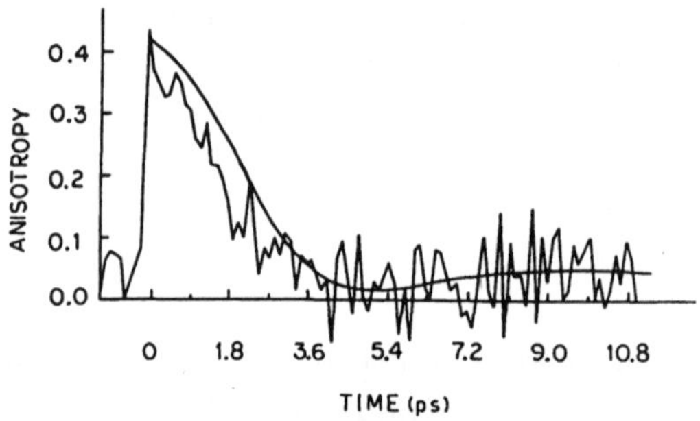

Fig.2. Experimental and calculated dependences of the absorption anisotropy on the delay time between exciting and probing pulses for POPOP molecules at T=600 K.

the dependence r(t) calculated on the basis of those results. Figure 2 exhibits also the dependence of anisotropy on time after excitation calculated by Blokhin [8] for the POPOP molecule. From this calculation value of τ is 4.5 ps from the value of the moment of inertia $1.68 \cdot 10^{-36}$ $g \cdot cm^2$.

As seen from Fig. 2 there is a good agreement between the experimental and theoretical dependences. Immediately after excitation we recorded maximum possible value of anisotropy r = 0.4. This is indicative, first, of parallel dipoles of $S_0 \rightarrow S_1^*$ and $S_1^* \rightarrow S_n^*$ transitions. Second, the maximum anisotropy can be observed only in the case when the pulse duration is much less than the measured decay time (the excitation by δ-function), and the excitation does not result in the saturation of $S_0 \rightarrow S_1^*$ transition ("weak" excitation). A stage of fast decay of anisotropy is occurring somewhat faster than predicted by the theory. This discrepancy may result from the following. The calculations were made under the assumption that a molecule had been absolutely rigid. However,one should not neglect turning of a part of an excited molecule relative to another, a respective changing of the transition dipole direction, and a decrease (or faster decay) of the anisotropy.

According to the calculations of Blokhin et al. [1,2], one should expect that the anisotropy achieves a stationary nonzero level (r=0.05 for POPOP) after fast decay to the minimal value. Such a deviation from zero is due to a nonchaotic rotation of free molecules. As seen from Fig. 2 this stage demonstrates a good agreement between experimental and theoretical data.

Thus, from the experimental results obtained it follows that we have observed, in a real time scale, the rotation of free molecules in the gas phase manifested as the decay of induced absorption anisotropy due to the dephasing of the molecules which rotate with different rates.

Acknowledgements. The authors would like to thank A.P.Blokhin, S.A.Tikhomirov, S.V.Kurbasov and A.P.Lutsenko for useful discussions and assistance in the experiment.

References

1. A.P.Blokhin, N.A.Borisevich, I.I.Kalosha, V.A.Tolkachev: Dokl.AN SSSR (Russian) **238**, 123 (1978)
2. A.P.Blokhin, V.N.Knyukshto, V.A.Tolkachev: Izv. AN SSSR (Russian) **42**, 359 (1978)
3. A.B.Myers, R.M.Hochstrasser: J.Chem.Phys. **85**, 6301 (1986)
4. P.M.Felker, A.H.Zewail: J.Chem.Phys. **86**, 2460 (1987)
5. N.F.Scherer, L.R.Khundkar, T.S.Rose, A.H.Zewail: J.Phys.Chem. **91**, 6478 (1987)
6. A.B.Myers,P.L.Holt, M.A.Pereira, R.M.Hochstrasser: Chem.Phys.Lett. **132**,585 (1986)
7. E.V.Khoroshilov, I.V.Kryukov, P.G.Kryukov, A.V.Sharkov: *Ultrafast Phenomena VI* (Springer, Berlin, Heidelberg, 1988) pp.22-23
8. A.P.Blokhin: Vest. AN BSSR (Russian) **4**, 77 (1986)

Rotation of Dye Molecules Induced by Optical Excitation

D.M. Gakamsky, N.A. Nemkovich, and A.N. Rubinov

Institute of Physics, BSSR Academy of Sciences, 220602 Minsk, USSR

Abstract. The results of time-resolved measurements of spectra and anisotropy fluorescence of 3-amino-N-methylphthalimide (3ANMP) in glycerol are presented. It is shown for the first time that the process of configurational relaxation of a solvate shell is accompanied by acceleration of the rotational movement of the dye molecule incorporated into the shell. The magnitude of the effect depends on the store of intermolecular interaction energy released in the process of the relaxation. This explains the observed specific dependences of fluorescence anisotropy spectra on the wavelength of excitation. It is also shown that the experimental data are well described by the continuum model of the polar solution if the inhomogeneous configurational broadening of dye electronic spectra is taken into account.

1. Introduction

The problem of chromophore rotation is urgent due to the wide application of the fluorescence probe method in biophysics as well as due to the important role of rotational diffusion in various photochemical reactions in solutions. The experimentally observable deviations from the Levshin-Perrin theory are associated, as a rule, either with the form of molecules or with the specific interactions in the solution. However, not a single existing theory takes into account inhomogeneous broadening of dye electronic spectra caused by the statistical spread of the microstructure of solution solvates.

On the basis of the model of polar solutions developed earlier [1], it is shown here that inhomogeneous spectral broadening may cause acceleration of rotational diffusion of dye molecules after their selective excitation in certain local microstructures of the solution.

2. Theory

2.1. Field Diagram

For clarity of further discussion, it is useful to briefly dwell upon the solution model [1]. This model considers the solution as a set of elementary subsystems - solvates. The diagram of the levels of solvate free energy, the so-called field diagram, is shown in Fig. 1. This diagram does not show the vibrational broadening of electronic levels. The local solvate field serves here as the generalized coordinate characterizing the state of the solvate. It changes due to the thermal fluctuations of the solution microstructure.

Analysis of the diagram shows that when the excitation frequency corresponds to ν_{ex} the process of solvate relaxation will be followed

Springer Proceedings in Physics, Vol. 49 Ultrafast Phenomena in Spectroscopy
Editors: E. Klose · B. Wilhelmi

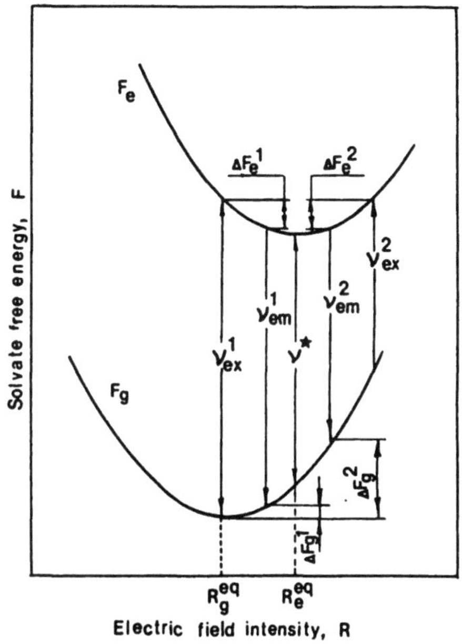

Fig. 1. Field diagram. The dependence of the free energy F_e on the solvate local electric field $\bar{R}$.

by a red shift of the fluorescence spectrum. But if the excitation frequency is equal to ν_{ex}^2 the temporal shift of spectrum will be to the blue region [2]. Between ν_{ex}^1 and ν_{ex}^2 there is a particular excitation frequency ν_{ex}^*, for which the solvates immediately appear at the equilibrium state and, therefore, spectral relaxation does not occur.

Such an approach makes it possible not only to qualitatively explain the experimentally observed effects, but also to give them a quantitative description. On this basis we can also predict a new phenomenon as follows. During relaxation in the solvate an excess free energy is released which can accelerate the chromophore rotation.

From analysis of the field diagram we can find the relationship between the excess free energy ΔF_e and the excitation ν_{ex} and emission ν_{em} frequencies:

$$\Delta F_e = \mathrm{Const}(\nu_{ex} - \nu_{em})(\nu_{ex} + \nu_{em} - \nu^*). \qquad (1)$$

From this it follows that only in one case, when $\nu_{ex} = \nu^*$, and therefore $\nu_{em} = \nu^*$, i.e. when the spectral relaxation does not take place, does no excess energy appear in a solvate. In all other cases, the excitation free energy is liberated. As the result of the intermolecular relaxation, this excess free energy is converted to kinetic energy of solvate molecules. Thus we may consider that in the solvate a heat source is functioning. To elucidate the role of this source in the process of fluorescence depolarization, it is necessary to know the law of spectral relexation, which can be found by analysing the experimental fluorescence spectrochronogram.

2.2. Analysis of the Spectrochronogram

In the above model it is assumed that the reactive field in the solvate affects only the frequency of the purely electronic

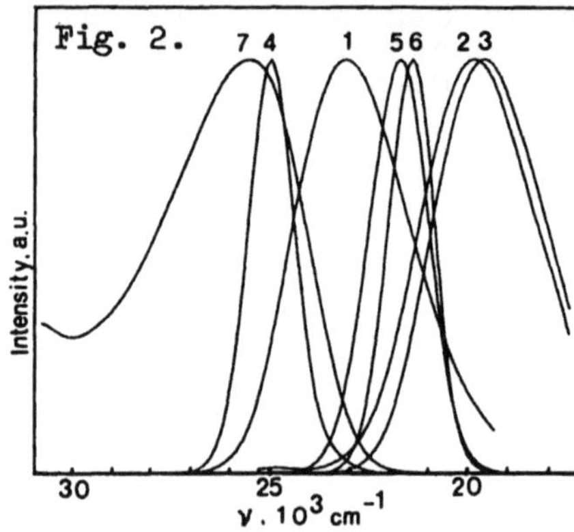

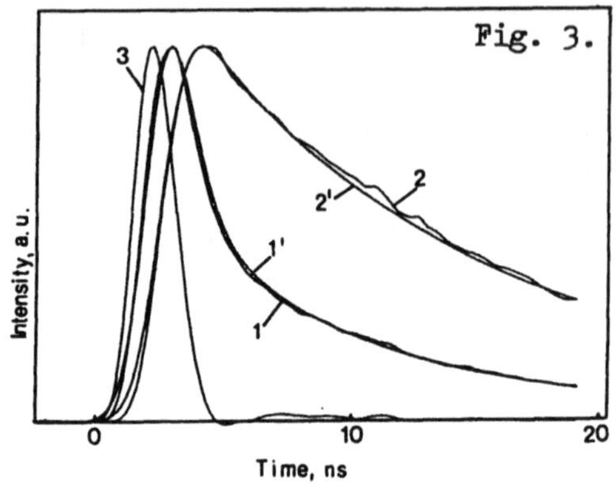

Fig. 2. Instantaneous fluorescence spectra (1,2,3), the IBF (4,5,6) and the absorption spectrum (7) of 3ANMP in glycerol. $C=10^{-6}$ M, $T=28^{\circ}$C, excitation frequency 25600 cm^{-1}.

Fig. 3. Experimental (1,2) and calculated (1',2') fluorescence pulses of 3ANMP in glycerol and the excitation pulse (3). The excitation frequency - 25600 cm^{-1}, the recording frequency - 22500 cm^{-1} (1,1'), and 18350 cm^{-1} (2,2').

transition. It means that the shapes of homogeneous absorption and fluorescence profiles of the chromophore are independent of the microstructure fluctuations and coincide with the chromophore spectra in a polar solution. Then the fluorescence spectrochronogram can be represented as a convolution of the homogeneous fluorescence spectrum $S_\bullet(\nu - \nu')$ and the function of the solvate distribution over the frequencies of purely electronic transitions $\varphi_\bullet(\nu',t)$:

$$I_\bullet(\nu,t) = i_o \exp(-t/\tau)\int_0^\infty \varphi_\bullet(\nu',t)S_\bullet(\nu - \nu')d\nu', \qquad (2)$$

(i_o is a constant and τ is a fluorescence lifetime). Thus, the process of spectral relaxation is fully determined by the evolution of the inhomogeneous broadening function (IBF) $\varphi_\bullet(\nu',t)$.

We have calculated IBF at different times for a glycerol solution of 3ANMP at room temperature and at Stokes excitation. Fig. 2 shows instantaneous fluorescence spectra (curves 1, 2, 3) and the corresponding IBF (curves 4, 5, 6) for three times - 0, 3, 16 ns. The calculations have shown that the centre of gravity of IBF at the zeroth time deviates significantly from the equilibrium position. We found the relaxation of the centre of gravity to equilibrium may be described by a two-exponential dependence. In this case the contribution of the fast process to the relaxation is about 88% and the fast and slow relaxation times differ by one order of magnitude (0.7 and 7 ns respectively). The shape of IBF fairly rapidly becomes Gaussian. Besides, the distribution halfwidth increases at the first stage of relaxation from 1100 cm^{-1} to 1700 cm^{-1}, and then decreases down to 1200 cm^{-1}. Having thus determined the IBF evolution, we may now describe the fluorescence kinetics at the blue and the red slopes of the emission spectrum. Fig. 3 shows good quantitative agreement between experimental and calculated curves. It should be noted that such a good coincidence has been obtained using the same set of a comparatively small number of physical parameters.

2.3. The Thermal Regime of the Solvate

Now we can calculate the thermal regime of the solvate. We are to analyse the balance of the heat $\delta Q(t)$ coming into the solvate in the process of its structural relaxation and leaving it due to its cooling. The intensity of the heat source functioning in the solvate during spectral relaxation must be determined by the time derivative of the free energy. We have found that during relaxation the instantaneous spectra are shifted by 3700 cm^{-1}. To this shift there corresponds an excess free energy of $0.3\ 10^{-19}$ J. The relaxation time constant of thermal fluctuations in the solution can be estimated by a relation known from thermodynamics:

$$\tau_T = (\rho\ C_p\ V\ L)/(\lambda\ S) \tag{3}$$

where ρ is the solvent density, C_p is the molar heat capacity of the solvent at a constant pressure, L is the average linear size of the cooling region, S and V are, respectively, the area of its surface and the volume, λ is the thermal conductivity of the solvent.

It turns out that the relaxation time of thermal fluctuations τ_T is much smaller than the short spectral relaxation time τ_1 (0.04 ns and 0.7 ns respectively). Therefore we can find from the balance equation

$$d/dt(\delta Q(t)) = (h\Delta\nu_s \tau_T / \tau_1)\exp(-2t/\tau_1) - \delta Q(t)/\tau_T\ , \tag{4}$$

that the temperature in the solvate during relaxation, except for its initial stage, is described by the following expression:

$$T(t) = T_0 + \Delta T\exp(-2t/\tau_1), \quad \Delta T = (h\Delta\nu_s\ /C^{solv})2\tau_T\ /(\tau_1)\ . \tag{5}$$

Here T_0 is the equilibrium solvate temperature and ΔT is the maximum temperature increase which is determined by the ratio of the times τ_T and τ_1, by the free excess energy and C^{solv} the solvate heat capacity.

Knowing the solvate temperature variation with time, we can calculate the Brownian diffusion time variation during relaxation:

$$\tau_{rot}(t) = (V^{chr}\ \eta(T))/(kT(t)), \tag{6}$$

where V^{chr} is the dye molecule volume and k is the Boltzmann constant. The variation of glycerol's dynamic viscosity with temperature can be given as [3]: $\log \eta$ (cP) = A' + B' /(T − T'); A' = −2.8834, B' = 997.86, T' = 128.481 K.

Then we can calculate the kinetics of fluorescence anisotropy [4]:

$$r(t) = r_0 \exp\left[-\int_0^t dt'/\tau_{rot}(t')\right]\ , \tag{7}$$

where r_0 is the anisotropy limiting value. The anisotropy kinetics calculated by this formula is shown in Fig. 4. It is seen that a significant change in the solvate temperature during relaxation (by 50 °C) leads to a noticeable deviation of the anisotropy kinetics from the exponential form, which means deviation from the Levshin-Perrin theory. We interpret this effect as light-induced rotation of the chromophore, since it is caused by the light quantum absorption and is determined by the absorbed light frequency. Of course, we understand that the direct reason for the acceleration of fluorescence depolarization is the action of accidental shocks caused by the microstructure rearrangement. The effect of light-induced rotation occurs mainly at the initial stage of relaxation. The solvate temperature rather rapidly becomes equal to the average temperature of the solvent, and the anisotropy kinetics becomes exponential.

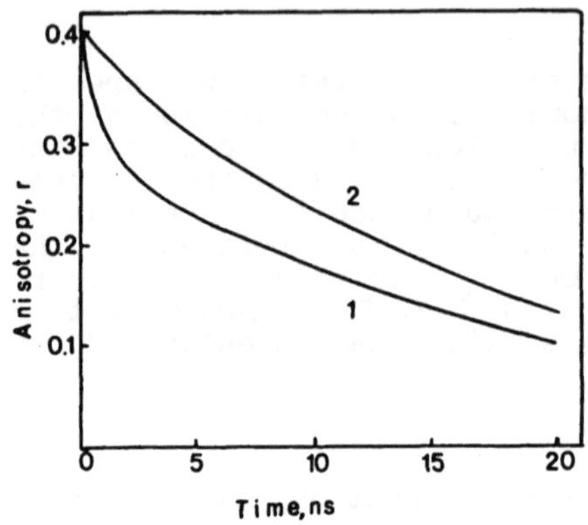

Fig. 4. The anisotropy kinetics calculated by the model taking into account the solvate temperature variation during relaxation (1) and the Levshin-Perrin model (the temperature is constant) (2).

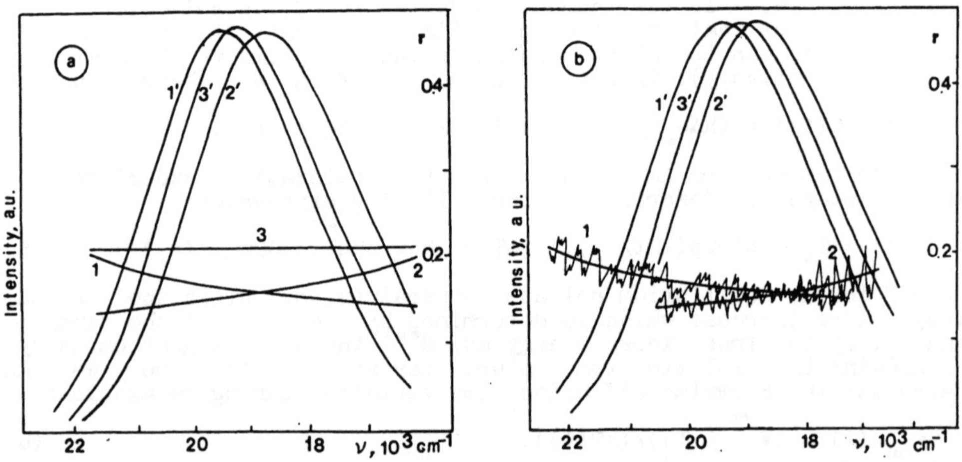

Fig. 5. Instantaneous spectra of fluorescence 3 ns after the excitation (curves 1',2',3') and instantaneous fluorescence anisotropy spectra 10 ns after excitation (curves 1,2,3) for different frequencies of excitation: 2,2' - ν_{ex} = 25600 cm^{-1}; 2,2' - ν_{ex} = 20800 cm^{-1}, 3,3' - ν_{ex} = 21700 cm^{-1}; a) calculation, b) experiment.

To compare the effect predicted by our model with experiment we must take into account that actually the instantaneous fluorescence spectrum is inhomogeneously broadened. And this means that both relaxed and nonequilibrium solvates will contribute to the fluorescence. For simplicity, we can assume that we have only two types of centres: nonequilibrium centres with the degree of anisotropy equal to r_0, and equilibrium ones with degree of anisotropy $r - \Delta r_0$. We also believe that the Brownian diffusion of equilibrium and nonequilibrium centres goes with the same speed. Then, using the rule of anisotropy additivity we calculate the shape of the instantaneous anisotropy spectra 10 ns after the onset of relaxation for three cases: Stokes excitation, excitation at ν^*, far anti-Stokes excitation (see Fig. 5a).

As seen from Fig. 5a, our model predicts the previously unreported dependence of the instantaneous anisotropy spectrum on the excitation wavelength. Under the excitation at the absorption maximum the anisotropy decreases with increasing fluorescence wavelength. Under excitation in the far anti-Stokes region we have the opposite picture: on the emission spectrum the instantaneous values of anisotropy are larger than on the blue one. And, finally, under excitation at the frequency ν^* there is no dependence of anisotropy on the fluorescence frequency at all.

3. Experimental Results

Experiments completely confirmed our predictions. Figure 5b shows experimental instantaneous anisotropy spectra of 3ANMP in glycerol at room temperature for two frequencies of excitation, located near the maximum of absorption (curve 1) and the anti-Stokes region (curve 2) 10 ns after the excitation. Curves 1',2' represent nonequilibrium instantaneous fluorescence spectra at the time 3 ns, corresponding to these two cases of excitation. Curve 3 represents the instantaneous fluorescence spectrum on completion of relaxation, which also corresponds to the instantaneous fluorescence spectrum under excitation (21700 cm^{-1}).

4. Conclusion

In conclusion it should be emphasized that since the cause of the light-induced rotation of dye molecules is the dipole-dipole interactions in solutions, this effect must be universal for all polar solutions. In particular, we observed this effect in studying rotational diffusion of the 1-phenylnaphthylamine probe in biological membranes [5]. Thus, when determining the microviscosity of biological objects from the probe rotational diffusion, we have to make a correction for the magnitude of this effect. We also believe that the light-induced rotation must contribute to the overcoming of the potential barrier of photochemical reactions in solutions.

5. References

1 A.N. Rubinov, V.I. Tomin: Inhomogeneous Broadening of Electronic Spectra in Solid and Liquid Solutions. Preprint No. 348 Inst. of Physics, BSSR Acad. Sci., Minsk, 1984) (Russ.).

2 N.A. Nemkovich, V.I. Matseiko, A.N. Rubinov et al.: Pisma Zh. Eksper. Teor. Fiz. **29**, 780 (1979).

3 R.C. Reeid, J.M. Prausnitz, T.K. Sherwood: The Properties of Gases and Liquids (New York, 1977).

4 J.R. Lakowicz. Bioph. Chem. **19**,13 (1984).

5 D.M. Gakamsky, A.P. Demchenko, N.A. Nemkovich et al.: Laser Kinetic Spectroscopy 1-AN Probe in Phosphatidilholin Membrane. Preprint No. 484 (Inst. of Physics, BSSR Acad. Sci., Minsk, 1987) (Russ.).

Fast Structural Transformations of Organic Molecules

N.A. Borisevich, N.A. Lysak, S.V. Melnichuk, S.A. Tikhomirov, and G.B. Tolstorozhev

Institute of Physics, BSSR Academy of Sciences, 220602 Minsk, USSR

Abstract. The mechanisms of photodissociation of aromatic disulfide molecules followed by charge transfer states formation have been studied using ultrafast spectroscopy methods. The thiyl radicals are found to be formed with the rate constant of 10^{12} s^{-1} by pre-dissociation mechanisms as a result of picosecond photolysis of di-sulfides. Fast geminate radical recombination is observed for diphenyldisulfide. Photodissociation of alkyl-substituted disulfides in polar media leads to thiyl radical formation in two different forms - "neutral" and "ionic". An increase in the medium viscosity leads to a decrease in the ionic form formation rate from 10^{11} to 10^{10} s^{-1}. It is assumed that the ionic form can represent a TICT state configuration.

1. Introduction.

One of the most fundamental types of structural transformations of organic molecules is molecule bond dissociation. Free radicals which are usually the products of such a reaction play an important, and often determining role in various photochemical reactions. Due to the presence of an unpaired electron their electronic states are doublet or quartet, unlike the ordinary organic molecules that have singlet or triplet states. For a long time free radicals have been the subject of comprehensive studies in physics, chemistry, biology. However, only the present state of ultrafast spectroscopy makes it possible to obtain qualitatively new information about detailed mechanisms of ultrafast relaxation processes in such systems. In this paper we present data concerning mainly two important stages in the free radical formation process. Kinetic studies of geminate recombination are expected to provide fundamental information on solvent cage effect in the bond dissociation reactions. The investigations of the charge separation and sequential stabilization processes taking place immediately after the bond cleavage are of great importance for the elucidation of the formed radicals reactivity.

Springer Proceedings in Physics, Vol. 49 Ultrafast Phenomena in Spectroscopy
Editors: E. Klose · B. Wilhelmi

2. Experimental

The details of our picosecond transient absorption spectrometer, based on Nd:glass passively mode-locked laser, have been given elsewhere [1]. Molecules under study are excited by 6 ps pulses of third and fourth harmonics of fundamental frequency. Broad-band picosecond continuum is used as a probe pulse. A double-beam optical arrangement was adopted, and absorption spectra in the 300-nm scanning region were recorded with an optical multichannel analyzer. The samples we have employed are R-Ph-S-S-Ph-R, where R = -H (I), $-NH_2$ (II), $-N(C_2H_5)_2$ (III). Solvents were of research grade and were used as received.

3. Results and discussion

When the hydrocarbonic solutions of the molecules under study (I, II, III) are excited by picosecond light pulses immediately after excitation the transient absorption spectra of the photolysis products are recorded (Fig. 1). In accordance with literature data [2,3] these transient spectra should belong to the phenylthiyl radical (curve 1), the aminophenylthiyl radical (curve 2) and the diethylaminophenylthiyl radical (curve 3), respectively. In other words, in all these cases the primary dissociation process is the S-S bond dissociation resulting in the corresponding radicals (radical pair) formation.

Fig. 2 shows the optical density kinetics in the absorption maximum of the phenylthiyl radical obtained after 4-th harmonic irradiation of the molecules I in isooctane (η=0.45 cP) and dodecane (η=1.49 cP). In isooctane the transient absorption kinetics has a peak at time

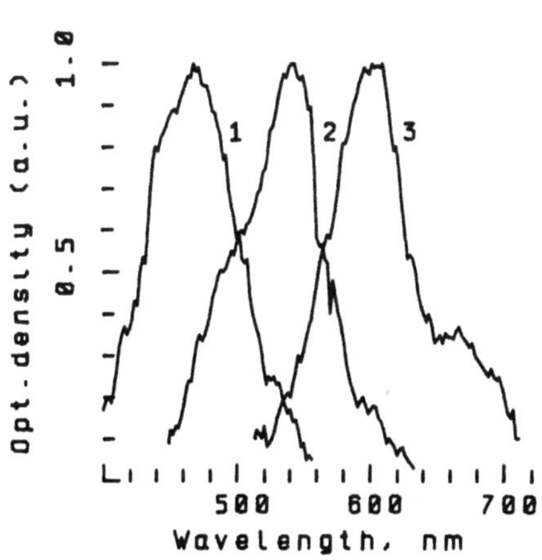

Fig.1 Transient absorption spectra of phenylthiyl (1), aminophenylthiyl (2), diethylaminophenylthiyl (3) radicals in heptane at 5 ps time delay

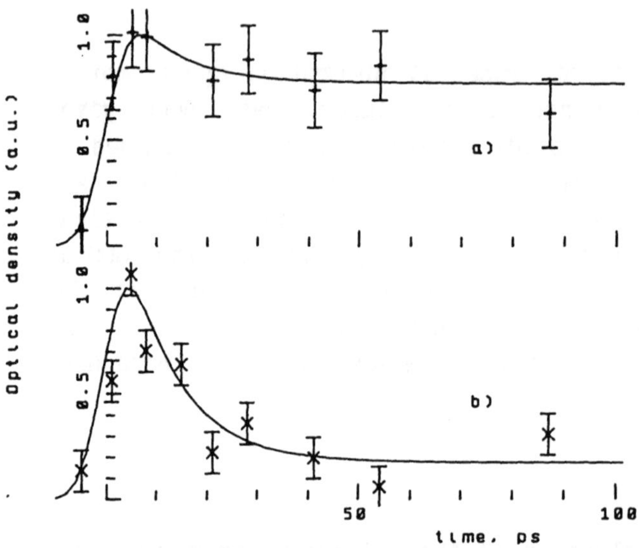

Fig.2 Time dependences of the optical density measured at 470 nm for phenylthiyl radicals in isooctane (a) and dodecane (b); solid lines are calculated curves

delay of 5 ps and then at 40 ps reaches a stationary value of 75% of the peak signal. For dodecane optical density has a pronounced peak at the same time delay of 5 ps, but the plateau signal has the intensity of only 15% of the transient absorption maximum value. Such transient absorption time evolution in non-polar solvents is typical for geminate recombination.

To obtain quantitative estimations of the photodissociation and geminate recombination kinetics parameters, the following scheme is used:

$$R{-}R \xrightarrow[k_d]{h\nu} R\text{- -}R \begin{cases} \xrightarrow{q} R\cdot\cdot R \xrightarrow{k_g} R{-}R \\ \xrightarrow{1-q} R\cdot+\cdot R \xrightarrow{k_v} R{-}R \end{cases},$$

where k_d is the dissociation rate constant, k_g is the geminate recombination rate constant, k_v is the volume recombination rate constant, q is the portion of geminate recombinating radicals. The kinetic balance equations were solved under the assumption of a Gaussian profile for excitating and probing pulses. To obtain the best agreement between experimental dependences and calculated curves the dissociation and recombination parameters k_d, k_g, k_v and q were

Table 1. Dissociation and recombination parameters derived from comparing experimental and calculated data

Solvent	η, cP	k_d, s^{-1}	k_g, s^{-1}	k_v, s^{-1}	q
isooctane	0.45	$>10^{12}$	$1-2x10^{11}$	$<10^9$	0.4
dodecane	1.49	$>10^{12}$	$1-2x10^{11}$	$<10^9$	0.9

varied. Fitting curves are presented in Fig. 2, and derived estimations are summarized in Table 1.

These data are in good correlation with those reported recently by T.W.Scott and S.N.Liu [4]. As far as molecules II and III are concerned, no geminate pair recombination was observed, perhaps due to localisation of electronic density on the S atom.

In previous studies of disulfide II picosecond photolysis we reported [5] two-band transient absorption of the aminophenylthiyl radical in polar solvents. Time-delayed long-wavelength band was attributed to absorption of the radical ionic form. To elucidate the nature of the ionic form the influence of the solvent viscosity and the nitrogen atom substituting groups volume on the ionic form risetime has been studied. Fig. 3 shows the transient absorption spectra of the diethylaminophenylthiyl radical in acetonitrile at different time delays. The delayed appearance of long-wavelength band is observed,while short wavelength absorption rises without any time delay. Optical density kinetics was measured for the maxima of these two bands and the data obtained are presented in Fig. 4. The solid lines are the result of model calculations performed in accordance with [5]. Data obtained for polar solvents of different viscosity are summarized in Table 2.

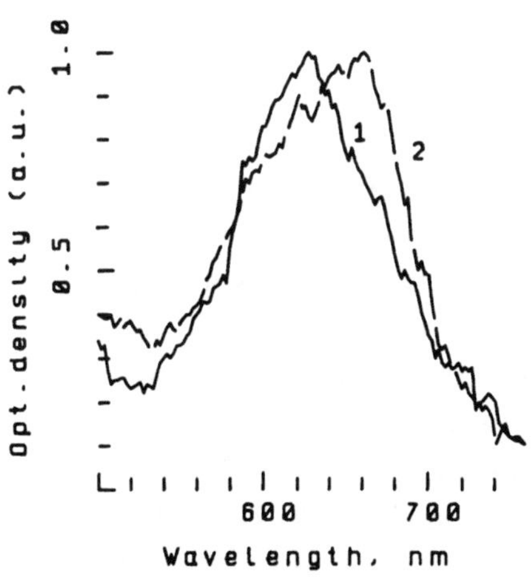

Fig.3 Transient absorption spectra of diethylaminophenylthiyl radicals in acetonitrile at time delays 5 ps (1) and 120 ps (2)

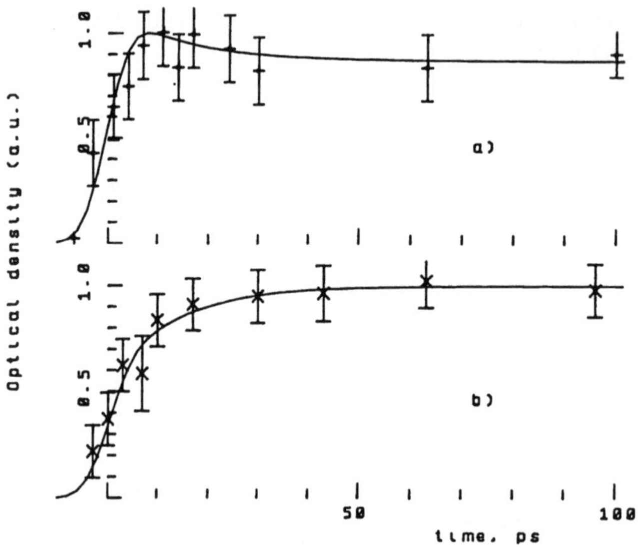

Fig.4 Time dependences of the optical density measured at 620 nm (a) and 680 nm (b) for diethylaninophenylthiyl radicals in ethanol; solid lines are calculated curves

Table 2. Substituent and viscosity effects on the ionic form formation time τ

Solvent	η,cP	ε	τ,ps[a)]	τ,ps[b)]
acetonitrile	0.35	37.4	10	-
pyridine	0.97	12.3	30	-
ethanol	1.20	25.2	40	25
buthanol	2.95	17.7	200	40

a) for the diethylaminophenylthiyl radical;
b) for the aminophenylthiyl radical

As we can see from Table 2, the characteristic time τ of the ionic form formation for the low viscosity range is a nearly linear function of viscosity. The increase in medium viscosity from 0.35 cP for acetonitrile to 2.95 cP for buthanol leads to an increase in the characteristic formation time of ionic form from 10 ps to 200 ps.

On the other hand, the comparison of data for the amino- and diethyiamino phenylthiyl radical derivatives in ethanol and buthanol solutions shows that an increase in the nitrogen atom substituting groups volume leads to a corresponding increase in the ionic form formation time τ. The characteristic time of the ionic form formation for the diethylaminophenylthiyl radical in ethanol is about 1.6 times greater than for the aminophenylthiyl radical. These results give an opportunity to suggest that a charge separation and radical ionic form stabilization is accompanied by nitrogen group rotation. This process in the radicals perhaps is similar to the formation of the twisted internal charge transfer states (TICT) [6]. In this case electronic density is transferred from nitrogen atom to the π-conjugated system. The reverse electronic density transfer is blocked by the turning of the planes of the donor aminoalkyl group and of the π-conjugated system to the mutually orthogonal position. To achieve more detailed understanding of mechanisms of the relaxation processes during ionic form formation further theoretical quantum chemistry investigations are necessary.

References

1. N.A.Lysak, S.V.Melnichuk, S.A.Tikhomirov and G.B.Tolstorozhev: Zh. Prikl. Spectr. 47, 267 (1987)
2. F.C.Thyrion: J. Phys. Chem. 77, 1478 (1973)
3. O.Ito, M.Matsuda: J. Phys. Chem. 88, 1002 (1984)
4. T.W.Scott, S.N.Liu: J. Phys. Chem. 93, 1393 (1989)
5. N.A.Lysak, S.V.Melnichuk, S.A.Tikhomirov and G.B.Tolstorozhev: Zh. Prikl. Spectr. 49, 949 (1988)
6. W.Rettig: Appl. Phys. B45, 145 (1988)

Photoisomerization Dynamics of DODCI

A. Penzkofer and W. Bäumler

Naturwissenschaftliche Fakultät II-Physik, Universität Regensburg,
D-8400 Regensburg, Fed. Rep. of Germany

Abstract. The photoisomerization dynamics of DODCI in ethylene glycol is studied by temperature dependent absorption and fluorescence measurements. A nearly quantitative picture of the S_0-S_1- potential energy curves versus the N-isomer (cis-cis) to P-isomer (all-trans) twisting angle is given.

1. Introduction

The polymethine dye DODCI (3,3'-diethyloxadicarbocyanine iodide) dissolved in ethylene glycol is the most widely used saturable absorber for the picosecond and femtosecond pulse generation in rhodamine 6G dye lasers [1]. The photoisomerization influences the mode-locking behaviour [2].

A nearly quantitative S_0- and S_1- potential energy surface diagram of DODCI in ethylene glycol versus the twisting angle from the normal cis-cis conformation [3] (N-isomer) to an elongated all-trans conformation (P-isomer) is obtained from the analysis of temperature dependent absorption and fluorescence spectra. The resulting potential energy diagram is shown in Fig.1.

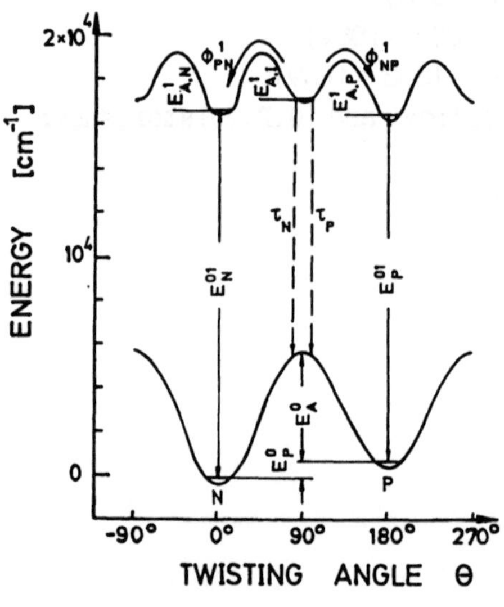

Fig.1: Potential energy diagram of DODCI in ethylene glycol. Parameters are explained in the text.

2. S_0 Ground-state Studies

The level position E_P^0 of the P-isomer ground-state is determined by temperature dependent absorption measurements in the long-wavelength absorption region [4]. The barrier height E_A^0 has been measured by Rulliere [5]. The absorption shoulder around 620 nm in Fig.2 indicates the thermal population of the P-isomer in the

Springer Proceedings in Physics, Vol. 49 Ultrafast Phenomena in Spectroscopy
Editors: E. Klose · B. Wilhelmi

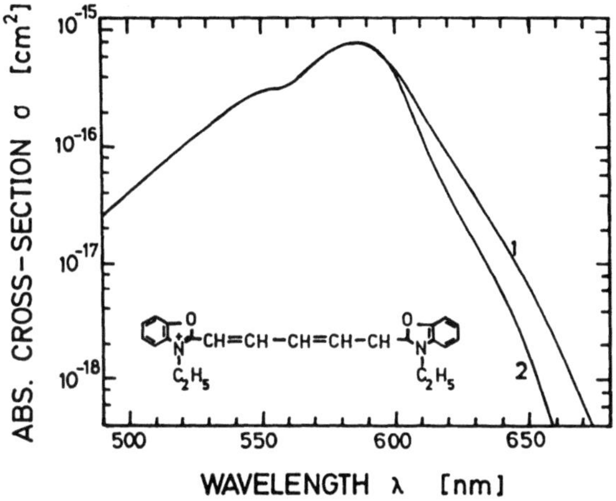

Fig.2: Absorption cross-section of DODCI in ethylene glycol. 1, ϑ = 95 °C. 2, ϑ = 2 °C. Structural formula of DODCI is included.

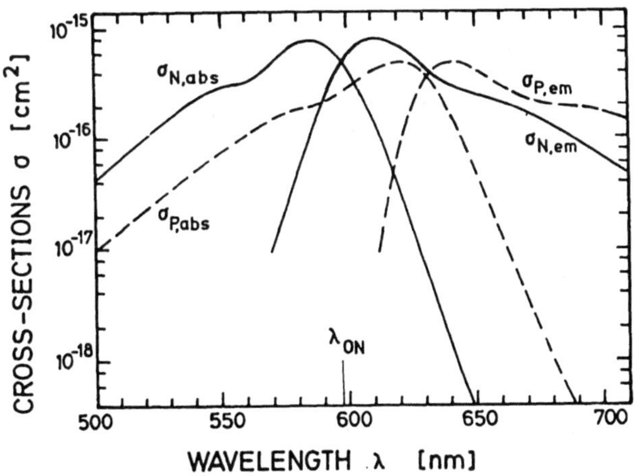

Fig.3: Absorption cross-sections and stimulated emission cross-sections of N-isomer and P-isomer of DODCI in ethylene glycol (ϑ = 22 °C).

S_0-ground-state. The separation of the shoulder from the N-isomer absorption spectrum allows the calculation of the P-isomer absorption cross-section spectrum. In Fig.3 the separated absorption cross-section spectra are displayed. The ratio of the absorption coefficients $\alpha_P/\alpha = \alpha_P/(\alpha_P+\alpha_N)$ of the P-isomer absorption to the total absorption is shown in Fig.4 for a temperature of ϑ = 22 °C. This ratio is equal to the population ratio $x_P^1 = N_P(S_1)/[N_P(S_1)+N_N(S_1)]$ in the S_1-state. In the long-wavelength region the P-isomer absorption dominates at thermal equilibrium.

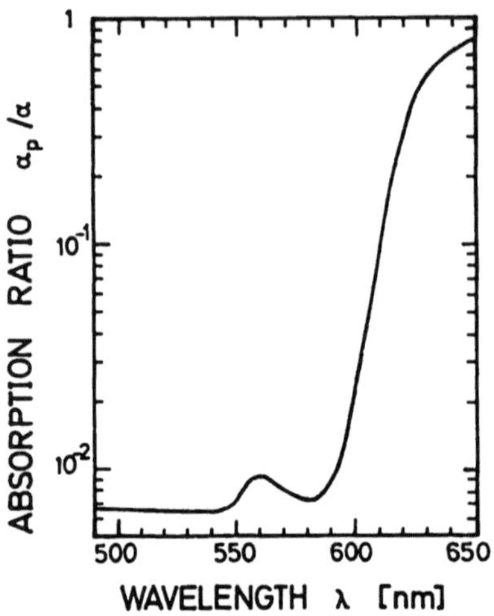

Fig.4: Ratio of absorption coefficients α_P/α for DODCI in ethylene glycol at 22 °C.

3. S_1-state Studies

The stimulated emission cross-section spectrum of the N-isomer is determined from fluorescence quantum distribution measurements at room temperature with an excitation wavelength near the S_0-S_1 zero-vibration wavelength λ_{0N} of the N-isomer absorption peak (see Fig.3)[6,7]. For long-wavelength excitation beyond the S_0-S_1-P-isomer absorption peak the P-isomer fluorescence dominates, a separation from the N-isomer contribution is possible, and the P-isomer stimulated emission cross-section spectrum may be calculated. The result is shown in Fig.3.

The S_1-level positions E_N^{01} and E_P^{01} are determined by the zero-vibration level wavelengths λ_{0N} and λ_{0P} where $\sigma_{i,abs} = \sigma_{i,em}$, i=N,P. The barrier heights $E^1_{A,N}$ and $E^1_{A,P}$ are derived from temperature dependent fluorescence quantum efficiencies q_i, i = N,P, by applying the Arrhenius-type deactivation relation

$$k_{nr,i}/k_{rad,i} = \mathrm{const}\ \exp(-E^1_{A,i}/k_B\vartheta) \approx q_i^{-1}-1. \qquad (1)$$

k_{nr} is the non-radiative S_1-state decay rate. k_{rad} is the radiative rate constant. k_B is the Boltzmann constant. q_N and q_P versus temperature are depicted in Fig.5.

The analysis of the spectral shapes of the fluorescence quantum distributions at different temperatures gives some information on the S_1-state N-isomer - P-isomer transfer efficiencies. In Fig.6 the normalized fluorescence quantum distributions are shown for the temperatures ϑ = -10 °C and ϑ = 90 °C. The excitation wavelength is λ_L = 597 nm. From the excess fluorescence around the P-isomer emission peak of 637 nm a S_1-state N-isomer to P-isomer transfer efficiency of $\phi^1_{NP} \approx 0.1$ may be deduced for ϑ = 90 °C [7].

The finite N-isomer to P-isomer S_1-state transfer rate sets an upper limit to the intermediate potential depths $E_{A,I}$ to $E_{A,I} \leq E^1_{A,N}$. The energy level positions E_N^{01} and E_P^{01} of the N-isomer and the P-isomer in the S_1-state and the activation energies $E^1_{A,N}$ and $E^1_{A,P}$ are compatible with a S_1-state double valley model ($E_{A,I}$ = 0) because the necessary condition of a double valley potential curve

$$E_N^{01} + E^1_{A,N} = E_P^0 + E_P^{01} + E^1_{A,P} \qquad (2)$$

is fulfilled within our experimental accuracy.

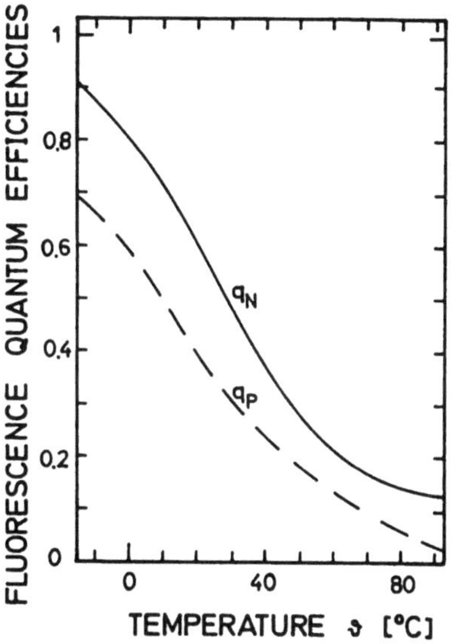

Fig.5: Fluorescence quantum efficiencies of the N-isomers, q_N, and the P-isomers, q_P, of DODCI in ethylene glycol.

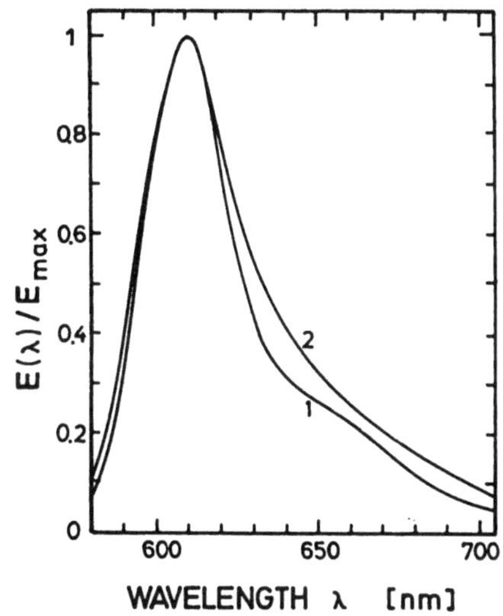

Fig.6: Normalized fluorescence quantum distribution $E(\lambda)/E_{max}$ of DODCI in ethylene glycol.
1, ϑ = -10 °C. 2, ϑ = 90 °C.

References

1. A. Penzkofer, Appl. Phys. B46, 43 (1988).
2. V. Petrov, W. Rudolph, and B. Wilhelmi, Rev. Phys. Appl. 22, 1639 (1987).
3. G.R. Fleming, A.E.W. Knight, J.M. Morris, R.J. Robbins, and G.W. Robinson, Chem. Phys. Letters 49, 1 (1977).
4. W. Bäumler and A. Penzkofer, Chem. Phys. Letters 150, 315 (1988).
5. C. Rulliere, Chem. Phys. Letters 43, 303 (1976).
6. S.J. Strickler and R.A. Berg, J. Chem. Phys. 37, 814 (1962).
7. O.G. Peterson, J.P. Webb, W.C. McColgin, and J.H. Eberly, J. Appl. Phys. 42, 1917 (1971).
8. W. Bäumler and A. Penzkofer, Chem. Phys., to be published.

Time-Resolved Resonance-CARS Spectroscopy of the First Excited Singlet State of Stilbene-3 and of Cyanine Photoisomers

W. Werncke, A. Lau, M. Pfeiffer, Kim Man Bok, and Tschoe Jong Tscholl

Zentralinstitut für Optik und Spektroskopie, Akademie der Wissenschaften, Rudower Chaussee 6, DDR-1199 Berlin, German Dem. Rep.

Abstract. CARS spectra of the first excited singlet state of stilbene-3 and of the photoisomers of bisdimethylaminopentamethineperchlorate (PMC) and bisdimethylaminoheptamethineperchlorate (BMC) are recorded. From the spectra we obtained information on structural changes of the short lived molecular states.

Time resolved excite-and-probe resonance CARS-spectroscopy is a very useful method to obtain vibrational spectra of short living molecules even in the case of a strong fluorescence caused by excitation of the excited states /1/.
Using pulses of a duration of about 10 ns, different excitation wavelengths λ_{exc}, pump wavelength λ_p and delays between excitation and probing time resolved CARS spectra of the first excited singlet state of stilbene-3 and of cyanine photoisomers have been recorded.

1. FIRST EXCITED SINGLET STATE OF STILBENE-3

CARS spectra of the first excited singlet state of stilbene-3 have been recorded and can be compared with the S_1-spectra of trans-stilbene (table 1).
S_1-spectra of stilbene-3 correspond to the S_1-trans stilbene spectra. However, there are minor frequency shifts to lower wave numbers in the C=C stretching region of stilbene-3 ($S_0 \rightarrow S_1$) than for trans-stilbene.
This can be explained by an enhancement of π-binding character of the C_0-C_0 bond in the central part of the stilbene-3 molecule.
Trans-stilbene S_1-Raman frequencies at 1567 cm^{-1} and at 1242 cm^{-1} are broadened contrary to the corresponding stilbene-3 lines. The latter feature has been assumed to be due to differently twisted conformers /2/ which probably do not occur in the stilbene-3 S_1-state.

Springer Proceedings in Physics, Vol. 49 Ultrafast Phenomena in Spectroscopy
Editors: E. Klose · B. Wilhelmi

TABLE 1. Comparison of the CARS-spectra of stilbene-3 (St-3) in the ground and in the first excited singlet state with the corresponding RRS spectra of trans-stilbene (T-St) /2/.

λ_{exc} = 347 nm, λ_p = 694 nm without delay

(Ø: phenylring; C_0: olefinic carbon)

S_1-state		S_0-state		
CARS/RRS /cm^{-1}/				assignment /2/
St-3	/ T-St	St-3	/ T-St	
1587 vs	/ 1567 vs	1620 m	/ 1641 s	C-C str olefin
		1593 vs	/ 1595 vs	C-C str ring
1563 w		1580 s	/ 1570 w	
		1558 w		
	1473 w	1495 w		C-C str
	1445 w	1447 w		C-C str + C-C-H bending
1335 m	/ 1333 m	1340 m		C_0-C_0-H bending
			1329 w	
		1280 m		
1199 s	/ 1242 s	1189 s	/ 1194 vs	C-H def olefin + C_0-Ø str
	1185 m		1181 vs	C-H def ring

2. PHOTOISOMERIZATION OF BMC AND PMC

CARS frequencies of the parent molecules of bisdimethylaminopentamethineperchlorate (PMC) (see Table 2) and -heptamethineperchlorate (BMC) /3/ and its transient photoisomers are obtained. The common feature of both molecules is the occurence of a very strong 1070 cm^{-1} line after photoexcitation.

In principle different photoisomers are possible (1,2 mono-cis, 2,3 mono-cis, etc.).

A changed spectral appearance can be explained by a lowering of the molecular symmetry. Canceling of the C2v-symmetry leads to a coupling of the former A1- and B2 vibrations resulting in a separation of close lying vibrational bands belonging to different species in the original molecule. Our selection from the different possible mono-cis configurations bases on a normal coordinate calculation for the different forms. From our calculations it follows that only for the 1,2 photoisomer a strong line at 1070 cm^{-1} is expected for BMC and for PMC as well.

TABLE 2. Parent molecule and photoisomer CARS spectra of PMC

photoisomer	parent molecule	assignment
λ_{exc} = 400 nm		
λ_p = 460 nm	λ_p = 460 nm	delay: 500 ns
CARS / cm^{-1} /	CARS / cm^{-1} /	
1660 vs	1660 vs	s C=N str
1574 m	1574 m	as C=C str
1487 w	1487 w	
1447 m	1442 m-s	
1420 vs	1410 vs	def CH_3
1362 w	1357 w	
1311 vs	1307 vs	def CH_3, C-Nstr
1214 m	1214 m	
1125 vs	1125 w	def CCC, C=Cstr
1082 m	1072 vs	def CH, C=Cstr, C=Nstr
832 m	832 m	

From the calculations we can conclude that after irradiation the 1,2 photoisomers are generated in both cases independent of the lengths of the methine chains.

References

/1/ W. Werncke, H.-J. Weigmann, J. Pätzold, A. Lau, K. Lenz, M. Pfeiffer; Chem. Phys. Lett. 61 (1979) 105

/2/ T. L. Gustafson, D. M. Roberts, D. A. Chernoff; J. Chem. Phys. 81 (1984) 3438

/3/ W. Werncke, A. Lau, M. Pfeiffer, H.-J. Weigmann, W. Freyer, Tschö Jong, Tscholl, Kim, Man Bok; Chem. Phys. 118 (1987) 133

Time-Resolved Absorption Spectroscopy of Polymethine J-Aggregates

K.-H. Feller[1], *R. Gadonas*[2], *and V. Krasauskas*[2]

[1]Sektion Chemie, Center of Optical Spectrosopy, Friedrich-Schiller-Universität Jena, DDR-6900 Jena, German Dem. Rep.
[2]Laser Research Center, Vilnius State University, 232054 Vilnius, Lithuania

The deactivation process of J-aggregates of DDEOCI (3,3′-diethyl-5,5′-diphenyl-9-ethyl-oxacarbocyanine iodide) was investigated by means of steady-state and time-resolved absorption spectroscopy.

1. INTRODUCTION

The photophysical properties of aggregated chromophores are of interest for several reasons. Several biologically active pigment systems, especially in photosynthesis, form dimers and higher aggregates or behave in a similar manner to dimers and aggregates. The fast energy as well as electron transfer to and from adsorbed dye aggregates plays a dominant role in spectral sensitization. For our investigations we used DDEOCI, which shows a very strong aggregation in methanol-water mixtures. First aggregates can be detected at dye concentrations as low as 10^{-6} mol/l. Whereas in pseudoisocyanine the J-aggregate band is uniform and the detected lifetime(s) belong(s) to only one aggregate structure, preliminary stationary investigations have indicated that the situation in DDEOCI is more complex. The first indices are changes in the absorption maximum of the J-band with the age of the sample (see Figure 1). But also at least two different maxima (or a shoulder) have been found, depending on the preparation conditions.

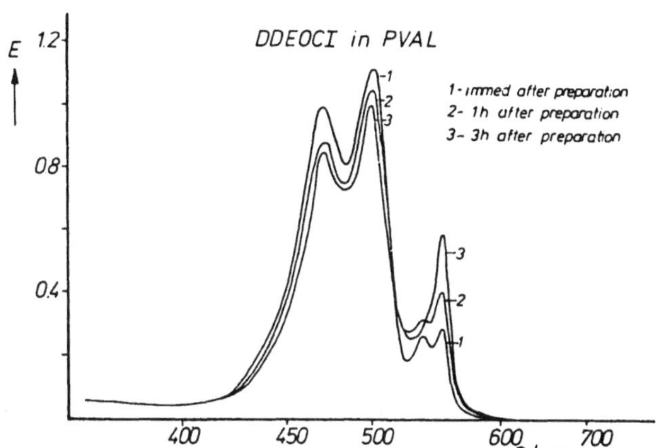

Figure 1: Changes in the absorption spectrum of DDEOCI depending on the age of the sample
a) immediately after preparation
b) 1 h after preparation
c) 3 h after preparation

Furthermore the DDEOCI J-aggregate band is less steep than that of PIC. From this point of view it should be neccessary to investigate different types of J-aggregate-forming cyanines to get additional information about effects determining the aggregate size. The extraordinary behaviour of DDEOCI also in this respect makes this dye very well-suited for detailed investigations of the J-aggregation process in polymethine dyes.

The time-resolved absorption investigations were done with a pump-and-probe pulse spectrometer which has been described in detail elsewhere [1]. The samples were excited by 532 nm 25 ps pump pulses and the absorption spectra at different time delays as well as the kinetics at different wavelengths were measured by means of a weak tunable probe beam. The pump pulse intensity was changed within the range $2 \cdot 10^{10} - 10^{15}$ phot/cm^2 per pulse by focusing or defocusing the irradiation spot.

The decay kinetics was fitted to the experimental curves in the case of the absorption investigations with the help of the ALAU program package [2] taking into account multiexponential as well as nonexponential (due to exciton-exciton annihilation, see Equ. (2)) decay kinetics. In this case the goodness-of-fit was checked by calculation of the standard deviation as well as the Durbin-Watson parameter.

2. RESULTS and DISCUSSION

Our single-photon counting experiments [3] have shown that the fluorescence decay of the J-aggregate band of DDEOCI cannot be described by only one lifetime component even at very low excitation intensities. Both the components detected by means of SPC investigations with subnanosecond time resolution are in the nanosecond region. The existence of different (at least 2) aggregate structures is assumed to be the reason for the detected biexponentiality of the decay curves. The non-uniform depletion of the ground state bleaching in DDEOCI J-aggregates indicates different relaxation of the two J-aggregate structures hidden in the common band.

First hints that the J-band is composed of different types of aggregates due to different isomers were given by Hada et al. [4] from the investigation of spectral changes of UV bands concerned with various J-aggregate types hidden in one J-band in the visible region. Furthermore, in the case of 5,5′-dichloro-3,3′-diethyl-9-ethyl-thiacarbocyanine chloride they found in aqueous solution three J-bands (at 619 nm, at 623 nm and at 642 nm). The band at 642 nm was shown to be unstable whereas both the other bands were stable. Hada et al. [4] were able to show that these spectral changes are due to different isomer-conformations in the ground state of the monomers building up different types of aggregates.

The expected aggregate structures should coincide with the conformer equilibrium detected in previous experiments [5] in the ground state of meso-substituted oxacarbocyanine dyes where, in accordance with results of Grahn and coworkers [6], an equilibrium of all-trans and 8,9 mono-cis isomers in the ground state have been found.

The results of the picosecond absorption investigations at medium excitation intensity ($2 \cdot 10^{10}$ phot/cm^2 per pulse) are depicted in Figure 2. It is seen that the deactivation kinetics should be fitted to an at least triple exponential decay (see Equ. (1))

$$\Delta D = A \exp(-t/\tau_1) + B \exp(-t/\tau_2) + C \exp(-t/\tau_3) \quad (1)$$

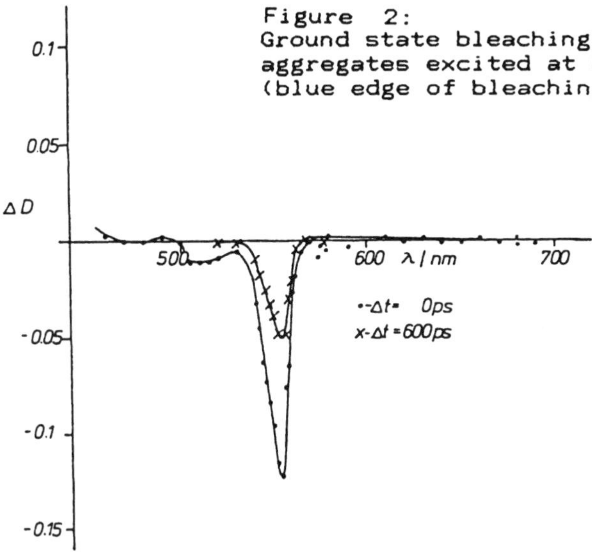

Figure 2:
Ground state bleaching kinetics of DDEOCI J-aggregates excited at 532 nm and probed at 553 nm (blue edge of bleaching curve)

with lifetime components of

$$\tau_1 = 50 \text{ ps}, \quad \tau_2 = 205 \text{ ps and} \quad \tau_3 = 3.50 \text{ ns}$$

DW parameter = 1.62

These lifetime components should be explained as belonging to different aggregate structures.
The investigation of the absorption spectra of DDEOCI J-aggregates at different delay times has shown that the bleaching curve in the ground state recovery region does not vanish uniformly. There is a short wavelength part of the bleaching curve which decays very fast whereas the long wavelength part decays more slowly (see Figure 3). As it was shown that the decay kinetics of polymethine aggregates depend on the excitation intensity [7,8] connected with exciton-exciton annihilation at high excitation intensities, we have measured the influence of excitation intensity upon the decay kinetics and the transient absorption spectra. Investigations of transient absorption spectra at different excitation intensities have been made. The higher excitation case (10^{15} photon/s cm^2) which is depicted in Figure 3 is connected with an absorption at 570 nm which decays with a very complicated deactivation rate law (see Figure 4). Furthermore there is a new very weak absorption band with a maximum at 680 nm which is nearly constant during the possible delay time changes of 1.5 ns. In the 510 nm region, where the J-aggregates have a weak absorption the bleaching of this ground state absorption is seen. In the lower excitation case ($2 \cdot 10^{10}$ photon/s cm^2) neither an absorption at 570 nm nor a weak red-shifted absorption is seen. The bleaching of the ground state absorption at 510 nm disappears within the used delay time of 600 ps. In both cases both sides of the J-aggregate bleaching curve do not relax uniformly. The pumping of the aggregates leads in both cases to a fast deactivation of the aggregates with a blue-shifted absorption spectrum compared with the aggregates with a red-shifted absorption spectrum. Even a transfer from one structure to the other has to be taken into account. Such a transfer is indicated by kinetic

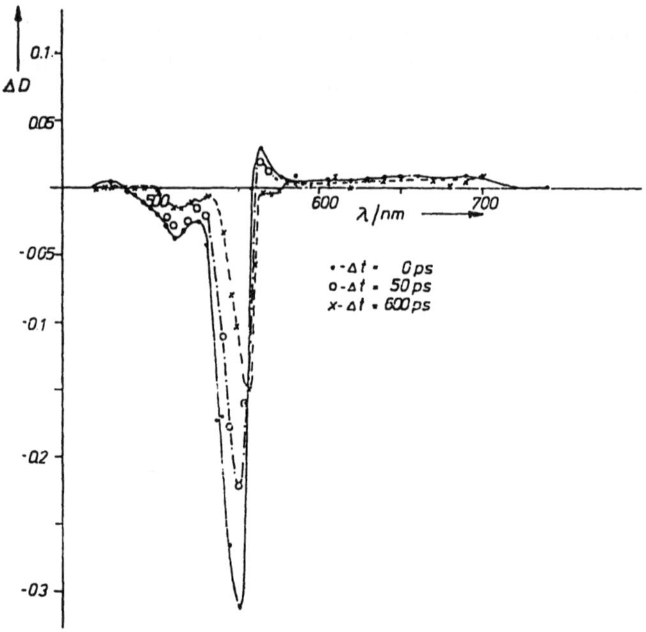

Figure 3: Picosecond absorption spectra of DDEOCI J-aggregates in water/methanol mixture at high excitation intensity excited at 532 nm
• - 0 ps delay, x - 50 ps delay, o - 600 ps delay

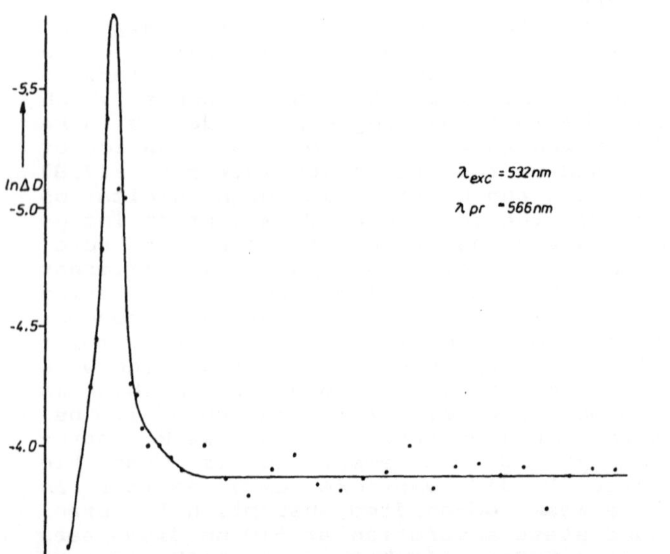

Figure 4: Picosecond transient kinetics of DDEOCI J-aggregates in the red-shifted transient absorption region
λ_{exc} = 532 nm λ_{probe} = 566 nm

measurements with excitation at 532 nm and probing at 568 nm where in addition to the slow decay of the bleaching a fast build-up process at short delay times appears.
Aggregation is also observed in nonaqueous solutions under circumstances where appropriate solvent structuring occurs [9].
J-aggregation of cyanine dyes at Langmuir-Blodgett films has been investigated with time-resolved methods by Lehmann [10] and Nakano et al. [11]. The latter have found two different J-bands of meso-substituted cyanine dyes in monolayer assemblies depending on the surface pressure. No explanation was given for the obtained surface pressure dependence. It can be supposed that also in this case different aggregate structures according to the surface pressure are the reason for the two observed J-bands.

The spectral properties of J-aggregates of cyanine dyes in solution and their absorption onto silver halides have been reviewed by Herz [12].
According to the exciton-exciton annihilation concept the decay kinetics should be described by the differential equation

$$\frac{d\Delta D}{dt} = - k_1 \Delta D - k_2 (\Delta D)^2 \qquad (2)$$

of which the solution is

$$\Delta D(t) = \frac{2 k_2 A_0}{\exp(k_1 t)(k_1 + 2k_2 A_0) - 2k_2 A_0} \qquad (3)$$

describing the decay of excited aggregates by competitive exciton-exciton annihilation and first-order decay [13].
Whereas in the case of uniform aggregates, as in the case of PIC, such a fit should lead to reasonable results, in our case, due to the involvement of different aggregate structures and their decay kinetics, we found a better fit of the decay curves by multiexponential fitting by equation (1).

REFERENCES

[1] R. Danielius, R. Gadonas, A. Piskarskas: Kvantovaya Elektronika 8, 669(1981)
[2] H. Schütz, E. Stutter, K. Geller, I. Petri: studia biophysica 104, 23(1984)
[3] K.-H. Feller, R. Gadonas, V. Krasauskas, V. Fidler, S. Vajda: Laser Chemistry (submitted)
[4] H. Hada, C. Honda, H. Tanemura: Photogr. Sci. Eng. 21, 83(1977)
[5] K.-H. Feller, R. Gadonas, V. Krasauskas: Laser Chemistry 8, 39(1988)
[6] R. Allmann, H.-J. Anis, R. Benn, W. Grahn, S. Olejnek, A. Waskowska: Angew. Chem. Suppl. 1147(1983)
[7] H. Stiel, K. Teuchner, S. Dáhne: J. Lumin. 39, 351(1988)
[8] V. Sundstróm, T. Gillbro, R. Gadonas, A. Piskarskas: J. Chem. Phys. 89, 2754(1988)
[9] K.-H. Feller, J. Siegel, D. Fassler, H. Bergner, F. Kerstan in: Teubner Texte zur Physik, Vol. 4, Proceedings of the Third Symposium Optical Spectroscopy (B.S.B. B.G. Teubner Verlagsgesellschaft, Leipzig, 1985) p. 216

[10] U. Lehmann, Thin Solid Films 160, 257(1988)
[11] A. Nakano, S. Shimizu, T. Takahashi, H. Nakahara, K. Fukuda: Thin Solid Films 160, 303(1988)
[12] A. H. Herz: Adv. Colloid. & Interface Sci.: 8, 237(1977)
[13] V. Brúckner, K.-H. Feller, U.-W. Grummt, Applications of Time-Resolved Optical Spectroscopy (Elsevier Science Publishers, Netherlands, 1989)

Light-Induced Intramolecular Proton-Transfer in a System in Which the Excited Donor is Separated from the Acceptor by a Flexible Alkyl Chain

C.J. Jalink, A.H. Huizer, and C.A.G.O. Varma

Department of Chemistry, Gorlaeus Laboratories, Leiden University, P.O. Box 9502, 2300 RA Leiden, The Netherlands

1. Introduction.

The rate constants of proton-transfer processes vary over many orders of magnitude. The reasons for this large variation are not well known. In a search for the factors which determine the mechanism and the rate constant of proton-transfer processes, Varma and coworkers [1-4] have previously concentrated their attention on the relatively small class of thermal and light-induced adiabatic intramolecular proton-transfer processes within aromatic molecules, having a proton-donor site and a proton-acceptor site as parts of the same conjugated system. If an intramolecular H-bond exists in such systems between the donor and acceptor sites, the transfer may proceed directly. When this H-bond is lacking the transfer may be catalysed by hydrogen bonding solvents, forming a cyclic structure in which donor and acceptor sites are incorporated.

In this study we have focussed our attention on a molecular system, quinacrine (QA, I in fig. 1), with an aromatic proton-donor separated from a proton-acceptor by a flexible alkyl chain. The time behaviour of the fluorescence of protonated forms of QA and related compounds have been studied previously [5-7]. The complex behaviour has been attributed to multiple excited state adiabatic proton transfer [6,7].

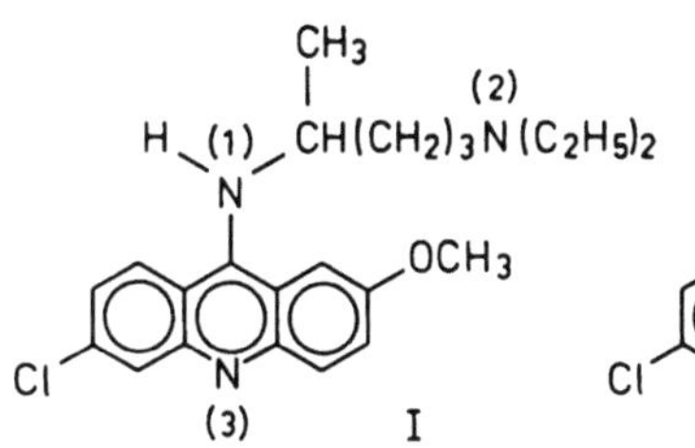

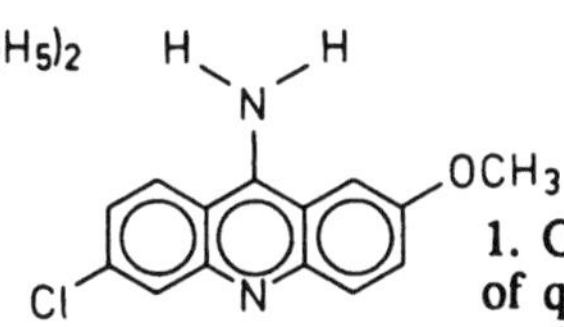

II

1. Chemical formula of the free base form of quinacrine (I) and the parent molecule 2-methoxy-6-chloro-9-aminoacridine (II).

2. Experimental.

Quinacrine dihydrochloride (Sigma Chemical Company) was converted with KOH to the free base form and purified by repeated crystalisation from solutions in n-hexane. Spectroscopic grade solvents were used as delivered. In the samples the concentration of the solute was kept below 2×10^{-4} M.

The time dependence of the fluorescence of QA has been studied by exciting the sample with a 265 nm laser pulse of 16 ps (FWHM) and detecting the fluorescence with a streak camera (Hadland, Imacon 500). The time axis of the camera has been corrected carefully and its response function R(t) for the excitation pulse has been recorded. The emission wavelength was selected by means of interference filters with a bandwidth of 10 Å (FWHM).

The presented time profiles were analysed numerically via a simulation procedure which represents the profiles as the convolution of the response function R(t) and a function I(t), a sum of exponentials. Attention has been paid to the effects of solvent polarity, protic or non-protic nature of the solvent, solvent's viscosity, temperature and deuteration of the proton-donor site on the time evolution of the fluorescence of QA.

3. Results and Discussion.

QA is known to have a single fluorescence band, independent of the type of solvent used, with a Stokes shift of about 3100 cm^{-1} relative to the first electronic absorption band. Time resolved spectroscopy reveals, for all the solutions, a biexponential decay at the blue edge of the fluorescence band (i.e. at 460 nm) with the short lifetime τ_S and the long lifetime τ_L. At the red edge (i.e. >580 nm) the fluorescence grows with the risetime τ_R and decays single exponentially with the lifetime τ_P (figs. 2a and 2b). In all cases the equalities $\tau_S = \tau_R$ and $\tau_L = \tau_P$ are found to hold within the experimental uncertainty of about 4% (table 1). This behaviour is characteristic for a system consisting of two types of excited species exhibiting a mother-daughter relation in which a blue emitting species (M^*) is transformed into a red emitting one (P^*), while the two fluorescence bands overlap strongly. Based on this picture and on the observed kinetics, the fluorescence band of QA in methanol is decomposed into the contributions from M^* and P^*, shown in fig. 3.

Excitation of a sample containing only one type of molecule, which exists in various conformations, may lead to a distribution of monoexponentially decaying fluorescing species

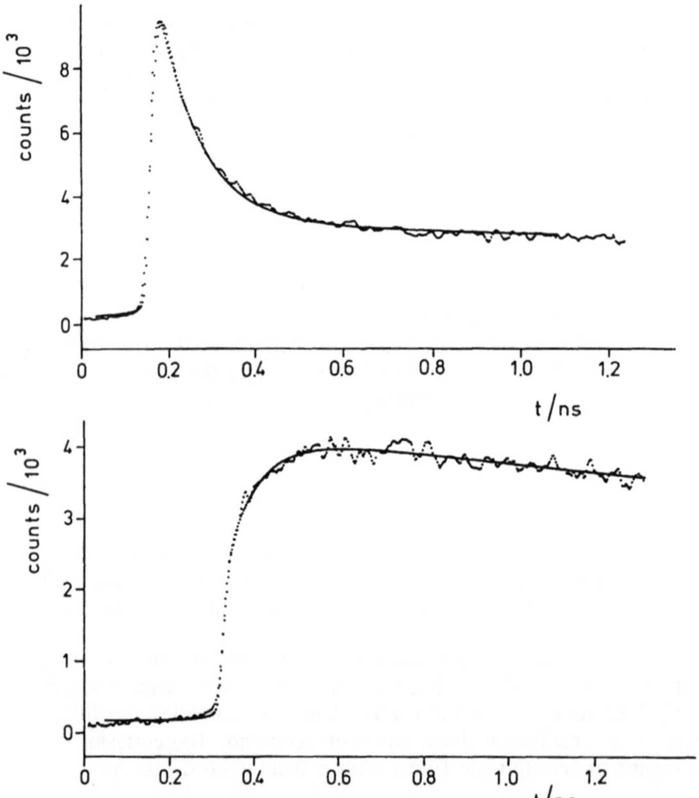

2. Recorded fluorescence of QA in methanol at 295 K. (A) emission wavelength 460 nm, short lifetime τ_S = 90 ps and long lifetime τ_L = 6000 ps. (B) emission wavelength 580 nm, risetime τ_R = 92 ps and lifetime τ_P = 6000 ps.

Table 1. *Fluorescence timeconstants of Quinacrine as a function of bulk properties of the solvent*

Solvent	η / cP	ε_D	τ_S / ps[1]	τ_L / ps[1]
methanol	0.60	32.6	90.	6000.
methanol-d			113.	7310.
ethanol	1.20	24.3	100.	6340.
1-propanol	2.26	20.1	105.	6500.
2-propanol	2.86	18.3	106.	6730.
1-butanol	2.95	17.1	110.	7900.
2-butanol	4.21	15.8	113.	7950.
1-hexanol	6.3	13.3	145.	9400.
1-octanol	10.6	10.3	165.	10300.
1-undecanol	17.2		247.	9930.
ethylene glycol	19.9	37.	269.	10500.
n-hexane	0.32	1.89	33.	1030.
liquid paraffin	68.81		138.	2450.
acetonitrile	0.44	37.5	97.	3420.
acetonitrile[2]			130.	4130.
undecanenitrile	2.	10.	137.	5450.
tetrahydrofuran	0.49	7.39	67.	3400.
1,2-dichloroethane	0.84	10.6	108.	3500.

1) $\tau_S = \tau_R$ and $\tau_L = \tau_P$ within experimental uncertainty (4%).
2) QA deuterated at N(1), see fig. 1.

corresponding to the different conformations. Under certain circumstances such a distribution may give rise to a total fluorescence intensity which appears to decay biexponentially [8]. However, the relation $\tau_S = \tau_R$, which is satisfied quite nicely, suggests that this does not happen here. The obtained value of τ_S has to be considered as the lifetime of M^* averaged over the distribution of its conformations.

There is a strong dependence of τ_S on viscosity (table 1). This indicates the involvement of a large conformational change in the alkyl chain during the transformation of M^* into P^*.

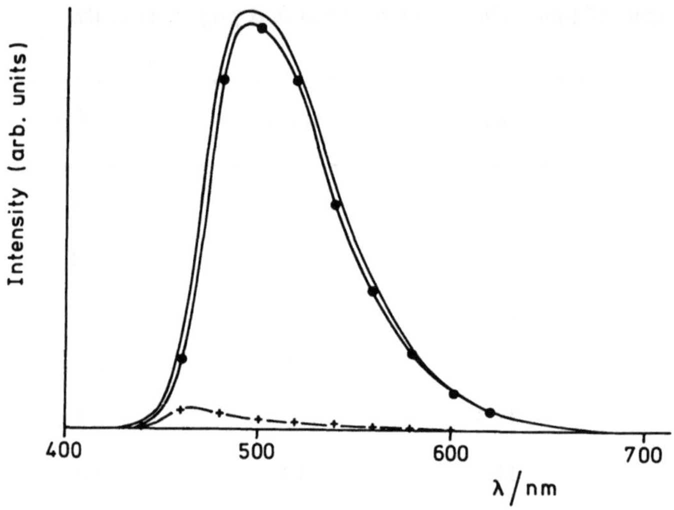

3. Deconvoluted fluorescence spectrum of QA in methanol at 295 K, showing the contributions of the blue emitting species M* (+) and the red emitting species P* (●).

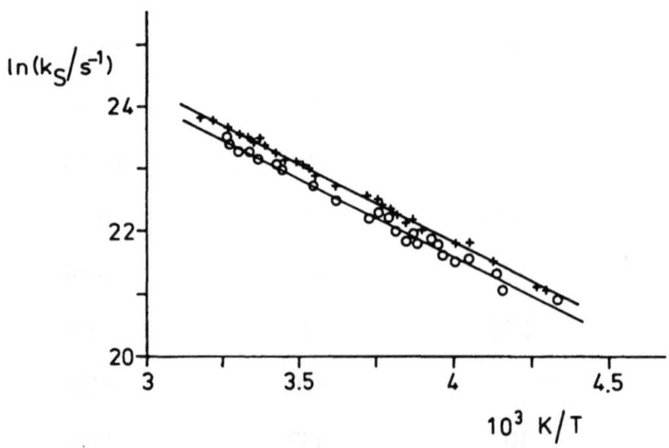

4. Arrhenius behaviour of k_s for the solvent composed of CH_3O^1H (+) and the solvent composed of CH_3O^2H (o). $k_s(H)/k_s(D) = 1.3$ and $E_{act} = 4.9$ kcal mol^{-1}.

The transformation could involve intramolecular hydrogen bonding (formation of an intramolecular exciplex) in apolar solvents or intramolecular proton-transfer, yielding a zwitter ion in polar solvents. The proton-donor site is to be identified with N(1) and the proton-acceptor site with N(2), see fig. 1.

Deuteration of the proton-donor site leads to a kinetic isotope effect (KIE) on the rate constant k_s (= $1/\tau_s$) for the decay of M*, amounting to $k_s(H)/k_s(D) = 1.3$ in the case of methanol (fig. 4) and in the case of acetonitrile as solvents (table 1). The temperature independence of the KIE is consistent with the following picture of the proton-transfer mechanism: a) An intramolecular or intermolecular hydrogen bonded complex with a special geometrical structure is formed upon encounter of donor and acceptor sites with rate constant k_C; b) Proton tunneling takes place prior to thermalization of the special complex, yielding a temperature independent probability α_P for tunneling per encounter. This amounts to a rate constant $k_s = \alpha_P k_C$.

Finally it should be noted that the rate constant for the photoreaction in excited Quinacrine is an order of magnitude larger than the rate constant for intramolecular exiplex formation [9,10] in molecules of type A-$(CH_2)_3$-D, in which A is an electron acceptor and D is an electron donor.

References.

1. J. Konijnenberg, A.H. Huizer, F.Th. Chaudron, C.A.G.O. Varma, B. Marciniak and S. Paszyc, J.Chem.Soc. Faraday Trans. 2., **83** (1987) 1475;
2. J. Konijnenberg, A.H. Huizer and C.A.G.O. Varma, J.Chem.Soc. Faraday Trans. 2., **84** (1988) 363;
3. J. Konijnenberg, A.H. Huizer and C.A.G.O. Varma, J.Chem.Soc. Faraday trans. 2., **84** (1988) 1163;
4. J. Konijnenberg, A.H. Huizer and C.A.G.O. Varma, J.Chem.Soc. Faraday Trans. 2., **85** (1989) 39;
5. A. Marty, P. Viallet, M. Bourdeaux and M. Dell'Amico, J.Photochem. **28** (1985) 71;
6. A. Andreoni, R. Cubeddu, S. de Silvestri and P. Laporta, Opt.Comm. **33** (1980) 277;
7. S. de Silvestri, F. Docchio, P. Laporta A. Longoni and F. Zaraga, Opt.Comm. **37** (1981) 20;
8. D.R. James and W.R. Ware, Chem.Phys.Lett. **126** (1986) 7;
9. Y. Wang, M.C. Crawford and K.B. Eisenthal, JACS, **104** (1982) 5874;
10. M. van der Auweraer, A. Gilbert and F.C. De Schrijver, JACS, **102** (1980) 4007;

Formation, Relaxation and Recombination of Polarons and Bipolarons in Poly(1,4-phenylene-1-phenylvinylene) (MP-PPV) in the Picosecond Time Region

J.P. Yang, S. Rentsch, M. Lenzner, and H.L. Li

Sektion Physik, Friedrich-Schiller-Universität Jena, Max-Wien-Platz 1, DDR-6900 Jena, German Dem. Rep.

Time resolved absorption, fluorescence and photoconductivity of MP-PPV in the picosecond regime show that the polaron and the bipolaron are the elementary excitations in photoinduced MP-PPV. The time difference of formation of bipolarons and polarons was found to be independent of the concentration of solution and of layer properties. It amounts to about 10 ps at all samples. In dilute solution (0.032 mg/ml) the formation of the negative polaron proceeds slower than the positive polaron formation. The mobility of the polarons and the bipolarons has been determined.

1. INTRODUCTION

Conducting polymers are currently attracting considerable interest as materials for a wide variety of applications. The poly(phenylene-vinylene) (PPV) derivatives have good material properties, including stability, solubility, photoconductivity and optical nonlinearity [1,2]. For the organic conducting polymers with non-degenerate ground state many studies [3,4,5] have demonstrated that polarons and bipolarons play a fundamental role in determining their electronic properties. We have studied the characteristics of the short-living photoinduced species of MP-PPV by time resolved spectroscopy.

2. TRANSIENT ABSORPTION MEASUREMENTS

The experiments were carried out by the excite-and-probe technique described in [6]. For the excitation of polymers the third harmonic of the Nd- phosphate glass laser was used with 6 ps pulses of (100 ± 40) µJ at 351 nm.

After photoexcitation a neutral bipolaron is rapidly formed [7]. This early process cannot be observed because of the time resolution limit of 6 ps. By interchain excitation the neutral bipolarons convert into polarons; the charged bipolarons are formed by the standard reaction

$$P^{+} + P^{+} \longrightarrow BP^{2+} \text{ and } P^{-} + P^{-} \longrightarrow BP^{2-}.$$

The stationary absorption spectrum of MP-PPV is shown in Fig.1.

The doped MP-PPV shows spectra in the visible region [8] similar to the transient spectra of undoped MP-PPV. The broad transient spectra (Fig.2) contain parts of different kinetics (Fig.3), R1-band (670 nm - 690 nm) and NIR1- band (780 nm - 810 nm), with a fast rise time and two bands, R2 (715 nm - 760 nm) and NIR2 (875 nm - 935 nm), with a delayed increase of about 10 ps. The R1-, NIR1- and NIR2- bands have a fast decay, whereas the R2- band shows a slower decay with two time components.

The bands R1 and NIR1 we assigned to E_2 and E_3 of the polarons, respectively, and the R2 band to E_2 of the bipolarons (Fig.4). The time decay of the NIR2 band shows an analogy to the R1 and R2 bands. The delay of about 10 ps in the increase of the NIR2 band in the dilute solution we explain as follows.

Springer Proceedings in Physics, Vol. 49 Ultrafast Phenomena in Spectroscopy
Editors: E. Klose · B. Wilhelmi

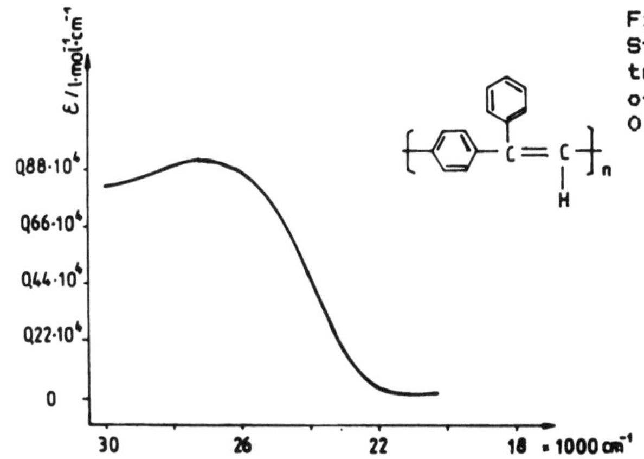

Fig.1
Stationary absorption spectrum and chemical structure of MP-PPV. E_g = 2.75 eV (at 0.1 of maximum extinction).

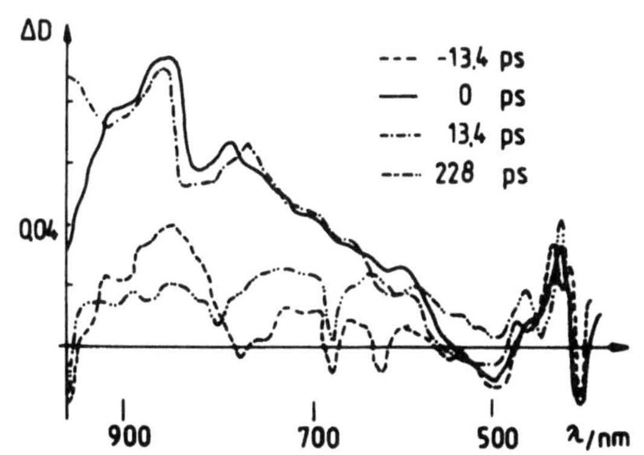

Fig.2 Transient optical density difference ΔD-spectra of MP-PPV in toluene (0.032 mg/l) after excitation (6 ps, 100 ± 40 μJ, 351 nm).

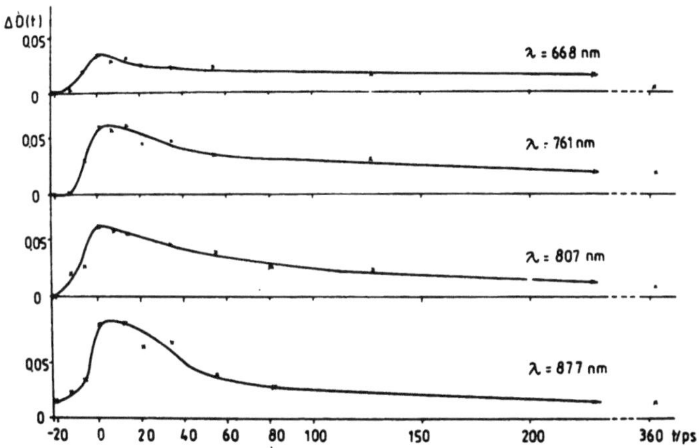

Fig.3
ΔD(t)-curves of MP-PPV in toluene after photoexcitation.

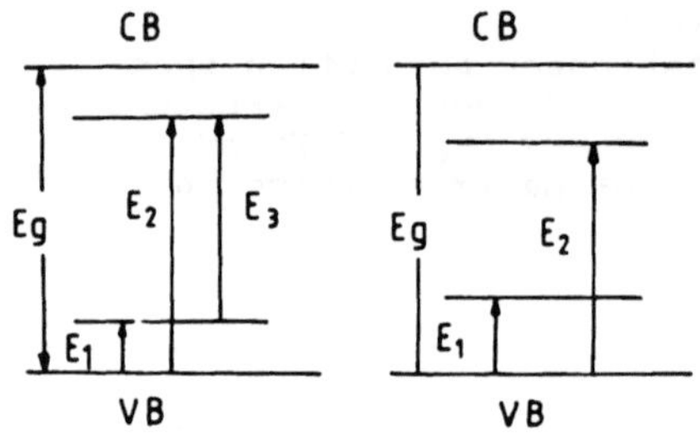

Fig. 4
Energy level schema and optical transitions for polarons and bipolarons.

Polaron (P^+) Bipolaron(BP^-)

We assigned the NIR2 band to the E_3 of P^-. In a dilute solution (0.032 mg/ml) the negative polarons may be formed after separation of an electron by a trapping process. Therefore the P absorption maximum appears later than that of P^+. This behaviour cannot be observed in solutions with higher concentration and in layers, because there the trapping proceeds faster.
In a preliminary fit of the curves with exponential functions an average lifetime of 170 ps for the polarons and 610 ps for the slow component of the bipolarons in the dilute solution was obtained. The time decay constants in the layer are 40 ps for the polarons and 980 ps for the slow component of the bipolarons. Compared with solutions there exists a broad absorption band (490 nm - 770 nm) of the bipolarons in layers and the polarons have a fast rise time.

3. TIME RESOLVED FLUORESCENCE MEASUREMENTS

The experimental measurements were carried out using time - correlated single photon counting equipment SPC - 100 (ZOS) [9]. The excitation was carried out by the second harmonic of a synchronously pumped dye laser.
Time resolved fluorescence spectra of MP-PPV in toluene (0.032 mg/ml) were measured under excitation with 5 ps pulses of 0.3 nJ at 350 nm. The kinetics of the fluorescence contains two exponential functions, the components of which are deduced from a deconvolution program [10]. These kinetic data are given in table 1.

Table 1.

λ/nm	500	550	600
τ_1/ps	166	222	266
τ_2/ps	535	634	771

By comparison with absorption measurements in dilute solution (0.032 mg/ml) in which were obtained a fast (170 ps) and a slow (610 ps) component, we assign the two fluorescence components (τ_1 and τ_2) to recombination of polarons and bipolarons, respectively. The observed spectral fluorescence maximum is located at 482 nm (2.57 eV) (Fig.5).
The fluorescence maximum energy (2.57 eV) is greater than the transition energy from the conducting band to the low energy levels of the gap states of the polarons and the bipolarons or from the upper energy levels of the gap states to the valence band, i.e. the fluorescence maximum does not come only from the recombination of the polarons or the bipolarons. The energy of the electron - hole pairs generated by the photo-excitation are situated near the bottom of the conducting band. Some of them can recombine fast [11, 12]. Therefore we can con-

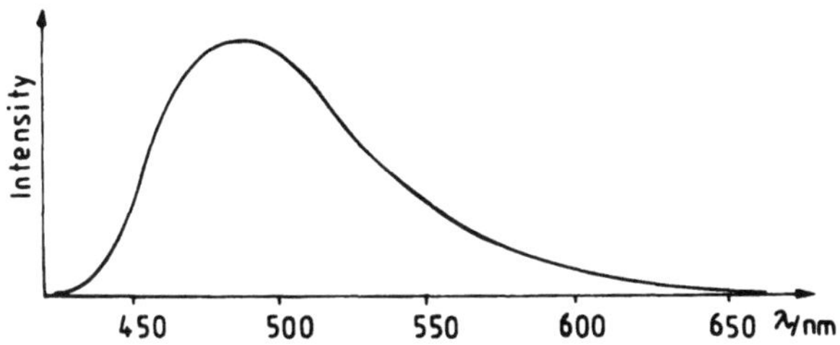

Fig.5
Fluorescence spectra of MP-PPV in toluene (0.032 mg/ml) after photoexcitation with 5 ps pulses of o.3 nJ at room temperature

clude that the fluorescence at the spectral maximum comes from the recombination of the electron - hole pairs. Practically the data in table 1 reflect that the fast recombination of electron hole pairs influences the time course of the whole fluorescence band. The decay times which we assigned to polaron and bipolaron recombination processes decrease with decreasing wavelength (600 nm - 500 nm). We assume this fact is caused by increasing influence of the very fast electron - hole pair recombination at shorter wavelengths.

4. TIME RESOLVED PHOTOCONDUCTION MEASUREMENTS

The transient photoconduction of a disordered MP-PPV layer was studied using the Auston switching technique [13] and the boxcar integrator with 220 ps time resolution. The sample was excited by a single pulse generated by the third harmonic of a passively mode - locked Nd:YAG laser (λ = 355 nm, T_i = 25 ps, about Nph = $2.7*10^{12}$ photons are absorbed per pulse).
The DC bias field applied on the gap was 10^4 V/cm. After the maximum of 63 µA the current decays with a non-resolvable fast (ps) component and a slower (ns) component. The decay time for the slower component can be directly obtained from the transient photoconductivity measurements and characterized by an exponential function with 3.6 ns. After deconvolution we obtained for the fast component 100 ps, which roughly corresponds to the fast decay of the polaron measured by picosecond-absorption experiments, whereas the slow one we assigned to the bipolaron. The integral of the current collected from the begin to 8 ns amounts to Ne = $1.2*10^8$ elementary charges. Corresponding to [14] we used Ne/Nph = $\eta \xi L/g$ with the assumption $\eta \xi$ = 1 in the ps experiment. For the gap distance g = 0.2 mm we obtained the mobility of the polarons as $4.4*10^{-3}$ cm^2/Vs and the bipolarons, $1.2*10^{-4}$ cm^2/Vs.

5. COMPUTER SIMULATION

According to our model the neutral bipolarons are formed at about 3 ps after excitation. The polarons are formed by interchain excitation and the bipolarons (BP^{2+} and BP^{2-}) will be formed through the combination of (P^+ + P^+) and (P^- + P^-).
The species vanish by different recombination and relaxation processes. We have described the kinetics of polarons by rate equations [15]. The results of computer simulation showed that the kinetics of bipolarons can only be understand as a superposition of both polaron and bipolaron absorption. The fast component was caused by polaron absorption. The formation time difference between polarons and bipolaron absorptions is also evident from the superposition of bands.

REFERENCES:

1. H.H. Hörhold, et al. Z. Chem. 27.Jg.(1087) Heft 4, 126
2. C. Bubeck, et al. (eds.) Nonlinear Optical Effects in Organic Polymers, 143-147, 1989 by Kluwer Academic Publishers
3. T.C. Chung, et al. Phys. Rev. B30 (1084) 702
4. G. Harbeke, et al. Solid State Commun. 55 (1985) 419
5. D. Bertho and c. Jouanin, Phys. Rev. B35 (1987) 626
6. T. Damm, et al. Exp. Technik der Physik 33 (1985) 409
7. A.J. Heeger, Rev. of Modern Phys. 60 (July 1988) 781
8. H.H. Hörhold et al. 3. Tagung Festkörperphysik, Dresden 1984, 87-111
9. W. Becker, et al. Proc. UPS-85, Teubner Leipzig 1985 Teubner Texte zur Physik Bd.10, p.77 -80
10. J. Opfermann, H. Reichel, Software Program, Jena 1988
11. M. Yoshizawa et al. Phys. Rev. B37 (1988) 8988
12. R.K. Yuan et al. Acta Physica Sinica (China) 36 (1987) 1161
13. D.R. Auston, Appl. Phys. Lett. 26 (1975) 101
14. Y. Yacoby et al. Solid State Commun. 47 (1983) 869
15. S. Rentsch, et al. to be published

Non-Markovian Effects of Vibronic Relaxation in Spectroscopy with Intense USP and In-Phase Modulation of USP in a Saturable Absorber

B.D. Fainberg and V.N. Nazarov

State Optical Institute, 199034 Leningrad, USSR

Abstract. The problem of calculating electronic transition non-linear polarization in a strongly broadened vibronic system with non-Markovian relaxation (a complex molecule, an impurity centre in a crystal) in the field of intensive ultrashort pulses (USP) has been solved. Generalization of theories of pump-probe femtosecond experiments for the case of sufficiently intense pump pulses is made using the obtained solution. The influence of non-Markovian effects of vibronic relaxation upon phase modulation of USP during transmission through a saturable absorber has been studied.

1. Introduction.

In this work the problem of calculating electronic transition non-linear polarization in a strongly broadened vibronic system with non-Markovian relaxation (a complex molecule, an impurity centre in a crystal) in the field of intense USP has been solved. The influence of non-Markovian effects of vibronic relaxation upon phase modulation and duration of USP during transmission through a saturable absorber is studied. The significance of such an investigation is due first to the presence of strong fields in USP dye lasers [1] and the use of intense USP in femtosecond spectroscopy [2], and second, to the importance of chirp in USP generation [3], and the commensurability of the duration $t_p \approx 50$-100 fs of USP generated in lasers with the vibrational relaxation time t_s of dyes [4].

2. Non-linear Polarization of Vibronic Transitions in the Field of Intense USP.

Consider a molecule with two electronic states $p=1,2$ ($E_2 > E_1$). It is possible to describe the influence of the vibrational subsystems of a molecule and a solvent on electronic transitions within the range 0 --> 1 related by high frequency optically active (OA) vibrations ($\approx$ 1000-1500 cm^{-1}) as modulation of this transition frequency by low frequency (LF) OA vibrations $\{\Omega_s\}$ with the correlation function $k_s(t)$. Within the "slow modulation" limit ($\sigma_{2s} t^2_s >> 1$) the following times are characteristic for the time evolution of the system under consideration:
$\sigma_{2s}^{-1/2} < T' << t_s$ [5]. Here σ_{2s} is the LF vibration contribution to the second central moment of an absorption spectrum, t_s is the characteristic decay time of $k_s(t)$, and $\sigma_{2s}^{-1/2}$ and $T' = (t_s / \sigma_{2s})^{1/3}$ mean the times of reversible and irreversible dephasing processes of the electronic transition, respectively.
The inequality $t_s >> T'$ suggests the possibility of describing vibrationally non-equilibrium populations in states 1 and 2 ($n_1(E/\hbar,t)$ and $n_2(E/\hbar,t)$) by balance equations for the intense pulse ($t_p > T'$)

Springer Proceedings in Physics, Vol. 49 Ultrafast Phenomena in Spectroscopy
Editors: E. Klose · B. Wilhelmi

excitation. This procedure allows one to solve the problem for strong fields.

In the case of $h\Omega_s << kT$ and $k_s(t) = \exp(-|t|/t_s)$ the problem reduces to the integral equation for the population difference $n(\Omega_{21}-\Omega,t) = n_1(\Omega_{21}-\Omega,t)-n_2(\Omega_{21}-\Omega,t)$ [6]:

$$n(t) = 1-\sigma'(\Omega_{21}) \int_0^t dt' J(t') n(t') R(t-t'), \qquad (1)$$

$$n(t) = n(\Omega_{21}-\Omega,t)/n(\Omega_{21}-\Omega,0).$$

Here Ω_{21} is the frequency of the Franck-Condon transition 1->2, $J(t)$ is the power density of the exciting radiation, Ω is its frequency, and $\sigma'(\Omega_{21})$ is the cross section at the absorption band maximum.

$$R(t) = (\sigma(t)/\sigma_{2s})^{-1/2} \sum_{j=a,e} \exp[-(\Omega-\Omega_j(t))^2/(2\Omega(t))]$$

describes the contributions of the induced absorption (j=a) and the induced emission (j=e) to n(t).
$\sigma(t) = \sigma_{2s}(1-k_s^2(t))$ is the "instantaneous" second moment, $\Omega_j(t) = \Omega_{21} - \delta_{je}\Omega_{st} + (\Omega-\Omega_{21} + \delta_{je}\Omega_{st})k_s(t)$ are the "instantaneous" central frequencies of the induced changes of absorption ($\Omega_a(t)$) or of emission ($\Omega_e(t)$), Ω_{st} is the Stokes shift of the equilibrium absorption and the luminescence spectra, δ_{je} is the Kronecker symbol. The solution to eq. (2) by Pade approximant [0/1] is

$$n(t)=[1 + \sigma'(\Omega_{21}) \int_0^t dt' J(t') R(t-t')]^{-1} . \qquad (2)$$

This solution (curve 1, fig.1) does not practically differ from the exact one (3, fig.1) even at a comparatively large saturation parameter $\sigma'(\Omega_{21})J_{max}t_p=0.88$, when the perturbation theory (curve 2) does not hold. As a result, for non-linear polarization we obtain

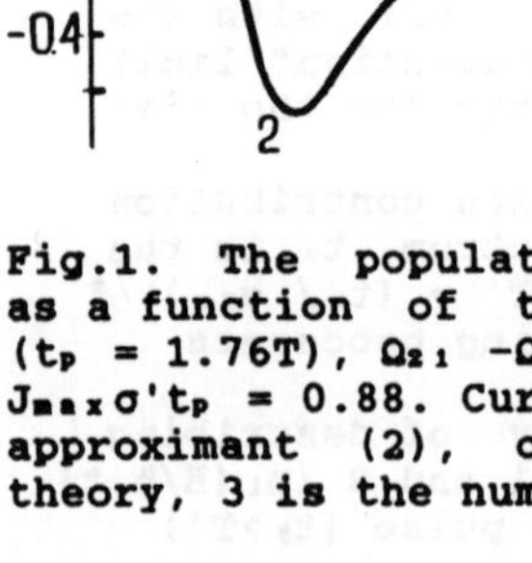

Fig.1. The population difference normalized to $n(\Omega_{21}-\Omega,0)$ as a function of t for the exciting pulse $J(t) = J_{max} / ch^2(t/T-3)$ ($t_p = 1.76T$), $\Omega_{21}-\Omega = 0$, $\Omega_{st}(2\Omega_{2s})^{-1/2} = 1.2$, $t_s/T = 1$, $J_{max}\sigma' t_p = 0.88$. Curve 1 is the solution to eq.(1) by Pade approximant (2), curve 2 is the calculation by the perturbation theory, 3 is the numerical solution to eq.(1).

$$P^{+nl}(t) = \int_0^t dt' J(t-t')n(t-t') \int_0^t dt'' \, E'(t-t'') \sum_{j=a,e} \exp[-\sigma(t')t''^2/2-$$

$$-it''(\Omega_j \, (t')-\Omega)]. \qquad (3)$$

where $E'(t)$ is the slowly varying envelope of the electric field. Formulae (1)-(3) solve the problem of calculating electronic transition non-linear polarization in a strongly broadened vibronic system for intense USP whose intensity is confined by the condition : $\sigma(\Omega_{21})J_{max} \ll (T')^{-1}$. Generalization of the theories [3,7,8] of pump-probe femtosecond experiments for the case of sufficiently intense pump pulses is made using these formulae [6].

3. A study of USP Phase Modulation in a Saturable Absorber with Vibronic Relaxation Processes Being Considered.

One should take into account the processes of vibronic relaxation for t_p <100 fs. A change of slowly varying USP envelope during transmission through a thin sample is given by

$$E'(l,t_1)-E'(0,t_1) \approx P^+(0,t_1) \qquad (4)$$

where l is the sample thickness. Using eq.(4) the complex amplitude of the field may be calculated

$$E'(l,t_1) = |E'(l,t_1)|\exp[if(l,t_1)] \; .$$

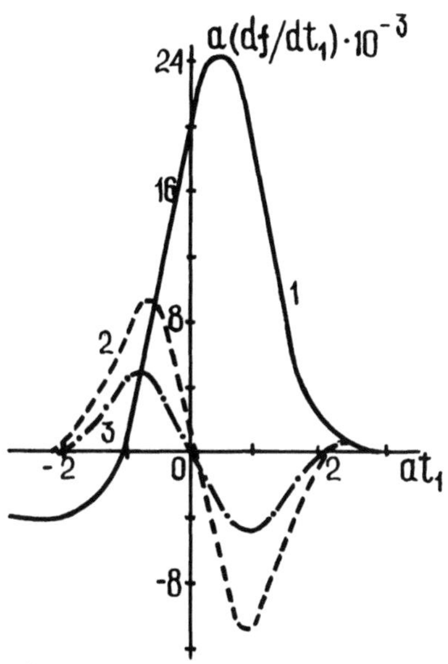

Fig.2. The dependence of df/dt_1 on time t_1 for $\sigma'Nl = 0.2$ in the cases of classical (1) and quantum (2,3) characters of OALF vibration system: $\sigma_{2s}^{1/2} / a = 12(1), \; 6(2)$ and $12.48(3)$; $(\Omega - \Omega_{21})(2\sigma_{2s})^{-1/2} = 0.6(1)$ and $0(2,3)$; $at_s = 0.8325(1)$; $t_p = 1.665/a$, $\sigma' J_{max} t_p = 0.4(1)$ and $0.33(2,3)$.

The chirp is given by df/dt_1. The effects of vibronic relaxation processes on the chirp are given in fig.2 for the Gaussian input pulse $E'(0,t_1) = E'\exp(-a^2 t_1^2/2)$. For the classical OALF vibration system ($h\Omega_s \ll kT$) df/dt_1 is an odd function of detuning $\Omega-\Omega_{21}$ for $|\Omega-\Omega_{21}| \ll \Omega_{st}$ and large Ω_{st}.

Now we consider the case when the OALF vibration system is a quantum one ($h\Omega_s \geq kT$). From fig.2 (curves 2 and 3) it will be obvious that the chirp arises even without detuning of the pulse carrier frequency with respect to the maximum of the steady-state absorption spectrum. The chirp originates from incompletion of the vibration relaxation processes during the pulse. Therefore, pulse absorption and pulse refraction occur in a vibrationally non-equilibrium system.

We have also studied the dependence of duration t_p' of the transmitted pulse $E'(l,t_1)$ on vibrational relaxation time t_s and detuning [9]. It has been shown that the dependence of t_p' on t_s has a mimimum for $at_s \approx 1$. Such a behaviour can be explained as follows. USP burns a hole in the absorption band for $at_s \leq 1$. A saturable absorber acts typically for $at_s < 1$ and for $at_s \gg 1$, sharpening the leading edge of the pulse. However, both the leading and the trailing edges of the pulse sharpen for $at_s \approx 1$ because the hole broadens and decreases during the pulse. Thus, it seems that the relationship $at_s \approx 1$ is the best for pulse shortening in a saturable absorber.

References

1. J.A.Valdmanis and Fork: IEEE, J.Quant. Electr. QE-22, 112 (1986).
2. P.C.Becker, R.L.Fork, C.H.Brito Cruz, J.P.Gordon, C.V.Shank: Phys. Rev. Letters 60, 2462 (1988).
3. J.Herrmann, B.Wilhelmi: Lasers for Ultrashort Light Pulses (Akad.-Verlag, Berlin, 1987).
4. B.D.Fainberg, I.N.Myakisheva, A.G.Spiro, S.V.Kulya: In Proceedings of the V International Symposium "Ultrafast Phenomena in Spectroscopy", ed.by Z.Rudzikas, A.Piskarkas, R.Baltramiejunas (World Scientific, Singapore, New Jersey, London, Hong Kong,1988) pp. 400-405.
5. B.D.Fainberg: In Abstracts of the XX All-Union Congress on Spectroscopy, pt.I (Kiev, 1988) p. 288(in Russian).
6. B.D.Fainberg: Opt.i spektr. 67, 241 (1989).
7. R.F.Loring, Y.J.Yan, S.Mukamel: J.Chem.Phys. 57, 5840 (1987).
8. W.Vogel, D-G.Welsh, B.Wilhelmi: Phys. Rev. A37, 3825 (1988).
9. B.D.Fainberg, V.N.Nazarov: Kvant. elektron. 16, 1715 (1989).

The Primary Steps of the Photosynthetic Process in Bacteria at Low Temperatures

G. Deinum[1], *T.J. Aartsma*[1], *R. van Grondelle*[2], *and J. Amesz*[1]

[1]Department of Biophysics, Huygens Laboratory, University of Leiden, P.O. Box 9504, 2300 RA Leiden, The Netherlands
[2]Department of Biophysics, Physics Laboratory, Free University, De Boelelaan 1081, 1081 HV Amsterdam, The Netherlands

Abstract. It is shown, by means of singlet-singlet annihilation measurements, that spectral inhomogeneities exist in the antenna of *Rhodospirillum rubrum* at low temperatures. These inhomogeneities limit the extent of energy transfer at 4 K.

1. Introduction

In photosynthetic systems one usually distinguishes between antenna pigments and reaction center pigments. This distinction refers to their different functions. The antenna pigments constitute the light harvesting component of the photosynthetic apparatus, while the reaction center pigments are part of the system where the chain of reactions of photosynthesis is initiated. The assembly of a reaction center with surrounding antenna pigments is called a photosynthetic unit which, for example, in the purple bacterium *Rhodospirillum rubrum* contains about 35 antenna bacteriochlorophylls (BChls). Photosynthetic units can be clustered together, forming aggregates. Such aggregates, which may contain as many as 10,000 pigment molecules, are referred to as domains. Optical excitations can move freely within a domain from one molecule in the aggregate to another by means of the Förster mechanism for energy transfer.

The antennae of many photosynthetic systems are heterogeneous, i.e. they contain different pigment pools, as identified, for example, by their absorption spectra. Optical excitations are rapidly transferred to the pigment pool with the lowest energy of the excited singlet state. The migration of the excitation energy towards the reaction center is therefore determined by the antenna component absorbing at long wavelength. Insight into this process, as well as into the size of the corresponding domains, can be obtained by singlet-singlet annihilation measurements. In such experiments the fluorescence yield of the antenna (bacterio)chlorophylls is measured as a function of the intensity of a laser excitation pulse. Singlet-singlet annihilation occurs when two optical excitations accumulate by energy transfer at one molecule, leading to higher excited states. Rapid decay by radiationless processes to the lowest excited state leads to dissipation of one of these excitations and a concomitant decrease of the fluorescence yield. The rate of the fluorescence quenching, and therefore the fluorescence yield, depends on the number of excitations per domain. An equation relating the fluorescence yield to the number of excitations per domain was given by Paillotin et al. (Eqn. 20 in [1]). This equation contains a parameter r which is the ratio of the rate of mono-excitation decay to the rate of annihilation per pair of excitations. Fitting experimental data to the theory of Paillotin

Springer Proceedings in Physics, Vol. 49 **Ultrafast Phenomena in Spectroscopy**
Editors: E. Klose · B. Wilhelmi

et al. [1] yields an explicit value for this parameter, and from that one may obtain the domain size.

An intriguing observation has been that the domain sizes calculated in this way show a strong temperature dependence. For example, in *R. rubrum* the calculated domain sizes varied from 150 BChl molecules at 4 K to larger than or equal to 1000 at room temperature [2]. Similar results were obtained for *Rhodobacter sphaeroides*. This temperature dependence was interpreted as arising from the existence of different pools of BChl 880 (the main BChl species in *R. rubrum*) which become disconnected (in terms of energy transfer) as the temperature decreases. It is difficult, however, to envisage a mechanism which differentiates between the temperature dependence of energy transfer within a BChl 880 pool and that among different BChl 880 pools. A possibility could be that the Förster-overlap integral is reduced at low temperature, inhibiting energy transfer between different BChl 880 pools, while still large enough to allow energy transfer within BChl 880 pools. It is more likely, however, that a spectral inhomogeneity may be responsible for the observed temperature dependence of annihilation. In particular, the existence of a long-wavelength antenna system, BChl 896, may have a profound effect on the annihilation behavior as a function of temperature.

2. Results and Discussion

To obtain information about the energy transfer within the antenna of *R. rubrum*, we examined the wavelength dependence of annihilation at low temperature. Typical results are shown in fig. 1. The measurements were performed at two wavelengths of detection (905 nm and 918 nm). The relative fluorescence yield as a function of the energy density differs for these wavelengths, leading to different domain sizes and r-values. This can only be explained if it is assumed that the antenna of *R. rubrum* is spectrally inhomogeneous. This spectral inhomogeneity in the annihilation curves was only noticeable below approximately 100 K.

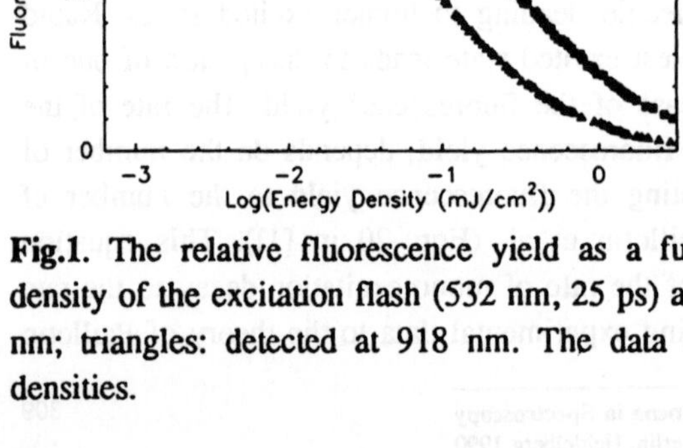

Fig.1. The relative fluorescence yield as a function of the incident energy density of the excitation flash (532 nm, 25 ps) at 4 K. Squares: detected at 905 nm; triangles: detected at 918 nm. The data are normalized at low energy densities.

It is useful to define the extent of annihilation as the ratio of the fluorescence yield at low excitation energy density (where no annihilation occurs) to that at high excitation energy density (where singlet-singlet annihilation is the dominant decay process). Experimentally, the extent of annihilation as a function of wavelength is obtained by dividing the fluorescence spectrum at low excitation energy density by the spectrum obtained at high excitation energy density. In the latter case we used 532 nm laser pulses with a duration of 25 ps and a typical energy density of 1 mJ/cm^2. The extent of annihilation for *R. rubrum* is shown in fig. 2. It can be observed that the extent of annihilation increases with wavelength until it reaches a constant level at 920 nm, indicating that the fluorescence above 920 nm originates from one type of pigment. If we assume that the fluorescence spectrum consists of two bands, the data in fig. 2 can be used to calculate the size and shape of these bands. The bands have a maximum at 918 and 911 nm, respectively. The ratio of the fluorescence yields of the two bands at low excitation density is 11.7 ± 0.8. These bands, which we call F918 and F911, presumably arise from BChl 896 and BChl 880, respectively.

Since BChl 880 is present in greater abundance, the excitations will reside predominantly in the BChl 880 pool after a short laser flash. A Boltzman equilibrium will be rapidly established between BChl 880 and BChl 896 by energy transfer. As noted already by Kramer et al. [3], this means that at 4K excitations will be transferred essentially irreversibly from BChl 880 to BChl 896. Annihilation will occur in both antenna components, but to a different degree, leading to a wavelength dependence as shown in fig. 2. As the temperature was increased from 4 K the wavelength dependence of the extent of annihilation diminished, until at 80 K it virtually had disappeared. Above this temperature the extent of annihilation is constant over the whole wavelength range of fluorescence, which indicates that excitations are completely delocalized over both antenna components, BChl 880 and BChl 896.

The theory by Paillotin et al.[1] is not simply applicable to this system, because it does not take a spectral inhomogeneity into account. When analyzing the fluorescence of F911 ar 4 K, the pigments emitting at 918 nm can be considered to act as traps and their presence is therefore reflected as a monoexcitation decay. Since energy transfer from BChl 896 to BChl 880 does not take place at 4 K [3], eqn. 20 in [1] remains applicable

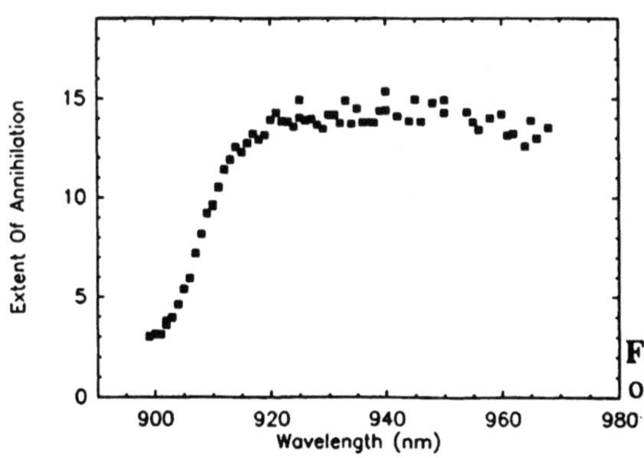

Fig.2. Wavelength dependence of the extent of annihilation (see text) at 4 K.

to F911. A correction, however, must be made for the fact that the annihilation curve (cf. fig. 1) recorded at 905 is not entirely due to F911: at low excitation intensity about 25 % of the fluorescence yield originates from F918, as can be calculated from the result in fig. 2. Applying the correction for the contribution of F918 at 905 nm a domain size of 75 ± 25 BChls was obtained. However, for F911 a correction must also be made for the pulse duration of 25 ps which is of the same order of magnitude as the lifetime of F911. At 77 K the lifetimes of F911 and F918 with closed reaction centers are approximately 20 and 160 ps, respectively [4]. If we take into account the increase of the total fluorescence yield by a factor of 2.5 to 3 and a ratio of the integrated fluorescence intensities of the two bands of 10 at 4 K, lifetimes of 50 ± 10 ps for F911 and of 500 ± 100 ps for F918 are calculated. Random walk simulations were used to study the influence of the length of the laser pulse. They showed that, when applying a 25-ps pulse the domain size is overestimated by a factor of about 1.3 in this particular case. Using this correction factor we determined the domain size for BChl 880 to be 55 ± 20 BChls.

Concerning the F918 fluorescence component we note that most of the excitations in the corresponding BChl 896 will result from energy transfer from BChl 880. Therefore the number of excitations present in BChl 896 has to be corrected for annihilation in BChl 880. In fact the annihilation curve of F911 determines the relative amount of energy transferred to BChl 896 and therefore the excitation density in BChl 896. On the assumption that this pigment constitutes about 15 % of the total antenna as in *Rhodobacter sphaeroides* [4] and receives 85 % of the excitations via BChl 880, an r-value of 0.5 ± 0.4 and a domain size of 21 ± 9 is found. Since probably only 5-7 BChls 896 are present per reaction center [4], this indicates that each reaction center is closely interconnected with one to four others.

Our results thus indicate that upon cooling to 4 K the antenna of *R. rubrum* is split up into separate BChl 880 and BChl 896 domains, due to inhomogeneities in energy transfer in the antenna.

3. References

1 Paillotin, G., Swenberg, C.E.,Breton, J. and Geacintov, N.E. (1979) Biophys. J. 25, 513-533.

2 Vos, M., Van Grondelle, R., Van der Kooij, F.W., Van de Poll, D., Amesz, J. and Duysens, L.N.M. (1986) Biochim. Biophys. Acta 850, 501-512.

3 Kramer, H.J.M., Pennoyer, J.D., Van Grondelle, R., Westerhuis, W.H.J., Niederman, R.A. and Amesz, J. (1984) Biochim. Biophys. Acta 767, 335-344.

4 Van Grondelle, R., Bergström, H., Sundström, V and Gillbro, T. (1987) Biochim. Biophys. Acta 894, 313-326.

Femtosecond Absorption Spectroscopy of Primary Processes in Bacterial Photosynthesis Reaction Centers

S.V. Chekalin, Yu.A. Matveyets, and A.P. Yartsev

Institute of Spectroscopy, USSR Academy of Sciences,
142092 Troitzk, Moscow Region, USSR

The reaction centers (RC) of purple bacteria include 6 molecules of a porphyrine nature: 4 bacteriochlorophylls (two of them being the monomers B_L and B_M and two forming the dimer P) and 2 pheophytins H_L and H_M [1]. To reveal the action mechanism of the RCs in the natural photosynthesis process, it is necessary to establish the role each RC component plays in energy and charge transfer and the sequence and rates of these processes. For this purpose, we have carried out femtosecond absorption spectrometry studies of intact and $NaBH_4$-modified RCs of Rhodobacter Sphaeroides [2-5] and demonstrated the following.

1. When exciting the RCs with a 300-fs pulse at a wavelength of 620 nm corresponding to absorption of the pigments P and B (around 10% of all the RCs getting excited), there occurs within 150±100 fs the fast energy transfer process $B^*_{600}P \rightarrow BP^*_{600}$ in those RCs where only the B pigment has undergone excitation. Thus, 0.5 ps after excitation, in all the RCs that have absorbed a quantum, there remains excited only the P pigment.

2. The relaxation of P^*_{870} (ΔA_{930} kinetics) and the bleaching of the H_L (755 nm) and B_L (805 nm) bands take one and the same characteristic time of around 3 ps.

3. The bleaching of the B_L band observed to take place during the first 3 ps after excitation is associated with the formation of the $P^+B_L^-$ radical and not with the electrochromatic shift of the B_L band.

4. The simultaneous bleaching of the B_L and H_L bands during the first 3 ps after excitation is due to the transfer of an electron from P^*_{870} to B_L and H_L.

5. The existing charge-transfer models (two-step transfer from P^* to B_L and thence to H_L [6], exciton mechanism involving the $B_L^+H_L^-$ state [7], or superexchange mechanism [8]) cannot explain the changes observed in the differential spectra. For explanation, we have suggested three schemes for the transfer of an electron from P^*_{870} [4] whereby the electron is distributed between B_L and H_L (Fig. 2).

Raising the excitation intensity to around 10^{16} ph/cm^2 causes all the kinetics to change substantially [2-3,5] compared to the case of linear excitation ($\sim 10^{15}$ ph/cm^2). Such a high sensitivity to excitation intensity is apparently specific precisely to the case of excitation at 620 nm, this being the wavelength at which P, B_L, and B_M absorb with comparable cross sections [2]. This circumstance complicates investigations into the details of natural photosynthesis, for even a slight increase in excitation intensity may cause simultaneous excitation of two or even

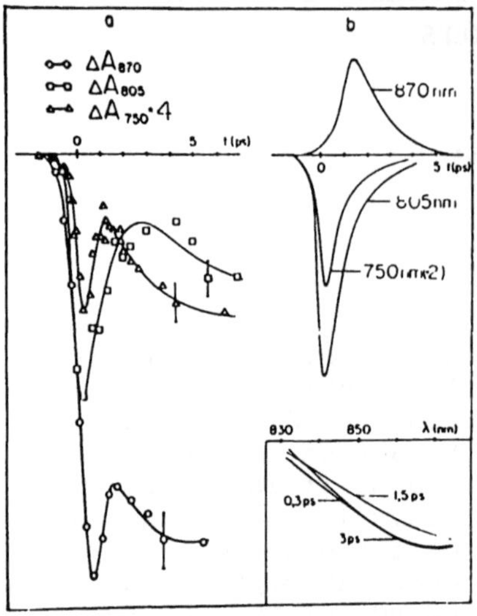

Fig. 1 RC absorption kinetics obtained at an excitation intensity of 10^{16} ph/cm^2: (a) real changes; (b) "additional" changes. Inset: P band shift at various delay times

three pigments in many RCs. On the other hand, however, simultaneous excitation of several pigments in RCs is of interest from the standpoint of studying superfast energy transfer and exciton annihilation processes in these unique objects where the distances between the excited pigments are no greater than the size of molecules themselves. What is more, these distances and the mutual arrangement of the pigments are known accurately enough [1], which allows for direct comparison between experiment and theory.

Figure 1 presents changes in the absorption kinetics in the P (870 nm), B(805 nm), and H(750 nm) bands obtained at an excitation intensity of 10^{16} ph/cm^2, giving rise to a large number of RCs containing simultaneously excited P and B pigments. As distinct from the case of linear excitation, the H band kinetics (ΔA_{750}) features a fast bleaching (on a time scale shorter than the exciting pulse duration), this bleaching then relaxing to give rise to a kinetics typical of the transfer of an electron to H_L under linear excitation [2,3]. The fast bleaching in the P band (ΔA_{870} kinetics) is observed to diminish slightly even during the exciting pulse. At the same time, the bleaching maximum in the 870-nm band is observed to shift toward the long-wavelength side. These changes relax on a time scale of 1-2 ps. The B band (805 nm) bleaches as instantaneously as in the case of linear excitation, but the amount of bleaching and its relaxation time are much greater. For clarity, Fig. 1b shows these "additional" changes separately. The fast bleaching of the H band has no relation to the direct excitation of the bacteriopheophytin, its absorption bands lying far enough from the exciting pulse wavelength. On the other hand, this phenomenon is characteristic exactly of the high-intensity excitation of RCs in the 600-nm region. For example, in[9], an instantaneous dip was observed to occur at the initial moment in the absorption curve near 545 nm, though no detailed absorption kinetics was obtained.

We believe that the fast bleaching of the H band is due to the transfer of excitation energy to the bacteriopheophytin as a result of the singlet-singlet annihilation of P* and B* by one of

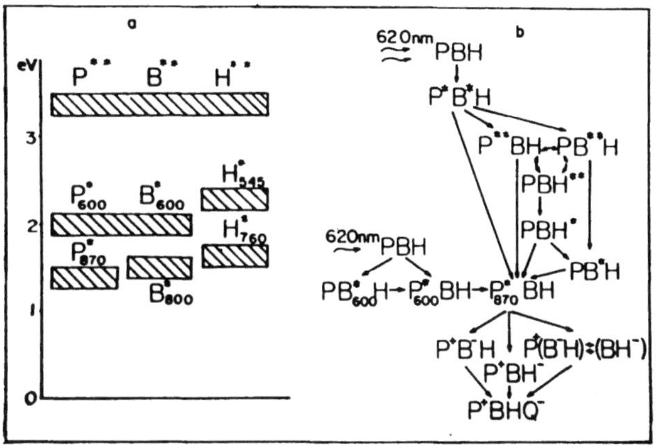

Fig. 2. (a) Simplified energy band diagram of RC pigments; (b) schemes of processes occurring in RCs at various excitation intensities

the schemes shown in Fig. 2. When only the pigment P is excited at 620 nm in an RC, there occurs the $PB^*_{600} \rightarrow P^*_{600}B$ energy transfer process which proceeds very effectively because the energy deficit between the states is small. The time of this process was measured as 100 fs [2]. The transfer of energy from the partially relaxed state B^*_{800} apparently proceeds not so effectively (according to [10], the time of this process amounts to 400 fs). When the P and B pigments in an RC are excited simultaneously, the B* decay channel $B^*P \rightarrow BP^*$ is closed. Obviously two versions of singlet-singlet annihilation of the excited states P^*_{600} and B^*_{600} are possible involving partial relaxation of one of the partners, namely, $P^*_{870}B^*_{600}$ and $P^*_{600}B^*_{800}$, because the sum of the excitation energies of the partners falls within the RC absorption band with a maximum at 365 nm. The 365-nm band is a combination of the absorption bands of the bacteriochlorophyll and bacteriopheophytin. For this reason, when one of the pigments is excited to a state corresponding to this band, there can take place a fast transfer of the excitation energy to other pigments, e.g., from B** to H**. This explains the rapid bleaching of the H band in our experiments. The band shift and diminishing of bleaching in the ΔA_{870} may be due to the absorption of B^*_{800} in this region. The lengthening (up to 1-2 ps) of the bleaching relaxation time of the B band compared to that in the case of linear excitation may be caused by the contribution to the relaxation kinetics from non-annihilating states (e.g., $P^*_{870}B^*_{800}$) or by the absorption of a third quantum by the RC. Obviously all these processes reduce the quantum yield of charge separation.

Further increase in the excitation intensity enhances the number of RCs in which the excitation of P causes no charge separation.

References

1. J.Deisenhofer, O.Epp, K.Miki, R.Huber, H.Michel: J.Mol.Biol., 180, 385 (1984)

2. Yu.A.Matveyets, S.V.Chekalin, A.P.Yartsev: Pis'ma Zh.Eksp.Teor. Fiz, 43, 545 (1986) (in Russian)
3. S.V.Chekalin, Yu.A.Matveyets, A.P.Yartsev, in: Ultrafast Phenomena V, ed. G.R.Fleming and A.E.Siegman (Springer, Berlin,1986), p.402
4. S.V.Chekalin, Yu.A.Matveyets, A.Ya.Shkuropatov, V.A.Shuvalov, A.P.Yartsev: FEBS Lett. 216, 245 (1987)
5. S.V.Chekalin, Yu.A.Matveyets, A.P.Yartsev: Rev.Phys.Appl. 22, 1761 (1987)
6. V.A.Shuvalov, A.V.Klevanik, A.V.Sharkov, Yu.A.Matveyets, P.G. Kryukov: FEBS Lett. 91, 135 (1978).
7. S.F.Fischer, P.O.J.Scherer:Chem.Phys. 115, 252 (1987)
8. W.W.Parson, A.Scherz, A.Warchel, in: Antennas and Reaction Centers of Photosynthetic Bacteria, ed.by M.E.Michel-Beyerle (Springer, Berlin, 1985).
9. D.Holten, C.Hoganson, M.W.Windsor, C.C.Schenck, W.W.Parson, A.Migus, R.L.Fork, C.V.Shank: Biochem.Biophys.Acta 592, 461 (1980)
10. J.Breton, J.-L.Martin, J.Petrich, A.Migus, A.Antonetti: FEBS 209, 37 (1986)

Evidence of Excited State Absorption of Strongly Interacting Chlorophylls Within Light-Harvesting Pigment Protein Complexes in Photosynthetic Systems

G. Kehrberg, J. Voigt, and T. Bittner

Humbold-Universität, Department of Physics, Division of Experimental Biophysics and Spectroscopy, Invalidenstr. 110, DDR-1040 Berlin, German Dem. Rep.

Abstract. The pumped absorption in the frequency domain and the nonlinear fluorescence in intact leaves, in chloroplasts, in light-harvesting complexes (LHC) and in photosystem I complexes (PS I) of higher plants in the region of the S_0-S_1 transition of chlorophyll have been studied. In LHC, chloroplasts and leaves a sharp dispersion of the test beam transmission was found in the nearest vicinity of the wavelength of the pump beam, which is absent in PS I. An excited state absorption process of the pigment protein subunit containing 6 or more strong interacting chlorophyll molecules have been suggested to be responsible for the pump beam induced absorption. A transverse relaxation time of about 1 ps for chlorophyll in LHC could be estimated by using this frequency domain method.

1. Introduction

In recent years the nonlinear optical properties of chlorophyll in variously isolated complexes of photosynthetic apparatus and in chloroplasts have been studied extensively to receive information about both the substructure of the photosynthetic apparatus and static and dynamic parameters of energy transfer. For investigating the energy pathway of the excitons from the absorbing antenna to the reaction center molecules and the structure of the pigment-protein complexes, time-resolved pumped absorption measurements and time-resolved fluorescence measurements were made /1-6/. The investigations of the nonlinear fluorescence and nonlinear transmission were performed to determine the number of interacting chlorophyll molecules in the various pigment-protein complexes /7-8/. Insights in the mechanisms of homogeneous and inhomogeneous broadening and in the electron phonon coupling were received by holeburning and site selection experiments /9-13/.

Still a great deal of uncertainty exists on the detailed organization of the pigment apparatus of higher plants and its function in the processes of the first steps of energy conversion following the absorption of photons. An important question not yet answered is whether the exciton states are predominantly localized as in disordered systems, or delocalized as in ordered systems, or whether there is a mixture of these in photosynthetically active systems. In systems with localized exciton states incoherent phonon-assisted hopping transfer (Förster mechanism) takes place whereas coherent transfer presumes delocalized exciton states /14/. The best way to answer this question is the knowledge of the transverse relaxation time of chlorophyll in vivo.

In the present work, we investigate the exciton-exciton interaction of chlorophyll molecules in pigment-protein complexes of LHC and PS I, respectively, and in chloroplasts and whole leaves with a two-beam "pump and probe" technique in the frequency domain

Editors: E. Klose · B. Wilhelmi

and with nonlinear fluorescence measurements. We find in LHC, chloroplasts and leaves a sharp dispersion of the test beam transmission in the nearest vicinity of the wavelength of the pump beam, which is absent in PS I. We discuss this result as an induced absorption, which is due to optical transitions from exciton to biexciton-like states of chlorophyll in pigment-protein subunits of LHC, and estimate for this chlorophyll a transverse relaxation time of about 1 ps.

2. Experimental set-up

The experimental arrangement for the pump-and-probe beam measurements is similar to the one described in ref. /15/ extended by a set-up for measuring the nonlinear fluorescence. The ns pulses of pump-and-probe beams, being variable in wavelength, are produced by two dye laser grating resonators (FWHM 4 ns pumped by N_2-laser). Both dye lasers have spectral halfwidths smaller than 0.05 nm. Pump-and-probe beams are incident perpendicularly on the sample cuvette. The diameter of the excited spot is about 0.15 mm, leading to a maximum excitation density of about $2 \cdot 10^{17}$ photons per cm^2 per pulse. The test beam is focused into a spot of o.1 mm in diameter inside the spot of the pump beam. The excitation density of the test beam is about $2 \cdot 10^{15}$ photons per cm^2 per pulse. The test beam, transmitted through the sample, is separated from the pump beam by a grating spectrometer with a linear dispersion of 0.36 nm/mm. The signal is detected by a photomultiplier and amplified by a boxcar integrator at fixed gate position.

We study with the test beam the change of transmission of the sample in the presence of an intensive pump beam. The wavelength of the pump beam is 620 nm in all measurements and is located on the high energy side of the absorption band of chlorophyll in vivo. At this wavelength the fluorescence of chlorophyll is very low and the error caused by this fluorescence can be neglected. The change of transmission is calculated by equation (1),

$$T^{rel} = \frac{I_{t+p} - I_p}{I_t}, \qquad (1)$$

where I_t and I_{t+p} are the transmitted intensities of the test beam through the unexcited and the excited sample, respectively. I_p is the intensity of the laser foot of the pump beam at the wavelength of the test beam, i.e. when the test beam is switched off. In the nonlinear emission experiments the integral chlorophyll fluorescence of the sample is detected by an avalanche silicon photodiode and amplified by a boxcar integrator. The exciting light is blocked by edge filters.

Intact primary leaves of wheat seedlings, chloroplasts of maize, and subchloroplast fragments of various structural organization are used in our experiments. Maize chloroplasts were isolated from leaves of 10-12-day seedlings on 0.35 M NaCl and 0.02 M Tricine-HCl buffer by a method of Hieke /16/. LHC fragments of monomeric form of the protein were prepared by ion-exchange chromatography from chloroplast membranes isolated by treatment with the detergent Triton X-100. The preparation of the LHC and its spectroscopic characterization is performed as described by Gulyaev et al. /17/. The isolation of the PS I complexes was performed by the method of Bengis and Nelson /18/. The polyacrylamide gel slices with LHC- and PS I-complexes are squeezed between two microscopic slides. All measurements are performed at a temperature of 2°C to ensure the stability of the samples within the time of measurement.

3. Experimental results

In Fig. 1 the changes of transmission of the test beam as a function of the wavelength difference between pump and test beam in chloroplasts and in intact leaves are shown. The pronounced transmission decrease of the test beam in the nearest vicinity of the wavelength of the pump beam is the striking feature of this figure. The halfwidth of this absorption band is about 0.24 nm. Fig. 2 shows the pump beam induced test beam transmission measured on LHC-complexes and PS I-complexes, respectively. Only in LHC-complexes are these strong transmission decreases around the pump beam wavelength established. The observation of these pumped absorption structures in chloroplasts, leaves and in LHC-complexes is restricted to pump beam intensities above about $5 \cdot 10^{16}$ photons per cm^2 per pulse. In PS I-complexes no transmission change occurs in the wavelength region investigated, within our experimental error.

At the pump beam photon flux density used in these experiments the exciton-exciton interaction is present in photosynthetic systems followed by exciton annihilation /19/. To prove the onset of exciton-exciton annihilation in the photosynthetic pigment-protein complexes and in chloroplasts, respectively, we measure the relative quantum yield of fluorescence in dependence on the laser intensity. The results are shown in Fig. 3 for chloroplasts, LHC-complexes and PS I-complexes, respectively. The excitation wavelength is in all cases 620 nm. It can be seen from Fig. 3 that there is no difference both in the onset and in the decline of nonlinear fluorescence for chloroplasts and LHC-complexes. The onset of the nonlinear fluorescence in PS I-complexes at about $3 \cdot 10^{16}$ photons per cm^2 per pulse is located at one order of magnitude of higher intensity in comparison with the onset of nonlinear emission in chloroplasts and LHC-complexes. These results are in good agreement with the results in /20/. Further-

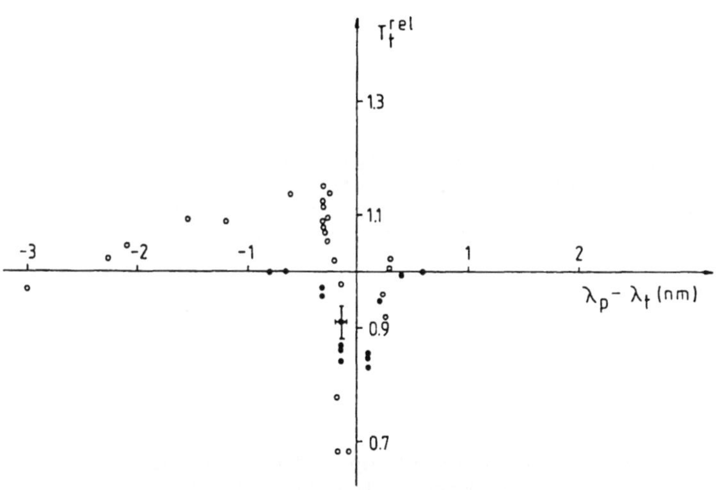

Figure 1. Pump beam-induced change of relative transmission versus wavelength difference between pump and test beam in leaves (o) and chloroplasts (•).

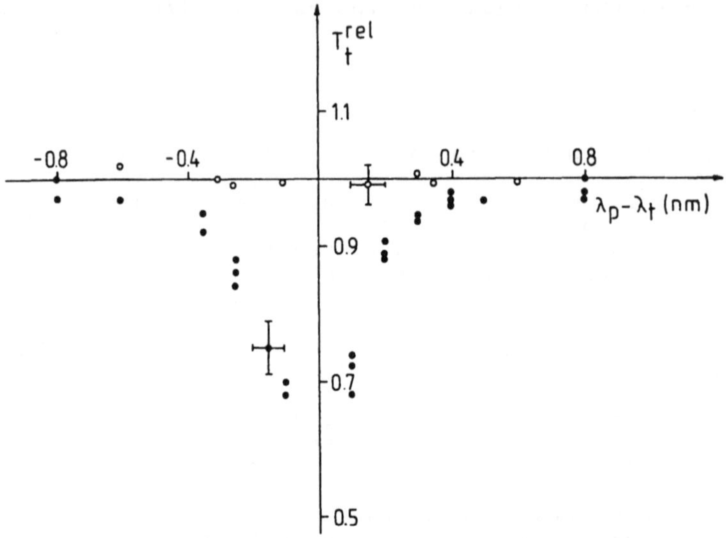

Figure 2. Pump beam-induced change of relative transmission versus wavelength difference between pump and test beam in LHC-complexes (•) and PS I particles (o).

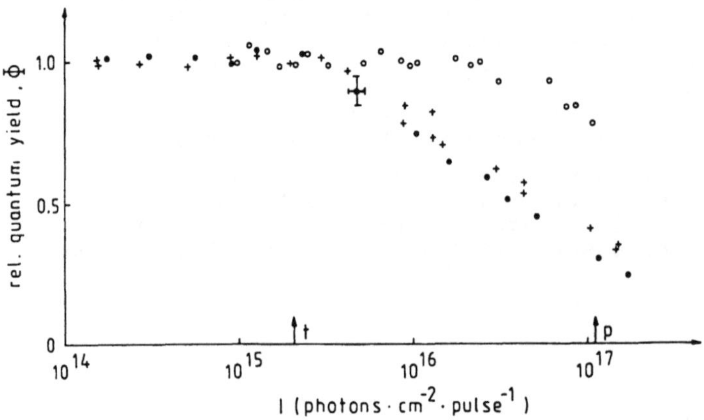

Figure 3. Relative fluorescence quantum yield as a function of the incident fluence measured in chloroplasts (+), LHC-complexes (•) and PS I particles (o). The arrows indicate the intensities of pump and test beam during the two-beam measurements.

more the dependence of the nonlinear course on the excitation intensity is weaker in PS I-complexes.

4. Discussion

The pronounced test beam transmission changes in chloroplasts, leaves and LHC-complexes in the nearest vicinity of the wavelength of the pump beam is measured for the first time. In leaves we find an absorption around the pump beam wavelength and a gain at a wavelength region of 0.3-2 nm. The halfwidth and the amount of

this absorption is nearly equal to that measured in chloroplasts and LHC-complexes. A very first interpretation of the pumped transmission curve of intact leaves was based on the interaction of the light fields with the pigment-protein subunits located in LHC, taking into account two-photon absorption, absorption saturation and three-wave mixing /15/. In this paper only the transmission decrease will be discussed. These pump beam-induced absorptions in chloroplasts, leaves and LHC-complexes, measured at nearly room temperature are a very surprising result:
(1) There is no bleaching of the absorption in the nearest vicinity of the pump beam but an increase, and (2) the pumped absorption structure is very sharp compared with the linewidths of subbands of chlorophyll in vivo of about 8 nm /21/. In principle, this transmission decrease can be explained both as pump beam induced scattering of the test beam or/and as an excited state absorption effect. Pump beam induced scattering processes can be excluded, because the transmission changes up to 30 % in LHC-complexes observed in this study would mean unusually high scattering efficiencies. We suggest therefore, that an excited state absorption process is responsible for the pump beam-induced transmission decrease. The appearance of the pumped absorption structure in the nearest vicinity of the pump beam wavelength means that the transition frequency of chlorophyll absorbing the second photon (test beam photon) is nearly the same as that of the chlorophyll absorbing the pump beam photon. Therefore a simple excited state absorption from S_1- to the most favoured S_3-excited state (Soret-band) of a single chlorophyll molecule with a transition energy of about 1.03 eV is not suitable to explain the absorption of the test beam with an energy of about 1.8 eV. Even in the case that excited state absorption from S_1 to S_n-excited states ($n > 3$) takes place, the low oscillator strengths would require a very large population of the S_1-level to receive transmission changes up to 30 %. On the other hand, by assuming an excited state (two particle) excitation of the whole pigment-protein subunit these discrepanciescan be avoided.

All chlorophyll molecules in LHC-complexes and the majority of chlorophylls in the photosynthetic units of chloroplasts are light-harvesting molecules. The light-harvesting chlorophylls are complexed with proteins in units containing clusters of six or more chromophores. The chlorophylls within such pigment protein subunits are strongly coupled, and electronic excitations exist as exciton states. These are delocalized over the pigment protein subunits /22/. Above a sufficient high excitation intensity excited state absorption occurs and forms a second exciton within the subunit. This second excitation is also delocalized over the subunit. Higher excited states of the subunit are assumed to be negligible at the intensities used in our experiments, so that it is reasonable to consider the subunit as an electronic three-level system /23/. The energy of this two-fold excited subunit is $E_{20}=2E_{10}-E_1$, where E_{10} is the energy of the single exciton and E_1 is the interaction energy between the two excitons. Due to the exciton-exciton interaction in the two-fold excited subunit, the transition frequency of the second exciton should differ from the first one. That means the absorption structure should not be symmetrical around the pump beam wavelength. It seems in LHC-complexes (Fig. 2) the absorption curve is weakly asymmetrical.

An important condition for explaining our pump beam induced absorption as two-fold excitation of a pigment-protein subunit is a relatively long life time of an exciton in a subunit. A first fit of the pumped transmission curves of intact leaves with formalism based on an effective $\chi^{(3)}$-theory allowed us to determine a transverse relaxation time of about 1 ps for the light-harvesting chlorophyll /15/. That would mean a small electron-phonon coupling in the antenna chlorophyll-protein subunits which was confirmed by Gillie et al. /12/ and Voigt and Stahl /21/ very recently.

Taking into account this model, the pumped test beam absorption is the result of the convolution of the single levels E_{10} and E_{21} of the subunit. Assuming LORENTZ distributions for these levels and nearly equal damping constants for both transitions, we estimate a transverse relaxation time of about 1 ps for chlorophyll in LHC by a fit of the convolution of two LORENTZ distributions to the measured absorption. From the position of the transition distributions, we determine a negative interaction energy of about 1 meV between two excitons within one subunit. The results of nonlinear fluorescence measurements on chloroplasts and LHC-complexes show that at a pump beam photon flux density of 10^{17} photons per cm^2 per pulse, being necessary for the transmission changes in two-beam investigations, a strong exciton-exciton annihilation proceeds. That is, the concentration of excitons in the pigment-proteins is so high that in the majority of pigment-protein subunits there is more than one exciton simultaneously. With respect to the results of PS I-complexes the situation is somewhat difficult. In PS I-complexes we don't find any pump beam induced absorption change. Also in PS I pigment-proteins the absorption is essentially determined by the core antenna chlorophyll a. The nonappearance of the pumped transmission structure may be caused by a smaller pigment-pigment interaction, which is the result of another structural arrangement of the antenna chlorophyll and/or of a lower chlorophyll concentration in the subunits. This hypothesis is supported by the results of nonlinear fluorescence measurements in PS I particles. The exciton-exciton annihilation in PS I particles starts at about $3 \cdot 10^{16}$ photons per cm^2 per pulse, one order of magnitude of intensity higher than in chloroplasts or in LHC-complexes.

Acknowledgements

The authors are grateful to V.Z. Paschenko and St. Lonunov for many useful discussions. We wish to thank V.A. Tetenkin and Mrs. Gebhardt for preparation of the chromatophores and chloroplasts, respectively.

5. Literature

/1/ A.J.Campillo and S.L.Shapiro: Photochem.Photobiol.28, 975(1978)
/2/ J.Deprez, A.Dobek, N.E.Geacintov, G.Paillotin and J.Breton: Biochim.Biophys.Acta 725, 444 (1983)
/3/ V.Z.Paschenko and L.B.Rubin: Kvantovaja Electronica (USSR) 8, 2569 (1981)
/4/ R.van Grondelle, H.Bergstrom, V.Sundstrom and T.Gillbro: Biochim.Biophys.Acta 894, 313 (1987)
/5/ A.M.Nuijs, H.J.van Gorkum, J.J.Plijter and L.N.M.Duysens: Biochim.Biophys.Acta 848, 167 (1986)
/6/ H.-J.Eckert, G.Renger, J.Bernarding, P.Faust, H.-J.Eichler and J.Salk: Biochim.Biophys.Acta 893, 208 (1987)
/7/ N.E.Geacintov, J.Breton, Ch.E.Swenberg and G.Paillotin: Photochem.Photobiol. 26, 629 (1977)

/8/ P.Hoffmann and D.Leupold: In Photosynthesis I, Photophysical Processes-Membrane Energization, ed. by G.Akoynoglou (Balaban International Science Service, Philadelphia 1981) pp. 79-88
/9/ R.Avarmaa, I.Renge and Mauring: FEBS Letters 167, 186 (1984)
/10/ S.R.Meech, A.J.Hoff and D.A.Wiersma: Chem.Phys.Letters 121, 287 (1985)
/11/ J.Hala, I.Pelant, L.Parma and K.Vacek: J.Lumin.24/25,803(1981)
/12/ J.K.Gillie, J.M.Hayes, G.J.Small and J.H.Golbeck: J.Phys.Chem. 91, 5524 (1987)
/13/ R.Jankowiak, D.Tang, G.J.Small and M.Seibert: J.Phys.Chem. 93, 1649 (1989)
/14/ R.S.Knox: In Primary Processes of Photosynthesis, ed. by J.Barber (Elsevier/North-Holland 1977)
/15/ G.Kehrberg, J.Voigt and V.May: Colloquia Pflanzenphysiologie, Humboldt-Universität Berlin 10, 37 (1986)
/16/ B.Hieke: Dissertation B, Humboldt-Universität Berlin (1980)
/17/ B.A.Gulyaev, V.A.Tetenkin and O.N.Pomeranceva: DAN SSR 248, 752 (1979)
/18/ C.Bengis and N.Nelson: J.Biol.Chem. 250, 2783 (1975)
/19/ J.Breton and N.Geacintov: Biochim.Biophys.Acta 594, 1 (1980)
/20/ T.Kolubayev, N.E.Geacintov, G.Paillotin and J.Breton: Biochim. Biophys.Acta 808, 66 (1985)
/21/ J.Voigt and U.Stahl: Z.phys.Chem., accepted for publ.
/22/ K.Sauer: In Bioenergetics of Photosynthesis, ed. by Govindjee (Academic Press, New York 1975)
/23/ J.Voigt: Photobiochem.Photobiophys. 12, 21 (1986)

Index of Contributors

MIX
Papier aus verantwortungsvollen Quellen
Paper from responsible sources
FSC® C105338

If you have any concerns about our products,
you can contact us on
ProductSafety@springernature.com

In case Publisher is established outside the EU,
the EU authorized representative is:
Springer Nature Customer Service Center GmbH
Europaplatz 3, 69115 Heidelberg, Germany

Printed by Libri Plureos GmbH
in Hamburg, Germany